Transforming Higher Education Through Digitalization

Demystifying Technologies for Computational Excellence: Moving Towards Society 5.0
Series Editors: Vikram Bali and Vishal Bhatnagar

This series encompasses research work in the field of Data Science, Edge Computing, Deep Learning, Distributed Ledger Technology, Extended Reality, Quantum Computing, Artificial Intelligence, and various other related areas, such as natural-language processing and technologies, high-level computer vision, cognitive robotics, automated reasoning, multivalent systems, symbolic learning theories and practice, knowledge representation and the semantic web, intelligent tutoring systems, AI and education.

The prime reason for developing and growing out this new book series is to focus on the latest technological advancements - their impact on the society, the challenges faced in implementation, and the drawbacks or reverse impact on the society due to technological innovations. With the technological advancements, every individual has personalized access to all the services, all devices connected with each other communicating amongst themselves, thanks to the technology for making our life simpler and easier. These aspects will help us to overcome the drawbacks of the existing systems and help in building new systems with latest technologies that will help the society in various ways proving Society 5.0 as one of the biggest revolutions in this era.

Data Science and Innovations for Intelligent Systems
Computational Excellence and Society 5.0
Edited by Kavita Taneja, Harmunish Taneja, Kuldeep Kumar, Arvind Selwal, and Ouh Lieh

Artificial Intelligence, Machine Learning, and Data Science Technologies
Future Impact and Well-Being for Society 5.0
Edited by Neeraj Mohan, Ruchi Singla, Priyanka Kaushal, and Seifedine Kadry

Transforming Higher Education Through Digitalization
Insights, Tools, and Techniques
Edited by S. L. Gupta, Nawal Kishor, Niraj Mishra, Sonali Mathur, and Utkarsh Gupta

A Step Towards Society 5.0
Research, Innovations, and Developments in Cloud-Based Computing Technologies
Edited by Shahnawaz Khan, Thirunavukkarasu K., Ayman AlDmour, and Salam Salameh Shreem

For more information on this series, please visit: https://www.routledge.com/Demystifying-Technologies-for-Computational-Excellence-Moving-Towards-Society-5.0/book-series/CRCDTCEMTS

Transforming Higher Education Through Digitalization

Insights, Tools, and Techniques

Edited by
S. L. Gupta, Nawal Kishor, Niraj Mishra,
Sonali Mathur, and Utkarsh Gupta

CRC Press
Taylor & Francis Group
Boca Raton London New York

CRC Press is an imprint of the
Taylor & Francis Group, an **informa** business

First edition published 2022
by CRC Press
6000 Broken Sound Parkway NW, Suite 300, Boca Raton, FL 33487-2742

and by CRC Press
2 Park Square, Milton Park, Abingdon, Oxon, OX14 4RN

CRC Press is an imprint of Taylor & Francis Group, LLC

Library of Congress Cataloging-in-Publication Data
Names: Gupta, S. L., editor. | Kishor, Nawal, editor. | Mishra, Niraj, editor. | Mathur, Sonali, editor. | Gupta, Utkarsh, editor.
Title: Transforming higher education through digitalization : insights, tools, and techniques / Edited by S. L. Gupta, Nawal Kishor, Niraj Mishra, Sonali Mathur, and Utkarsh Gupta.
Description: Boca Raton : CRC Press, 2022. | Series: Demystifying technologies for computational excellence | Includes bibliographical references and index.
Identifiers: LCCN 2021018918 (print) | LCCN 2021018919 (ebook) | ISBN 9780367676292 (hardback) | ISBN 9780367676308 (paperback) | ISBN 9781003132097 (ebook)
Subjects: LCSH: Education, Higher—Computer-assisted instruction. | Educational technology—Computer-assisted instruction. | MOOCs (Web-based instruction) | Distance education—Technological innovations. | Education, Higher—Aims and objectives.
Classification: LCC LB2395.7 .T73 2022 (print) | LCC LB2395.7 (ebook) | DDC 378.1/734—dc23
LC record available at https://lccn.loc.gov/2021018918
LC ebook record available at https://lccn.loc.gov/2021018919

ISBN: 978-0-367-67629-2 (hbk)
ISBN: 978-0-367-67630-8 (pbk)
ISBN: 978-1-003-13209-7 (ebk)

DOI: 10.1201/9781003132097

Typeset in Times
by KnowledgeWorks Global Ltd.

Contents

SECTION II Understanding Technology in Education

SECTION III Enhancing Teaching Quality in Digital Age

Preface

Amidst the backdrop of a dramatically changed world affected by COVID-19, the education sector has witnessed a massive transformation. From primary to the higher education levels, online teaching and assessment became mainstreamed. This has proved to be a boon in disguise as the process of education needed to be reinvented and aligned with the requisites of the fourth industrial revolution called Education 4.0. Digitalization and the adoption of technology lie at the core of this next generation of education. In fact, experts have opined that "education will be an entirely digital pursuit fortified by artificial intelligence and virtual reality".[1]

A simplistic definition of digitalization is that it is "the use of digital technology to teach students". It forms the core of Education 4.0 that aims to prepare graduates for a future that is more aligned and engaged with technology. The distinctive characteristics of Education 4.0 are the transformed education processes such as peer-to-peer learning, critical thinking skills, automated assessment methods, advanced data analytics and personalized learning.

The success of digitalization in higher education institutions is not just adopting new technologies or upgrading to the latest tools. It is rather a reflection of the process in creating a sustainable education model where the success accrues in different capability areas. It needs a curriculum overhaul aimed at imparting the right tools and skills for preparing *"the workforce of the future"*.

The program delivery is set to include both remote and face-to-face learning. Remote learning by the digitalized medium will enhance theoretical learning, whereas the latter medium will be useful for practical skills.

Innovative technologies such as Artificial Intelligence (AI), robotics, Big Data, Internet of Things, and social media are all bound to influence the skills required in the workplace. The new set of skills for the future includes digital skills as well as training in science, technology, engineering and mathematics (STEM).

The teaching and learning process is the pivot of educational transformation. It implies that higher education institutions across the world should gear up for a major overhaul of their curricula, assessment, learning delivery methods and learning outcomes.

Adopting technology in education should not just be aimed at meeting industry needs but also at enhancing students' learning experience and optimizing resource utilization. "Customizable degrees" are going to be a reality where students do not limit themselves to a few courses. They would instead be allowed to study modules from varied programs. Such complex permutations in program management would only be possible through digitalization.

Students' assessment methods must be altered so that the results reflect the skills learned by students and not a mere reflection of numerical grades. There will be a continuous assessment based on practical and experiential activities. Digitalization will play an important role in assessing the performance of each student throughout

[1] https://www.plm.automation.siemens.com/global/en/our-story/glossary/digitalization-in-education/25307

his/her "learning journey". Technologies that store student data would help in "optimizing learning strategies".

The need for higher education institutions to embrace digitalization for aligning with Education 4.0 motivated us to offer this book. We aim at providing critical insights into the role of digitalization in achieving a holistic transformation of higher education institutions.

The book covers various facets of transformation and the challenges associated with them. These challenges include those associated with sustainability, delivery and outcomes.

The book has been organized into three main sections and 18 chapters. The first section covers the issues and challenges in digitalization of education. The second section includes the various technologies and applications that are currently being used or will soon be embraced by higher education institutions. The role of digitalization in teaching and learning is the area covered in the third section. A brief summary of the contents of each chapter is presented below:

Chapter 1 covers "Determining Sustainability of Online Teaching: Issues and Challenges".
Chapter 2 discusses "Effectiveness of Online Learning and Face-to-Face Teaching Pedagogy".
Chapter 3 contains "Issues and Challenges Faced by College Students in Online Learning during the Pandemic Period".
Chapter 4 describes "Teacher's Perception towards Online Higher Education Teaching during COVID-19".
Chapter 5 covers "Challenges Faced by Faculty and Students in Online Teaching and Learning: A Study of Higher Education Institutions in Oman".
Chapter 6 describes "Ramifications of Digitalization in Higher Education Institutions with Reference to Indian Educators: A Thematic Analysis".
Chapter 7 discusses "Digital Education and Society 5.0".
Chapter 8 explains "Moodle: Learning Management System".
Chapter 9 describes "Digital Transformation of Higher Education: Opportunities and Constraints for Teaching and Learning and Research".
Chapter 10 includes "'Rubrics' as a Tool for Holistic Assessment: Design Considerations and Emerging Trends".
Chapter 11 comprises "A Systematic Review of Barriers to Scientific Crowdsourcing in Higher Education".
Chapter 12 discusses "Effects of Technology-Based Feedback on Learning".
Chapter 13 "Storyboarding: A Pedagogical Tool for Digital Learning" is discussed.
Chapter 14 describes "Online Social Capital and Its Role in Students' Career Development".
Chapter 15 explains "Upskilling and Reskilling in the Digital Age: The Way Forward for Higher Educational Institutions".
Chapter 16 consists "Strengthening the retention rate of Massive Open Online Courses through Emotional Intelligence and Intrinsic Motivation".
Chapter 17 covers "Digitalization of Higher Education: Issues and Challenges".

Chapter 18 discusses "Creating a Sustainable Future with Digitalization in Online Education: Issues and Challenges".

With the enduring support and encouragement of our respective families, we believe we have compiled a book that should serve as an important resource for both scholarly and practical needs. It explores the challenges associated with digitalization, the trends in education technology and how higher education institutions should prepare to achieve the goals of Education 4.0.

S.L. Gupta
Mesra, Ranchi, India

Nawal Kishor
New Delhi, India

Niraj Mishra
Mesra, Ranchi, India

Sonali Mathur
New Delhi, India

Utkarsh Gupta
Connecticut, United States of America

Editors

S. L. Gupta is the Director, Birla Institute of Technology, Noida. Before that he was the Dean of Waljat College of Applied Sciences. Prior to joining WCAS, he had been the Director of Birla Institute of Technology (Deemed University – Ranchi, India) Patna Campus. He brings with him a rich experience of 27 years in academia. His professional qualifications include an Executive Programme in Retail Management from IIM-Kolkata and a PGDBM (Marketing) from CMD Modinagar, India and M.Com from University of Rajasthan, Jaipur, India. His fields of specialization are Sales and Distribution Management, Marketing Research, Marketing of Service, Retail Management, and Research Methodology. He has to his credit many publications in national and international journals. He has published eight books, which are internationally recognized and recommended in many universities and colleges, and research papers on his area of specialization.

Nawal Kishor is a Professor and former Director, School of Management Studies, Indira Gandhi National Open University, New Delhi, India. He earned PhD, PGDIM and M.Com. He is the Managing Editor of the leading journal, *The Indian Journal of Commerce* and the Editor of *Indian Journal of Open Learning*. He has been engaged in teaching, training, research and other academic and administrative activities for the last 30 years. He has been involved in the development of B.Com, M.Com, PGDIBO, BBA (Retail), M.Phil and PhD programs along with the conduct of Orientation Programs, Workshops, Refresher Courses, Faculty Development Programs, etc. He has published more than 60 research papers in international and national reputed journals. He has presented more than 15 research papers at international and national conferences. He has been actively involved as Keynote Speaker, Technical Session Chairman, Guest of Honor, Resource Person in various international and national conferences. His areas of interest are International Business Management, International Marketing, Marketing Management, Consumer Behavior, General Management, Organizational Behavior, Human Resource Management, Foreign Trade, Export Import Procedures and Documentation, Retail Management and Distance Education. He has visited the USA, Canada, Australia, France, Germany, UK, Netherlands, Italy, Switzerland, Ethiopia, Singapore, Hong Kong, UAE, Nambia, among others for academic purposes. He received the Best Researcher Silver Medal for the "Second Best Research Paper Award" in 2014.

Niraj Mishra is an Assistant Professor in the Department of Management at Birla Institute of Technology, Ranchi. He was formerly working as Head in the Department of Management at Waljat College of Applied Sciences, Muscat, Oman. Dr. Mishra completed his MBA degree from Birla Institute of Technology, Mesra, India and PhD in Management from BR Ambedkar Bihar University, India. Dr. Niraj Mishra has published several research papers in areas of service marketing, e-services and quality management. Dr. Mishra has participated in many national and international conferences and presented research papers. He has also received the best paper award for one of his papers presented at an international conference. He is also guiding PhD scholars in various universities in India in the area of e-services. Dr. Niraj Mishra has served as Deputy Head – Department of Management, Quality Coordinator (Academic) at Waljat College of Applied Sciences) and is the Chairman of Risk Management Committee of the college. Dr. Mishra has played an active role in preparation of strategic plans and operational plans, risk register and various HR policies of the college.

Sonali Mathur is associated with JSS Academy of Technical Education, Noida, since 2010, working as Assistant Professor in Computer Science & Engineering Department. She earned her BE in Computer Science and Engineering from M.J.P. Rohilkhand University, Bareilly, M.Tech. in Information Technology from Guru Gobind Singh Indraprastha University and she is pursuing her PhD in Data Warehouse Testing and Security from Birla Institute of Technology, Mesra, Ranchi. She has more than 16 years of teaching and research experience and published research papers in various journals and conferences. Her areas of interest include Data Warehouse, Data Mining, Security and Testing.

Utkarsh Gupta is currently pursuing his MBA in Marketing and Business Analytics from the University of Connecticut, United States. He is also serving as the President of UConn Graduate Consulting, where he manages a team of 50 consultants. He also has extensive work experience with developing marketing and sales strategy for startups. He had worked as Decision Scientist in Mu Sigma delivering data-based strategy solutions to Fortune 500 companies. He had a brief stint with The Coca-Cola Company's global headquarters in Atlanta developing marketing strategies.

Contributors

Irum Alvi
HEAS Department
Rajasthan Technical University
Kota, Rajasthan, India

Manju Amla
University Business School
Panjab University
Chandigarh, India

Soumitro Chakravarty
Department of Management, BIT Mesra
Off-Campus Lalpur
Ranchi, India

Richa Chauhan
FMS-WISDOM,
Banasthali Vidyapith
Rajasthan, India

Kavita Chavali
College of Commerce and Business Administration
Dhofar University
Sultanate of Oman

Gopu J.
University of Technology and Applied Sciences
Salalah, Sultanate of Oman

Mukta Goyal
Guru Nanak Dev Institute of Technology
New Delhi, India

S. L. Gupta
Birla Institute of Technology, Mesra,
Ranchi, India

Najmul Hoda
Department of Business Administration
College of Business
Umm Al-Qura University
Makkah, Saudi Arabia

Abduroof Ahmed Ismail
University of Technology and Applied Sciences
Salalah, Sultanate of Oman

Karishma Jain
Institute of Management Studies
Banaras Hindu University
Varanasi, India

Biswa Mohana Jena
NSCB College
Sambalpur, Odisha, India

Lokesh Jindal
Atal Bihari Vajpayee School of Management and Entrepreneurship
JNU
New Delhi, India

Pooja Kansra
Mittal School of Business
Lovely Professional University
Punjab, India

Rajni Kansra
Sidana Institute of Education
Punjab, India

Regina Lenart-Gansiniec
Jagiellonian University
Krakow, Poland

Nidhi Maheshwari
Delhi Technological University
Delhi, India

Viju Mathew
Department of Scientific Research
University of Technology and Applied Sciences
Salalah, Sultanate of Oman

Niraj Mishra
Department of Management
Birla Institute of Technology, Mesra
Ranchi, India

Arun Mittal
Department of Management
Birla Institute of Technology, Mesra
Ranchi (Off Campus – Noida), India

Kumar Mukul
KLE Society's Institute of Management Studies and Research
Hubli, Karnataka, India

V. Padmaja
Dept. of HRM
Ramaiah Institute of Management
Bengaluru, India

Prajna Pani
School of Management
Centurion University of Technology and Management
Parlakhemundi, Odisha, India

Umesh Prasad
Department of Computer Science, BIT Mesra
Off-Campus Lalpur
Ranchi, India

Twinkle Sanghavi
Department of Sociology
Maniben Nanavati Women's College
Mumbai, India

Shouvik Sanyal
College of Commerce and Business Administration
Dhofar University
Sultanate of Oman

Gautam Shandilya
Department of Hotel Management & Catering Technology
Birla Institute of Technology, Mesra
Ranchi, India

Swasti Singh
Faculty of Management Studies
Dr. Harisingh Gour University
Sagar – Madhya Pradesh, India

Abhaya Ranjan Srivastava
Department of Management, BIT Mesra
Off-Campus Lalpur
Ranchi, India

Abhishek Srivastava
Department of Information Technology
Gopal Narayan Singh University
Rohtas, India

Praveen Srivastava
Department of Hotel Management and Catering Technology
Birla Institute of Technology, Mesra
Ranchi, India

Shelly Srivastava
Department of Management
Birla Institute of Technology, Mesra
Ranchi, India

Section I

Transformation of Education

1 Determining Sustainability of Online Teaching
Issues and Challenges

Arun Mittal

CONTENTS

1.1 INTRODUCTION

The world has taken an abrupt turn due to the spread of the virulent disease COVID-19. The pandemic has enforced social distancing and mobility restrictions, which are now a part of the "new-normal." Every sector is going through a major setback and is trying to implement innovative ways to operate efficiently in this situation.

DOI: 10.1201/9781003132097-1

Educational institutions all over the world are facing a similar fix and have taken the assistance of digital technology to maintain regularity in classes. However, this system is not easily accepted by all nations due to many glitches. According to Dhawan (2020), the Indian educational system has always had a conventional approach. Technology was not a thing for Indian educators and learners as long as it was not a compulsion. Even though several learning institutions have amalgamated digital into the curriculum, the majority of studies were based on traditional courses. The new transformation was not only a new dawn on the horizon of education but also a challenging episode for tutors who have inadequate knowledge about e-learning. The thought of accepting the virtual mode of education was not easy for both the teachers and the tutors. Even though children nowadays are much exposed to technology, teachers with a constrained mindset lack in this area. Technology came in handy during the crisis as the educational management had no other option to conduct classes with utmost safety. Higher educational institutions all over the country have adopted virtual classes. The different applications that are implemented for e-learning are Zoom, Google Hangouts, Skype Meetup, social media, etc. A lot of institutions have also conducted motivational programs for teachers to help them feel efficient (George, 2020). However, many tutors feel that there is a thin line of difference between the traditional and virtual modes of education as one can still present PPTs, show videos, and use the board and marker to address the class (Moll and Nielsen, 2017). Kebritchi et al. (2017) found that the issues pertaining to online learning are expectations, readiness, identity, and participation in online courses. The issues for instructors are changing faculty roles, transitioning from a face-to-face mode to online, time management, and teaching styles. At the same time, the faculty members also have to face issues such as the integration of multimedia in content, role of instructional strategies in content development, and considerations for content development while they develop the content for their lectures.

Shenoy et al. (2020) found that there was a lack of infrastructure with the teachers like "configured laptops," "Internet," and "microphones" that are all required for online classes. Since there is a lack of technical assistance, teachers were unable to solve technical problems like "connectivity issues," "system failure," and "bandwidth issues" while conducting online sessions. The teachers were annoyed due to the unavailability of "hardware devices," "software," "Internet connectivity," and "power backup," which disrupts online teaching and assessments. Gratz and Looney (2020) conducted a study in Los Angeles and found that due to lack of experience and skills of online teaching, they were not really willing to teach online and they were also showing their resistance to change since they had a lack of time to prepare online course material. Some of the teachers also share their views that their subject and the course are not suitable for online teaching and learning as well. As per the survey-based estimate given by the International Association of K-12 Online Education, around 1.5 million students have taken one or more courses online since 2010 (Wicks, 2010). In the courses, students received all or part of the instructions online and took classes online with the help of peers, teachers, digital learning, etc. Some states such as Florida, Alabama, and Michigan have helped in making online education a part of the requirements of graduation.

E-education has gained popularity due to its perceived potential for providing a flexible access to instruction as well as content by:

- Increasing availability of the learning experience for people who can't attend or prefer to not attend the traditional school.
- Assembling as well as disseminating the instructional content effectively.
- Increasing the student:faculty ratio while also achieving outcomes of learning equal to that of the traditional classroom teaching. Some of the proponents consider technology to have the potential beyond raising the efficiency in the instructional delivery, for instance, by offering the community of students to assist in the understanding of the knowledge which is extremely crucial.

Online technological advancements can expand and help these kinds of communities with the promotion of the participatory model of education rather than just changing the delivery models of education (Riel et al. 2004; Barab et al. 2000).

1.2 REVIEW OF EXISTING LITERATURE

1.2.1 Framework of Online Teaching

Sangrà et al. (2012) stated that online learning is a method of sending and receiving information with the application of an educational framework. It can only be executed with the help of digital instruments that can accelerate communication and interaction and allow the absorption of contemporary methods of education. Sharma and Kitchens (2004) illustrated that e-learning involves the use of Internet-based training amenities such as electronic-based learning institutions and classroom settings that facilitate the digital conglomeration that supported online learning. In order to make effective use of online learning, both the teacher and the student should have great Internet facilities and sufficient digital instruments. In spite of the assortment of descriptions, e-learning has a major role in the pedagogical expansion of many countries as it caters to the prospect of the countries in order to develop their educational progression.

Though the list of existing online learning platforms is not exhaustive; however, online learning may be divided into two major categories, namely recorded lectures and live sessions. Both forms have their own advantages and limitations. The live sessions are more interactive; however, they are not available 24/7. In addition to that, live online learning is much costlier than pre-recorded content, which is not only available 24/7 but also easy to store and reuse. The live sessions are being used as an alternative to classroom learning. E-learning is an alternative arrangement to physical classroom settings; however, our institutions do not seem to be fully prepared for the same. This caused discomfort, anxiety, and unease to a large number of teachers. The teaching is less interactive because students do not actively participate in the process. It has been found in the extant literature that students' participation and engagement has been limited during online classes and establishing this discussion and writing is very important (Romiszowski and Mason, 2004; Vonderwell and

Zachariah, 2005). In the case of group assignments and presentations, the interaction among group members remains limited and the purpose of assigning a group activity remains unfulfilled (Graham and Misanchuk, 2004; Jaques and Salmon, 2007). The importance of online learning has been rapidly increasing. This shift has gradually taken an inclination toward the adoption of online learning in the long run (McCallum and Price, 2010).

1.2.2 Technical Constraints

Shenoy et al. (2020) found that teachers are facing problems in adapting to the technology for online classes. They are facing personal challenges in teaching and assessments since they are not familiar with the new technology and have no experience in taking online classes. Despite their lack of knowledge, institutions pressure the teachers to teach online, which increases their negative attitude. There are teachers who are using the "institutional support" and have a good knowledge of the technology but do not feel motivated to take online classes due to the reason that it is a very "tiring," and "time-consuming" activity. Teachers were facing problems with the subjects having numerical studies and experiments and where personal interaction is needed. In online classes teachers find difficulty in solving the problems and clearing the doubts as compared to physical classes.

Hatlevik and Hatlevik (2018) define that the hardships related to Information and Communication Technology (ICT) can be a major blocking factor for teachers who are technologically incapable, and it involves a lot of time and effort to figure out its operations. The process of teaching and learning, with the intrusion of ICT, becomes more complex than the traditional face-to-face method of education. The sundry multifaceted matters are included in the incorporation of ICT into the process. Educators must analyze how to work with several applications and should synchronize their time and slots with the other elements of education required in the process. It can lead to uncertain, inundated, apprehensive, and anxious teacher functionalities. In a longer span of time, it can hamper work performance.

1.2.3 Limited Interaction and Bounded Teaching

According to Singh and Thurman (2019), the respective educational institutions and educators should lend mental support that can help students cope with the distress and embrace the new normal in a timely manner. Simones et al. (2015) state that the analysis done on tutors' approach toward students by taking into consideration the technical information like note names in association to the pitch or rhythm or even inquiring about a child's perspective of their conduct. Crawford et al. (2020) state that the research done on online studies shows that educators spent less time in questioning students who were present in face-to-face communication, whereas digital classes have levels of questioning that can be an equivalent alternative to face-to-face teaching. In total, there is slight evidence to depict that online teaching has a strong impact on the teachers who can spend time questioning students. Bao (2020) stressed the relevance to offer psychological assistance to the university group. In this exploration, students have elaborated on the pros and cons of e-learning. There has been an incidence of perplexity, disparity, anxiety, and gloominess about learning in isolation. These thoughts need to be addressed by an ultimate superior.

Ungar and Baruch (2016) are of the opinion that ICT is not anticipated to cater to groundbreaking, student-centric learning ambience until the teacher takes the initiative to utilize the full potential of the Internet and allow curriculum segregation. The tutors should play an active part in training students, irrespective of any medium. Introducing a student-based online learning system needs an alteration from the traditional framework of pedagogy into an atmosphere that facilitates both online and offline learning techniques (Mooij and Smeets, 2001).

Aboagye et al. (2020) concluded that the teachers were also facing problems because they have no experience in giving online classes, and since there is a lack of eye contact in online teaching, they are not able to understand the behavior of the students. They were also facing problems in synchronizing their class because of the disturbance created by the students sitting in their homes. "Administrative issues," "academic skills," "social interactions," "technical skills," "learner motivation," "time and support for studies," "cost, and access to the Internet," and "technical problems" were some of the challenges that were faced by the teachers, students, and the facility providers during online learning.

The process of e-learning is often stereotyped as being a degraded version of face-to-face learning. The immediate transfer of the educational system into a completely online learning process can help in healing the concept and increasing confidence in it, whereas in reality, no one making the switch to e-learning under the given situation will be actually able to utilize the full benefit of the chances provided by the online version (Hodges et al. 2020).

1.2.4 Ineffective Evaluation

Kebritchi et al. (2017) recommended that efforts should be made to improve the system of taking online exams. Every institution should follow the online exam proctoring method, in which a student will be under strict supervision during the exam, therefore increasing the authenticity of these evaluations and helping in validating the success of educators. Along with the "resources," "staff readiness," and "confidence," the motivation and the accessibility of the students also play a very important role in ICT-integrated learning. The study suggests that the faculty members should use the "technology" and the "technological gadgets" to increase the process of teaching and learning during this time of exceptions. It is observed that there not only "technical issues" in the online learning environment but also "pedagogical," and "instructional challenges" (Ali 2020).

Christian et al. (2020) found that though the students actively collaborate in the online learning process, which definitely assists the educators in accepting the new normal, the students were seen to be more effective in a conventional classroom setting. Their effectiveness during online classes was not on par with that during face-to-face interaction. The mock examinations are taken keeping all strict measures in a classroom setting; there is a higher possibility that students can cheat during the exams (MacPhail et al. 2019).

Chang and Fang (2020) describe that online education makes education possible anytime, anywhere. The research elaborates on the learning itself and its mechanisms. The outcome shows that even though the majority of the educators try to organize the elements of instruction, nevertheless, it is not a rosy way to judge and modify the pupils' educational approach within a small span of time. There has been a projection of insinuations for practice and notions related to upcoming explorations.

1.2.5 Sustainability of Online Learning

Various empirical studies have been done for examining the quality of e-education from different perspectives. Various studies have examined as well as identified crucial problems affecting the quality of e-education such as technology, communication, pedagogy, management of time, and assessment. There are certain organizations such as Online Education Consortium as well as Quality Matters that focus on the improvement of the quality of e-education in higher education with the help of resources and opportunities to collaborate on the development of the curriculum. However, the reports relating to e-education require a literature review that synthesizes as well as integrates empirical studies further and offers an integrated report on the existing challenges of teaching online courses. Online educators should go through the daunting tasks of moving through the increasing expansion of literature for identifying problems for themselves (Limperos et al. 2015). Further, due to continuous reports regarding the high rate of dropout as well as achievement issues in the online courses (Morris and Finnegan 2008), conducting these kinds of investigations and offering the desired result is increasingly becoming critical for informing the educators regarding the changes as well as considerations needed to improve the quality of e-education. The main purpose of these studies is to inform the educators regarding the issues as well as strategies that affect the quality of online courses of higher education. These research reviews have been examined for identifying the major challenges and issues faced during teaching online various courses of higher education. The reviews also address the issues provided under the topical segregation and offer certain suggestions for addressing the problems for the online educators.

Sustainable education mainly focuses on the students and the innovative pedagogies that bring them close to the social reality as well as its primary conflicts. The aim of the students is to understand the environment better for the intended profession. It is important for the university to be the driving force that teaches the students to know about sustainability as well as changes implied. Thus, it is the priority to work for the values that cause reflection as well as critical thoughts and for integrating the theme of sustainability in the contents of the subject material (Takala and Korhonen-Yrjänheikki 2019). For achieving the competencies, it is important to consider sustainability as a continuous process.

Regarding academic requirements, the higher education community is tagged as the most vulnerable group who are pumped up with increasing stress and apprehensions (Lee 2020). Practical steps were taken to support the staff and students to prepare them for online learning. The digital means were particularly adopted not to cause disruption in education but to avoid delay in graduation (Flores 2020).

1.3 RESEARCH GAPS

There are a good number of studies on topics pertaining to issues and challenges of online learning. Learning through online sources has many limitations. The extant literature has also identified that e-learning cannot replace face-to-face and classroom teaching; however, during the pandemic, it was the only resort to mitigate the students' study loss. Despite the fact that it requires many resources including laptops/phones and Internet data, overall, it is still cheaper than face-to-face learning. A student saves multiple forms of expenses, and the most important among all is transportation cost. However, in normal situations, classroom learning is always preferred by teachers as, in physical settings, they do not require specific knowledge for conducting online classes and they ensure better classroom engagement and fair evaluation.

The extant literature has limited discussion on challenges in a comprehensive way, and there is hardly any measurement scale on the issues and challenges faced by faculty in the context of online teaching. The major objective of this study is to explore the various challenges faced by faculty members while teaching online and develop a measurement scale in this context. Further, the study also determines the impact of these challenges on the sustainability of online teaching.

1.4 RESEARCH QUESTIONS

- What are the various issues and challenges that determine the sustainability aspects of online teaching?
- What is the influence of the issues and challenges of online teaching on its sustainability aspects?

1.5 OBJECTIVES, THEORETICAL FRAMEWORK, AND HYPOTHESES

- To determine the various issues and challenges that determine the sustainability aspects of online teaching.
- To ascertain the influence of the issues and challenges of online teaching on its sustainability aspects.

1.6 THEORETICAL FRAMEWORK

The present study mainly focuses on how various issues and challenges affect the sustainability aspect of online teaching. The theoretical framework in the above context has been portrayed in Figure 1.1.

The whole world seems to be fascinated by online teaching and learning. The online setting is producing instant results as well. However, with the passage of time, the effectiveness of online learning has been questioned due to many important reasons such as classroom interaction and engagement, challenges in conducting exams, performing continuous evaluations, and technical constraints. The feeling of the existence of a boundary in teachers' minds while conducting classes, because they do not feel free to share in the way they would in a physical classroom setting,

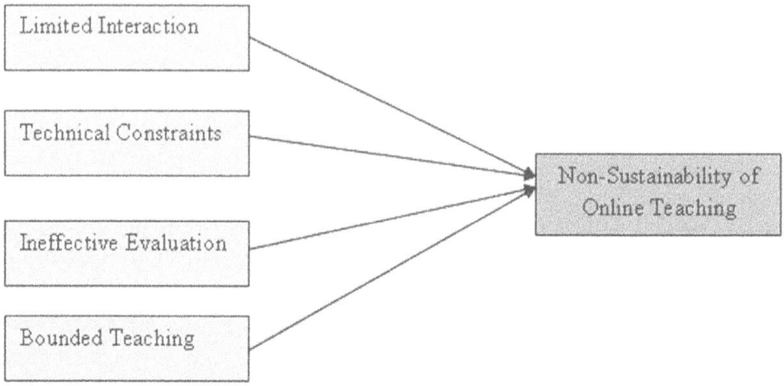

FIGURE 1.1 Theoretical framework of the study.

is also a challenge. Such challenges question the sustainability of online teaching as a full-fledged format of teaching. Hence, the theoretical framework of this study shows the causal relationship between the various issues and challenges of online teaching and the sustainability of online teaching by taking teachers' perspectives into account.

1.7 HYPOTHESES

Ha1: Limited Interaction influences the Non-Sustainability of online teaching.
Ha2: Technical Constraints influence the Non-Sustainability of online teaching.
Ha3: Ineffective Evaluation influences the Non-Sustainability of online teaching.
Ha4: Bounded Teaching influences the Non-Sustainability of online teaching.

1.8 MATERIALS AND METHODS

The study explores the issues and challenges of online learning experienced by faculty members and mainly contains variables such as the knowledge of technology, evaluation of students, classroom engagement, freewheel learning, and sustainability of online teaching. The study focuses on the exploration of the factors around the abovementioned areas and the causal relationship between the various factors and the sustainability of online teaching for the future. The present study is exploratory-cum-descriptive in nature. A structured questionnaire was designed for the collection of data, in which the responses were captured on a 5-point Likert scale ranging from 1 to 5, where 1 means strongly disagree and 5 means strongly agree. The sample size for the study was 192 teachers who teach courses in various higher education programs in the areas of commerce, management, and humanities. In this study, only teachers who do not teach any laboratory courses were approached for the collection of data. The statistical tools applied in this study are exploratory factor analysis (EFA), confirmatory factor analysis, and structural equation modeling.

1.8.1 Justification of Sample Adequacy

Kelloway (1998) suggests that the least proper sample for structural equation modeling should be not less than 200 observations. The final sample size is 192, which is very close to the recommended sample size. According to Hair et al. (2006), the smallest sampling size should be not less than five times the number of observed variables during factor analysis. The number of observed variables in this research is 20. This means that the minimum sampling size required for the analysis is 100 observations to meet the requirement.

1.9 DATA ANALYSIS AND INTERPRETATION

1.9.1 Exploratory Factor Analysis

EFA gives a variety of outputs that include the measure of sample adequacy, variance explained by the various factors, reliability of the respective factors, and component matrix with factor loading. The Kaiser-Meyer-Olkin (KMO) value was found to be 0.857, which is more than the recommended value of 0.5, implying that the requirement of sample adequacy was fulfilled. Five factors were obtained from the 20 variables. The total variance explained by the five factors was 84.331.

The first factor was named "Limited Interaction," which explained 22.387% of the total variance. Factor 2 explained 17.538% of the variance and was named "Technical Constraints." The third factor was named "Ineffective Evaluation" and explained 16.499% of the variance. The fourth factor explained 15.261% of the variance and it was named "Bounded Teaching." The last factor was named the "Non-Sustainability" of online teaching with 12.646% of the variance explained. Table 1.1 presents the factor loading of the individual statements and the construct reliability and composite reliability of the factors obtained. Both the values of Cronbach's alpha and composite reliability (McDonald's Omega) were found to be satisfactory.

1.9.2 Confirmatory Factor Analysis

Table 1.2 specifies the model fit indices for the confirmatory factor analysis (measurement model). It may be seen from the table that all the model fit criteria have been satisfied as per their desired values mentioned in the table, which proves that model is fit for further analysis. Figure 1.2 represents the measurement model for sustainability of online teaching.

Table 1.3 shows the validity measures of the measurement model. It is found from the table that composite reliability is above 0.7, which establishes the convergent validity. Similarly, the ASV (Average Shared Variance) is above 0.5, which also determines the convergent validity. However, the AVE (Average Variance Estimate) is above the MSV (Maximum Shared Variance) and ASV (Average Shared Variance), which establishes the discriminant validity.

TABLE 1.1
Factors Extracted by EFA

Factors and Items	Factor Loading	Cronbach's Alpha	McDonald's Omega
Limited Interaction		0.965	0.958
LI_1	0.939		
LI_5	0.935		
LI_3	0.917		
LI_4	0.905		
LI_2	0.899		
Technical Constraints		0.949	0.949
TC_2	0.921		
TC_3	0.905		
TC_1	0.903		
TC_4	0.871		
Ineffective Evaluation		0.923	0.923
IE_1	0.883		
IE_3	0.858		
IE_2	0.845		
IE_4	0.837		
Bounded Teaching		0.888	0.888
BT_4	0.891		
BT_3	0.852		
BT_2	0.841		
BT_1	0.818		
Non-Sustainability		0.938	0.942
NSUS_1	0.871		
NSUS_2	0.858		
NSUS_3	0.836		

TABLE 1.2
Model Fit Indices – Social and Cultural Determinants

Model Fit Indices	Values as per Model	Criteria	Criteria Fulfilled
CIMIN/DF	1.426	≤3.00	Yes
CFI	0.983	≥0.95	Yes
GFI	0.899	≥0.90	Yes
AGFI	0.866	≥0.80	Yes
RMSEA	0.047	≤0.10	Yes
P Value	0.000	≤0.05	Yes

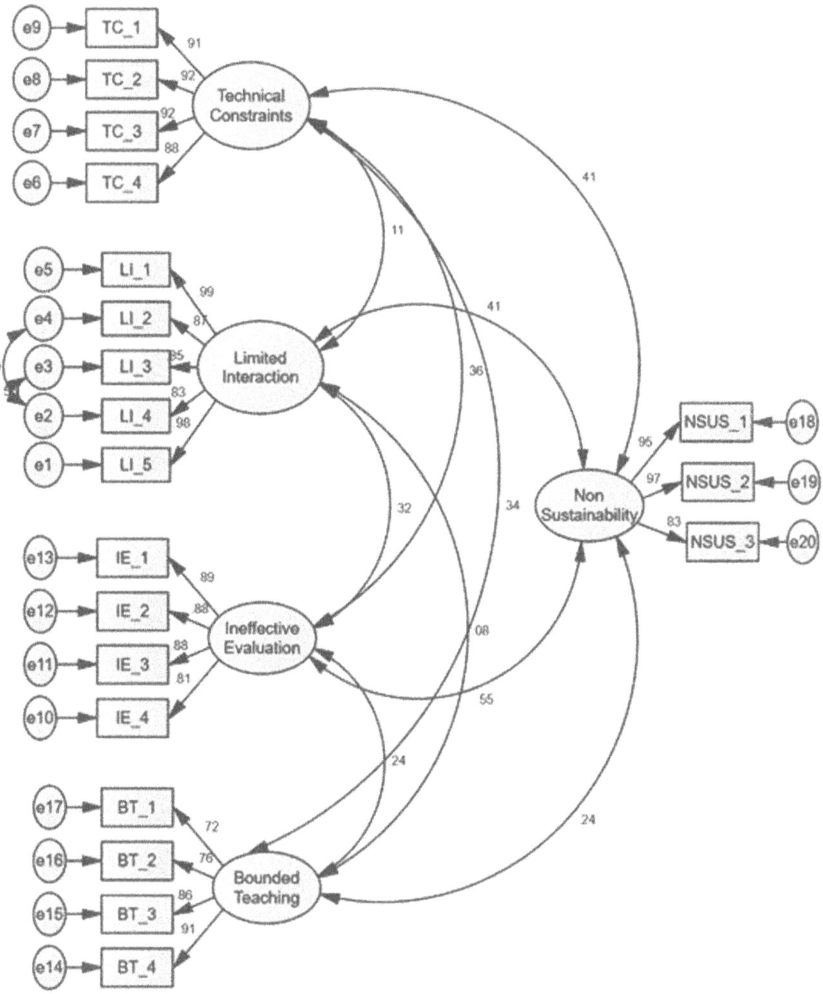

FIGURE 1.2 Measurement model for sustainability of online teaching.

TABLE 1.3
Constructs of Issues and Challenges of Online Teaching

Constructs of Issues and Challenges of Online Teaching	Reliability (CR)	Average Variance Estimate (AVE)	Maximum Shared Variance (MSV)	Average Shared Variance (ASV)	Convergent Validity	Discriminant Validity
Technical Constraints	0.950	0.826	0.165	0.105	Yes	Yes
Limited Interaction	0.959	0.824	0.166	0.073	Yes	Yes
Bounded Teaching	0.890	0.671	0.114	0.059	Yes	Yes
Ineffective Evaluation	0.924	0.753	0.298	0.148	Yes	Yes
Non-Sustainability	0.940	0.840	0.298	0.171	Yes	Yes

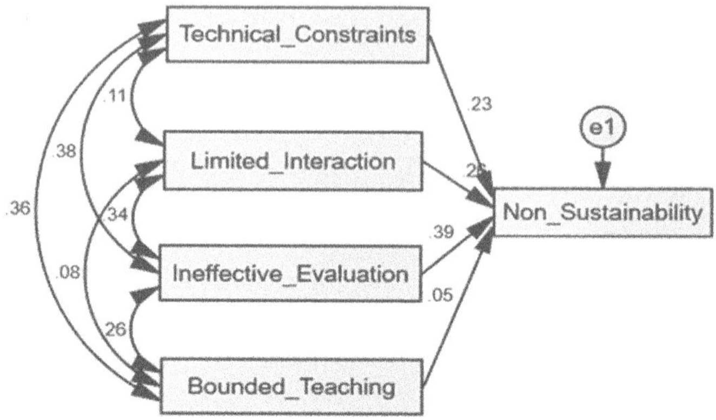

FIGURE 1.3 Impact of online learning challenges on the non-sustainability of online teaching.

1.9.3 Structural Model

Figure 1.3 shows the structural model for the impact of online learning challenges on the non-sustainability of online teaching.

Table 1.4 shows the results of structural equation modeling regarding the impact of the various issues and challenges of online teaching on the Non-Sustainability of online teaching. It is found that there is a significant impact of Technical Constraints, Limited Interaction, and Ineffective Evaluation on the Non-Sustainability of Online teaching as the P values of the mentioned relationships are below 0.05. Briefly, it may be concluded that these factors are contributing significantly if online teaching is found non-sustainable tomorrow. However, there is no significant impact of "Bounded Teaching" on the Non-Sustainability of online teaching (the P value is 0.429, which is more than 0.05). In the opinion of teachers, the bounded nature of teaching is not significantly correlated with the non-sustainability of online teaching.

TABLE 1.4
Impact of Online Learning Challenges on the Non-Sustainability of Online Teaching

Causal Relationships	Estimate	S.E.	C.R.	P
Non-Sustainability <--- Technical Constraints	0.236	0.063	3.739	***
Non-Sustainability <--- Limited Interaction	0.232	0.052	4.465	***
Non-Sustainability <--- Ineffective Evaluation	0.405	0.065	6.226	***
Non-Sustainability<--- Bounded Teaching	0.048	0.061	0.790	0.429

Note: Estimate – standardized beta, SE – standard error, CR – critical ratio (t – value in regression).

1.10 CONCLUSION

Technology is used as a magic charm to heal many sectors in the present world; however, India lacked its infusion in the educational ground as people have always followed the traditional path for it. Unlike other countries, India has relied on a conventional approach to deliver quality education. In spite of the introduction of the Internet, online education has not been used to provide education in a full-fledged manner (Toquero 2020). Technological momentum has altered all sectors of operations. The academic fraternity across the world is adopting e-learning. Education, being a mandatory area, was shifted to adopt a digital process in order to maintain consistency and ensure academic outcomes. Even when other nations are easily grasping it, India has difficulties in adapting to e-learning (Quennerstedt 2019).

The outcome is based on the results derived from an empirical study, in which teachers who use online platforms to teach students were the respondents. Such platforms include Zoom, Google Meet, Microsoft Teams, JioMeet, and Cisco Webex. It has been found from the study that if, in the future, online teaching becomes non-sustainable, then the major factors contributing to it would the Technical Challenges, Limited Interaction, and Ineffective Evaluation of the students. Though the teacher also feels bounded while teaching online, this factor will not influence the sustainability of online teaching.

The extant literature has revealed that online learning has been a great alternative to a physical mode of learning. Even though it cannot completely replace the physical touch and feel of it, it is highly economical if students are offered adequate digital instruments. The biggest constraint that blocks its smooth functioning is the inadequate technical competency of teachers who fail to evaluate students' doubts effectively. It makes assessment and validation difficult during exams as there is a high incidence of cheating. It also restricts students and teachers from communicating freely as they would in a physical setting. Teachers feel isolated, as it is difficult to implement engaging strategies to get the attention of students, virtually. Also, it destroys the creativity that could have otherwise been portrayed lucidly in a physical environment.

Apart from all these factors, Internet errors and electricity issues can hamper online classes and finally affect the suitability of online teaching. Therefore, there is a long way to go for teachers and students in the country to master the art of online education and accept its presence formally. To ensure that the online classrooms are interactive and the learning is student centric, the educators need to create an ambience that can promote learning and can be adjusted accordingly to the requirements, capabilities, and interests of each student, therefore motivating students to be dynamic, to associate and create more room for online learning.

1.11 IMPLICATIONS OF THE STUDY

There has been an exponential rise in the implementation of online modes of teaching and training in the past few years. Online teaching has a wide continuum that ranges from recorded lectures to live online interaction. Teachers and educational institutions have put in their best efforts to make online teaching the closest alternative to physical classroom teaching. However, it is not easy to claim that online teaching can replace the normal teaching format. The findings of the present study clearly reflect that there are many important aspects that may cause problems to the sustainability of online teaching. The policymakers must keep in mind that an online teaching format requires many preparations to avoid some serious issues and challenges. The technical glitches must be removed first and the stakeholders should be trained for the technology used while teaching online. The participants in online teaching must be more agile and find alternative ways of evaluation and exams that are more dynamic than typical online quizzes and assignments. Further, it is also recommended that the stalwarts of the education section give teachers all the rights and privacy while teaching and ensure that their classes are not recorded without their permission. This will enable teachers to conduct their classes with a free mind.

There is a need to ensure the best interaction between teachers and students by removing all technical barriers and to set up a robust evaluation system; otherwise, online teaching will only remain as an assistant to physical classroom teaching and will never become a full-fledged alternative of it.

1.12 LIMITATIONS OF THE STUDY

The present study has captured the responses from teachers only, and the conclusions have been derived based on the responses from the teachers. The study is based on primary data only that was collected with the help of a structured questionnaire; hence, to some extent, the biases of respondents may not be avoided. The study has been conducted in the context of courses and for those teachers who only teach the subjects that do not have laboratory requirements; hence the results of the study may not be generalized for all types of courses and programs.

1.13 SCOPE FOR FUTURE RESEARCH

Online teaching and learning is a vast area of research. Future studies that extend the present study may be conducted on the impact of Internet speed and availability on teachers' online classroom experience. Further, the studies may be conducted to explore the best and robust evaluation system for students that is a proximate alternative to the pen-paper exam. In addition to this, studies may also be carried out on comparing the sustainability aspect for various types of courses – pure theory, semi-practical, and fully practical. Future researchers may also compare online and offline teaching in the light of affordability, convenience, and effectiveness.

REFERENCES

Aboagye, Emmanuel, Joseph Anthony Yawson, and Kofi Nyantakyi Appiah. 2020. "COVID-19 and E-Learning: The Challenges of Students in Tertiary Institutions." *Social Education Research* 2 (1): 1–8. https://doi.org/10.37256/ser.212021422, http://ojs.wiserpub.com/index.php/SER/article/view/ser.212021422/282.

Ali, Wahab. 2020. "Online, and Remote Learning in Higher Education Institutes: A Necessity in Light of COVID-19 Pandemic." *Higher Education Studies* 10 (3): 16–25. https://doi.org/10.5539/hes.v10n3p16. http://www.ccsenet.org/journal/index.php/hes/article/view/0/42784.

Bao, Wei. 2020. "COVID-19 and Online Teaching in Higher Education: A Case Study of Peking University." *Human Behavior and Emerging Technologies* 2 (2): 113–115. DOI: 10.1002/hbe2.191.

Barab, Sasha A., Kurt D. Squire, and William Dueber. 2000. "Supporting Authenticity Through Participatory Learning." *Educational Technology Research and Development* 48 (2): 37–62. https://sashabarab.org/article/supporting-authenticity-through-participatory-learning/

Chang, Chiu-Lan and Ming Fang. 2020. "E-Learning and Online Instructions of Higher Education During the 2019 Novel Corona Virus Diseases (COVID-19) Epidemic." *Journal of Physics* 1574: 012166. DOI: 10.1088/1742-6596/1574/1/012166.

Christian, Michael, Edi Purwant, and Suryo Wibowo. 2020. "Techno Stress Creators on Teaching Performance of Private Universities in Jakarta During Covid-19 Pandemic." *Technology Reports of Kansai University* 62 (6): 2799. https://www.researchgate.net/publication/343230929_Technostress_Creators_on_Teaching_Performance_of_Private_Universities_in_Jakarta_During_Covid19_Pandemic/link/5f1ea8b892851cd5fa4b268c/download.

Crawford, Joseph, Kerryn Butler-Henderson, Jurgen Rudolph, Bashar Malkawi, Matt Glowatz, Rob Burton, Paola A. Magni, and Sophia Lam. 2020. "COVID-19: 20 Countries' Higher Education Intra-Period Digital Pedagogy Responses." *Journal of Applied Teaching and Learning* 3 (1): 1–20. DOI: https://doi.org/10.37074/jalt.2020.3.1.7.

Dhawan, Shivangi. 2020. "Online Learning: A Panacea in the Time of COVID-19 Crisis." *Journal of Educational Technology Systems*: 1–18. doi:10.1177/0047239520934018.

Flores, Maria Assunção. 2020. "Preparing Teachers to Teach in Complex Settings: Opportunities for Professional Learning and Development." *European Journal of Teacher Education* 43: 1–4. https://doi.org/10.1080/02619768.2020.1771895.

George, Marcus L. 2020. "Effective Teaching and Examination Strategies for Undergraduate Learning During COVID-19 School Restrictions." *Journal of Educational Technology Systems*: 1–26. https://doi.org/10.1177/0047239520934017.

Graham, Charles. R., and Melanie Misanchuk. 2004. "Computer-mediated learning groups: benefits and challenges to using group work in online learning environments." In T. S. Roberts (Ed.), *Online collaborative learning: theory and practice*: 181–202. Hershey, PA: Idea Group.

Gratz, Erin and Lisa Looney. 2020. "Faculty Resistance to Change: An Examination of Motivators and Barriers to Teaching Online in Higher Education." *International Journal of Online Pedagogy and Course Design* 10 (1): 1–14. DOI: 10.4018/IJOPCD.2020010101.

Hatlevik, Ida K. R. and Ove E. Hatlevik. 2018. "Examining the Relationship Between Teachers' ICT Self-Efficacy for Educational Purposes, Collegial Collaboration, Lack of Facilitation and the Use of ICT in Teaching Practice." *Frontiers in Psychology* 9: 1–8. https://doi.org/10.3389/fpsyg.2018.00935.

Hodges, Charles, Stephanie Moore, Barb Lockee, Torrey Trust, and Aaron Bond. 2020. "The Difference between Emergency Remote Teaching and Online Learning." https://er.educause.edu/articles/2020/3/the-difference-between-emergency-remote-teaching-and-online-learning.

Jaques, David and Gilly Salmon. 2007. *Learning in groups: a handbook for face-to-face and online environments*. Abingdon, UK: Routledge.

Kebritchi, Mansureh, Angie Lipschuetz, and Lilia Santiague. 2017. "Issues and Challenges for Teaching Successful Online Courses in Higher Education." *Journal of Educational Technology Systems* 46 (1): 4–29. DOI: 10.1177/0047239516661713.

Lee, Joyce. 2020. "Mental Health Effects of School Closures During COVID-19." *The Lancet Child & Adolescent Health* 4 (6): 421. DOI: 10.1016/S2352-4642(20)30109-7.

Limperos, Anthony M., Marjorie M. Buckner, Renee Kaufmann, and Brandi N. Frisby. 2015. "Online Teaching and Technological Affordances: An Experimental Investigation into the Impact of Modality and Clarity on Perceived and Actual Learning." *Computers and Education* 83: 1–9. http://dx.doi.org/10.1016/j.compedu.2014.12.015.

MacPhail, Ann, Deborah Tannehill, and Zuleyha Avsar. 2019. *European physical education teacher education practices: initial, induction, and professional development.* Maidenhead: Meyer & Meyer Sport.

McCallum, Faye and Deborah Price. 2010. "Well Teachers, Well Students." *The Journal of Student Wellbeing* 4 (1): 19–34. DOI: 10.21913/JSW.v4i1.599.

Moll, Rachel and Wendy Nielsen. 2017. "Development and Validation of a Social Media and Science Learning Survey." *International Journal of Science Education, Part B: Communication and Public Engagement* 7: 14–30. DOI: 10.1080/21548455.2016.1161255.

Mooij, Tom and Ed Smeets. 2001. "Modeling and Supporting ICT Implementation in Secondary Schools Computers and Education" 36: 265–281. DOI: 10.1016/S0360-1315(00)00068-3.

Morris, Libby V. and Catherine L. Finnegan. 2008. "Best Practices in Predicting and Encouraging Student Persistence and Achievement." *Journal of College Students Retention* 10 (1) 55–64.

Quennerstedt, Michael. 2019. "Physical Education and the Art of Teaching: Transformative Learning and Teaching in Physical Education and Sports Pedagogy." *Sport, Education and Society* 24 (6): 611–623. DOI: 10.1080/13573322.2019.1574731.

Romiszowski, A. and Mason, R. 2004. "Computer-mediated communication." In D. H. Jonassen (Ed.), *Handbook of research for educational communications and technology*: 397–431. New Jersey, NJ: Lawrence Erlbaum. http://teresarichards.weebly.com/uploads/1/1/6/2/1162870/research_article_computer_mediated_communication.pdf.

Sangrà, A., Dimitrios Vlachopoulos, and Nati Cabrera. 2012. "Building an Inclusive Definition for E-Learning: An Approach to its Conceptual Framework." *The International Review of Research in Open and Distance Learning* 13 (2): 145–159. https://www.researchgate.net/publication/268980149_Building_an_inclusive_definition_for_e-learning_an_approach_to_its_conceptual_framework.

Sharma, Sushil K. and Fred L. Kitchens. 2004. "Web Services Architecture for M-Learning." *International Journal of Mobile Communications* 2 (1): 203–216. https://silo.tips/download/web-services-architecture-for-m-learning.

Shenoy, Veena, Sheetal Mahendra, and Navita Vijay. 2020. "COVID 19 Lockdown Technology Adaption, Teaching, Learning, Students Engagement and Faculty Experience." *Mukt Shabd Journal* IX (IV): 698–702. http://shabdbooks.com/gallery/78-april2020.pdf.

Simones, Lilian, Franziska Schroeder, and Mathew Rodger. 2015. "Categorizations of Physical Gesture in Piano Teaching: A Preliminary Enquiry." *Psychology of Music* 43 (1): 103–121. https://doi.org/10.1177/0305735613498918.

Singh, Vandana and Alexander Thurman. 2019. "How Many Ways Can We Define Online Learning? A Systematic Literature Review of Definitions of Online Learning (1988–2018)." *American Journal of Distance Education* 33 (4): 289–306. https://doi.org/10.1080/08923647.2019.1663082.

Takala, Annina and Kati Korhonen-Yrjänheikki. 2019. "A Decade of Finnish Engineering Education for Sustainable Development." *International Journal of Sustainability in Higher Education* 20 (1): 170–186. https://doi.org/10.1108/IJSHE-07-2018-0132.

Toquero, Cathy Mae. 2020. "Challenges and Opportunities for Higher Education Amid the Covid-19 Pandemic: The Philippine Context." *Pedagogical Research* 5 (4): 1–5. https://doi.org/10.29333/pr/7947.

Ungar, Orit Avidov and Alona Forkosh Baruch. 2016. "Perceptions of Teacher Educators Regarding ICT Implementation." *Interdisciplinary Journal of e-Skills and Life Long Learning* 12: 279–296. http://www.ijello.org/Volume12/IJELLv12p279-296Ungar2793.pdf.

Vonderwell, Selma and Sajit Zachariah. 2005. "Factors that Influence Participation in Online Learning." *Journal of Research on Technology in Education* 38 (2): 213–230. DOI: 10.1080/15391523.2005.10782457.

Wicks, Matthew. 2010. "A National Primer on K-12 Online Learning." https://aurora-institute.org/wp-content/uploads/iNCL_NationalPrimerv22010-web1.pdf.

APPENDIX

Kaiser-Meyer-Olkin and Bartlett's Test

KMO of Sampling Adequacy		0.857
Bartlett's Test of Sphericity	**Approx. Chi-Square**	3959.400
	df	190
	Sig.	0.000

Total Variance Explained

	Initial Eigenvalues			Extraction Sums of Squared Loadings			Rotation Sums of Squared Loadings		
Component	**Total**	**% of Variance**	**Cumulative %**	**Total**	**% of Variance**	**Cumulative %**	**Total**	**% of Variance**	**Cumulative %**
1	7.257	36.286	36.286	7.257	36.286	36.286	4.477	22.387	22.387
2	3.935	19.673	55.958	3.935	19.673	55.958	3.508	17.538	39.926
3	2.347	11.737	67.695	2.347	11.737	67.695	3.300	16.499	56.425
4	1.996	9.981	77.676	1.996	9.981	77.676	3.052	15.261	71.686
5	1.331	6.655	84.331	1.331	6.655	84.331	2.529	12.646	84.331
6	0.470	2.352	86.683						
7	0.406	2.029	88.712						
8	0.323	1.615	90.327						
9	0.300	1.500	91.827						
10	0.253	1.265	93.092						
11	0.244	1.219	94.311						
12	0.203	1.017	95.328						
13	0.173	0.864	96.192						
14	0.168	0.842	97.034						
15	0.149	0.745	97.779						
16	0.135	0.674	98.452						
17	0.116	0.581	99.034						
18	0.103	0.516	99.550						
19	0.069	0.343	99.893						
20	0.021	0.107	100.000						

2 Effectiveness of Online Learning and Face-to-Face Teaching Pedagogy

Biswa Mohana Jena, S. L. Gupta, and Niraj Mishra

CONTENTS

DOI: 10.1201/9781003132097-2

2.1 INTRODUCTION

Online teaching is having a significant impact on the teaching sector with adaptation to various technological developments. The teaching techniques and technologies involved in the teaching are taking a step forward with the implementation of online teaching. Online teaching is not new, and it has already been part of various curriculums across the world, which has got the significance of its own. The technological developments have taken online teaching to a whole new level, which has a variety of tools to adapt and make the students attentive in classes. Online teaching is making its wings spread over the education system across the world, which makes the dissemination of knowledge at ease. The new era has come into being with the evolution of online teaching in the country. The working of online teaching is dependent on various technological factors, and it has to be available with the recipients to make full use of the online teaching.

"In higher education, especially in the format of blended learning, e-learning is gaining more and more influence, and this new style of traditional teaching and learning can be practiced in many ways. Many studies have compared face-to-face teaching to online learning and/or hybrid learning to try to decide which of the formats provides, for instance, the best learning outcome, produces the most satisfied students, or has the highest rate of course completion. However, these findings also suggest that teaching and learning are influenced by more than just the teaching style. Many variables play significant roles, and this literature review will further discuss some of them. In the literature examined, the dimensions that are reported to have a significant impact on student learning in professional programs delivered by integrated or online formats on educator roles and relationships include the educator's role in establishing a strong presence of educators in online settings and in fostering positive relationships in online learning communities" (Nortvig, Petersen, and Balle 2018).

Face-to-face teaching has been the structure that has been followed for the dissemination of knowledge in the colleges and universities in the country. This technique has been considered the significant one, and it has been followed across the world with high success rates. This school of education has been long standing in the field of teaching for imparting knowledge. Face-to-face teaching has its pros and cons, with the effectiveness measured with assessment. This teaching method is vastly used at all educational institutions, which makes them prominent in the field of teaching. Face-to-face learning has been long standing and needs to be reformed with the advent of online teaching.

The blended teaching method with the involvement of online teaching and face-to-face teaching is taking shape in developed countries for making more effective dissemination of knowledge than the existing one, namely face-to-face learning.

The need for blended teaching can be substantiated only with the studies that are proving the beneficial effects of the online teaching methods. The steps for making things work in favor of online teaching to bring it into mainline teaching can be succeeded based on the effectiveness of online teaching from the student's points of view, including collaborative learning, research projects reflection, and quiz during classes.

"The necessary credits to obtain a post-secondary degree are intimately linked to this cost and postsecondary education issue. Traditionally, students have to earn most of an institution's college credits before they obtain bachelor's degrees at that institution. The point of disagreement is how online classes will play a role in granting credits or certificates, and with some online classes, many educators linked to online learning expect that there will be credit equivalence. For example, for some online courses, he collaborated with the American Council on Education to suggest credit equivalency. This initiative aims to increase the graduation rate, reduce the time needed to obtain a degree, reduce postsecondary education costs, and provide non-traditional students with more access. Five online courses for college credit were accredited by the American Council on Education as of 2013. There is, however, uncertainty about whether the recommendation would be approved by colleges and there is also concern about the dilution of a typical degree due to the change. Last but not least, there is the possibility that online learning can provide anybody, wherever, and anywhere with world-class education as long as they have access to the Internet. Khan Academy, Udacity, edX, and Coursera are some of the most prominent websites and businesses that are founded on this concept, and many well-respected scholars and entrepreneurs have high standards and expectations for online learning, particularly for large open online courses. The success of the online format in educating students is fundamental to this unique gain. If online learning is usually less efficient than the traditional face-to-face format, some of the alleged statements and advantages of online learning listed above are highly suspect. The root of the issue lies in this, the fundamental issue of online learning and the subject of this paper: the efficacy of the online format relative to the conventional format in educating students. The good, negative, and mixed, and null effects of the efficacy of online learning compared to the conventional format will be analyzed to resolve this issue" (Nguyen 2015).

2.2 REVIEW OF LITERATURE

"Online learning has reshaped instruction, giving learners, despite their physical position, the ability to achieve their academic objectives. Currently, 6.4 million students take a minimum of one distance education course, representing 32% of all teaching enrolments (Allen et al. 2017) of substantial significance is that 36% of the upper education population and 31% of the undergraduate online population serve the community college sector (McFarland et al. 2017). Recently, with President Obama's need for community colleges to graduate another five million students by 2020, the relevance of community colleges has been brought to the forefront" (The White House Summit on Community Colleges 2011).

2.2.1 Online Learning

The use of online learning has continued to grow globally in higher education institutions (HEIs). The early forms of distance education focused on correspondence style courses, video conferencing, and educational television programs were surpassed by online learning in the 21st century. Instead, today's higher education technology consists of internet courses, such as Massive Online Open Courses (MOOCs) that offer Internet courses.

Increasingly, college and university teachers integrate online instruments into face-to-face teaching methods, so that blended teaching is expected to become "the new traditional model" However, less than 5% of higher education mixing scholarships pursue the academic practice. The findings of a systematic review of literature on the adoption and usage of online resources for face-to-face training by faculty members are reported in this discussion. Six literature-wide influences are identified: interactions of faculty members with technology, academic workload, institutional environment, student interactions, attitudes and beliefs of the teacher, and professional development opportunities. Literature strengths and constraints and future directions for research on socio-technical instructional systems are identified. Over the past two decades, the use of the internet and information communication technology (ICT) has become a common norm in all areas of higher education. Higher education institutions are increasingly relying on ICT and Internet media to support their leadership and administrative functions. In higher education, emerging technologies such as online tools complicate organizational practices such as instructional work.

2.2.2 Global Access

Global access to higher education courses, web-based applications, interactive programs, and more developed virtual learning environments such as Moodle or Blackboard are being used on a large-scale. This rapid spread in higher education online learning methods has led to various gaps in delivery and uptake. Educators are now faced with a variety of teaching techniques to choose from, such as online, face-to-face, or mixed learning. Several literature reviews are available that consider the subject of higher education online and blended learning (Mccutcheon et al. 2015).

More recently, 12 studies conducted in 2013–2014 that compared learning in an online or hybrid format to learning in conventional or face-to-face environments were evaluated by meta-analytic information collected by Means et al. (2013). The findings of this study reveal similar conclusions for the U.S. Study of the Department of Education; students in online and hybrid formats performed better than students in conventional versions of equivalent courses. In addition to online humanities classes, Wu suggested several required issues for future research, as well as calculating additional reliable short- and long-run performance overall and finding results. Typically, students taking part in online and conventional courses face entirely different goals and characteristics, making it more difficult to verify the actual results of the two types of courses. At this point, long-run learning effects such as the impact on the field are still unknown, especially in the fields of social sciences and humanities.

2.2.3 Teaching Pedagogy

"In particular, students indicated that the web class was more or more challenging due to the conventional or face-to-face version of the course and that the success of the students within the course led to their willingness to register for future online courses in the program" (Daniel et al. 2016).

"Even if study and practice in the pedagogy of online linguistics have been quite slow to evolve, in most universities today, online teaching has become a major mode of teaching. About 30 percent of college undergraduates already engage in their online coursework, and this figure will certainly continue to grow. An increasing body of research on online pedagogy has been conducted, much of which focuses on the ongoing debate about the quality of online versus F2F courses. Online courses provide material that students can explore and learn, accompanied by video lectures and podcasts that can be replayed as needed; those students ask and answer questions with the teacher and each other archived in online forums for later reference, and students take asynchronous self-assessments and teacher-graded assessments that help them evaluate their learning. An empirical study has been carried out to compare the efficacy of the same course delivered across specific disciplines in both online and F2F formats. Therefore, it is not clear if the needs of similar student populations are fulfilled by F2F and online courses. It is not certain that the sort of provocative back-and-forth that sometimes occurs during in-class debates is accomplished through asynchronous conversation. And it is uncertain if some disciplines and subjects within disciplines are better suited for one mode of delivery over another, mainly German for the topic of this paper. However, to decide whether such reluctance regarding the online distribution of certain subjects is warranted, empirical research is required" (Johnson and Palmer 2015).

2.2.4 E-learning Opportunities

"In one research, pre-service lecturers experiencing a lot of merging and e-learning opportunities in their teacher training program indicated that they felt encouraged in their learning during the job with targeted instructor support and scaffolding into the web modules. However, a small proportion of students tended to complain during this same study that they did not like online formats and required more transparency by actually engaging in a face-to-face category (Chigeza and Halbert 2014). This is also a critical factor because it means that students who are most relaxed and benefit most from online learning experiences are students who are comfortable and choose to participate in online courses over older versions" (Jaggars 2014).

The authors conclude that online learning inside this teacher education program is an associate equally effective platform for getting ready teacher candidates because the F2F degree when measuring candidate-learning outcomes using edTPA (Petty, Good, and Heafner 2019). In this study, researchers found that teachers who had graduated from an online program were as or more capable than teachers who had graduated from traditional programs as measured by their students' achievement on a literacy assessment. Further exploration into the impact of online programs upon teaching practice is needed to examine this potential.

2.2.5 Online Learning Platforms

"With the use of online learning platforms and social networks, various tertiary institutions have implemented digital learning. The research on the effectiveness of such platforms, however, is confused, as is the field itself, partly because of the rapidly changing technology and also because of a lack of clarification on what constitutes a platform for learning. The role of blended and online learning, including 'flipped learning' as well as the use of different systems or 'platforms' of learning management in tertiary education is the subject of many studies among scholars. There is growing interest in how online or cloud-based resources and pedagogy applied to help such tools could generate greater engagement and interaction between students in tertiary education and between students and their teachers. Although some scholars are avowed 'techno-optimists' in that they conclude that the use of such technologies would revolutionize higher education by increasing student interaction, other scholars are more reluctant to democratize access and enhance learning, proposing only minimal or even no use of social media tools in the classroom due to the limited effect on student learning, the potential for spreading district learning. Various perspectives on the efficacy of online learning have contributed to greater confusion on how these resources should be deployed by higher education institutions (or even whether they should be deployed at all) and how educators might use the appetite of students for social media to boost learning. In order to capitalize on these changes, it is necessary to carefully examine how pedagogical practices need to shift as tertiary institutions rely more heavily on digital media to structure students' learning experiences. While studies of these pedagogies are still in their infancy, we believe that the above framework can provide a guide for assessing the effectiveness of a learning platform and how best to use it in a tertiary environment. Teacher educators should also be motivated to be self-reflexive and focused to investigate their use of technology and its effect on students" (Heggart and Yoo 2018).

"In order to complete the project, students in the online segment were required to use email, Google channels, message boards, or other chosen interaction methods, but it was never a necessary expectation for students to physically meet to discuss the project. In contrast to conventional ways of learning, future research could benefit from exploring peer contact and collaboration in both synchronous and asynchronous online environments (Daniel et al. 2016), (Petty, Good, and Heafner 2019). All other rating score metrics were reasonably consistent between the two iterations of the class, including contact with the course teacher, course material, and feedback."

"Online education is one of the newest modalities for distance learning that helps more students to access higher education (Allen et al. 2017) and its growth continues to outpace total enrollment in higher education (Allen et al. 2017)."

2.3 MOOC AND ONLINE LEARNING

A couple of years ago, the Massive Online Open Course (MOOC) wave declared the demise of conventional universities. This has not occurred. Now several voices are declaring that MOOCs are gone because that hope has not been fulfilled. We assume that the MOOC format has a lot of potential, even more, profound than other online teaching pedagogies as MOOCs are used with short videos, immersive experiments and simulations, and social interaction. One form of taking advantage of flipping the classroom is content MOOCs. Students view lectures online in a flipped classroom and use touch time with teachers for interaction that cannot be replaced by a computer: conversations, misconceptions solved, work checked, etc. But the only way to continue is the flipped classroom model. There are also other ways that technology can be paired with face-to-face technology. Finding the best ways to integrate face-to-face and digital learning in meaningful, efficient, and engaging ways is the great promise of the use of technology in education. We have sought to progress in the right direction with these interactions (Delgado Kloos et al. 2015).

"As both respected pedagogical approaches to teaching are online learning and peer teaching, it was agreed to merge the two approaches for one blended learning course. To substitute several compulsory face-to-face tutorial sessions and a written group evaluation assignment, this strategy was adopted. Based on the alphabetical listing of student names, all students enrolled in this course were assigned to small online peer learning groups manually. In groups of 6–8 students, students were allocated so that they could complete exercises in community online forum sessions and evaluation assignments. The Moodle course management system was used as the technology framework in which peer learning activities and other required course material were accessed by the students. Interestingly, students showed a preference for mixed learning, where they used both face-to-face and online learning as learning methods together. It has been shown that learners learn in various ways, with several learners choosing to learn in different ways, using all learning modalities. This poses challenges for scholars who need to ensure that teaching methods address the needs of a wide variety of learners, while still covering the material needed to train eligible graduates. One way of achieving diversity in teaching approaches to account for diversity in learning needs is to use various teaching methods, including online learning" (Raymond et al. 2016).

2.4 FACE-TO-FACE TRAINING

"Under these conditions, the temptation to equate online learning to face-to-face training would be great. In reality, an article in the Higher Education Chronicle has already called for a 'grand experiment' to do just that. However, this is an extremely controversial idea. The politics of any such discussion must, first and foremost, be remembered. Online learning can become a politicized concept that depending on the argument that someone wishes to move forward, can take on any number of meanings. Without paying due attention to the fact that institutions will make different choices and spend differently, the concept of blended learning was dragged into political agendas, resulting in widely varying solutions and outcomes from one institution to another. We aim to advance some careful distinctions with some of that hindsight as wisdom, which we hope will guide the assessments and reflections that will surely arise from this mass move by colleges and universities. Despite studies suggesting otherwise, online learning holds a stigma of being of poorer quality than face-to-face learning. Such rushed online steps by too many institutions at once could seal the image of online learning as a poor choice because, in fact, under these circumstances, no one making the transition to online teaching would be designed to take full advantage of the online format's affordances and possibilities" (Hodges et al. 2020).

In 2000, at least one distance education course was taken by 8% of community college students, doubling to 16% in 2003, 20% in 2007 (Lewis and Parsad 2009), and 32% in 2016 (McFarland et al. 2017).

2.5 RESEARCH GAPS

The review of literature has given a critical insight into the existing aspects of online teaching. The variety of studies in relation to online teaching has an explanation of the benefits and drawbacks of online teaching. Numerous studies discuss the working of online teaching and various technologies that have been the foundation for the operation of online teaching. The research studies of the past have a lot of concentration on the technological tools that are in favor of online teaching and how they have an effect on the students in listening in the classes. Online teaching helps to make the students attentive to the class with the adoption of technological tools from the teacher's side. The working of online teaching has been completely documented in the literature survey. On the other hand, the effectiveness of face-to-face learning has been assessed in various studies with numerous combinations of the variables. The studies in the literature have failed to capture the comparative assessment regarding which teaching method has more effectiveness in delivering the content to the students. There is also another missing link on the comparative assessment to identify in which areas online teaching is more effective in comparison with face-to-face learning. The current study focuses on the research gap of making a comparative assessment in understanding the effectiveness of online teaching over the existing pattern of face-to-face teaching. The key challenges of higher education online are alliances, classroom experience, experiments in labs, faculty as role models, and research on student interests.

2.6 SIGNIFICANCE OF THE STUDY

The current desperate times have rendered the need for assessing the various modes of delivering education to students. Face-to-face learning has practical difficulties at times when there is a lack of personal contact between the students and teachers (Ben-David, Kushilevitz, and Mansour 1997). The most prominent area among the various sectors that contribute to the growth of online education is the technological aspects. The pandemic has given a new path for the growth of online education with a minimal classroom atmosphere for the learners (Singh, Rylander, and Mims 2012). The path of online education will be on the rise in the coming decades based on the various technological developments and atmosphere-oriented health concerns. The working of online education will have a significant impact on the offline mode of education, and the differences have to be balanced out by making an exact assessment of facts (Singh 2003). Some mutual pros and cons have to be taken care of in implementing online education and phasing out offline education. These assessments regarding the various pros and cons as well as implementation criteria have to be determined in examining the impact of adopting online education (Lim and Han 2020). The study, on that note, helps to examine the effectiveness of online teaching in the process of being an effective teaching model for imparting education to replace face-to-face learning. There is also a need to know these impacts from the students' points of view to assess the real effectiveness of online education, which is being a substitute for face-to-face learning in tough times.

2.7 RESEARCH QUESTIONS

a. What are the areas of effectiveness that make online teaching more effective than the existing pattern of conventional teaching?
b. What are the various problems from the viewpoint of students that affect the efficiency of online learning?

2.8 OBJECTIVES OF THE STUDY

a. To examine the effects of attributes of online teaching available with the students on the factors of online learning.
b. To comparatively identify the dimensions of online learning that prove to be beneficial for the students.
c. To identify the various dimensions of problems that affect the efficiency of online learning.

2.9 RESEARCH METHODOLOGY

The study adopts the applied research method, which aims to give a solution to the applied problem of identifying a better source of learning. The study assesses the problem from the students' points of view and gives a scientific solution for the research problem. The research methodology has adopted a step-by-step process using various methods for testing hypotheses to identify the relationship between the attributes of online teaching and the factors of online teaching. It also focuses on making a comparative assessment of the effectiveness of online teaching with that of face-to-face teaching.

2.10 SAMPLING DESIGN

Undergraduate and post-graduate students and research scholars comprised the population of the study. The sample was selected from a pool of students from the area of Delhi. The number of samples was fixed to be 313 students based on the infinite sampling calculator. The sample of the study was selected using the non-random snowball sampling technique. The reason for the adoption of a non-random technique is that the total number of students has the feature of being fixed, which rules out the option of using a random sampling technique. Therefore, a non-random sampling technique was used for the selection of the sample for the study.

2.11 TYPE OF DATA AND DATA COLLECTION INSTRUMENT

The study has been conducted with the help of primary data that was collected using a mailed questionnaire. The data collection instrument was divided into three parts: the first part consisted of demographic information; the second part consisted of comparative assessment; and the third part consisted of an analysis of the problems of online teaching. The questionnaire built for the study was tested for reliability based on the pilot survey, which consisted of 60 students. The reliability of the study was found to be highly satisfactory, and the metrics for the same are given in the following table:

Reliability Statistics			
Comparative Assessment		Problems of Online Teaching	
Cronbach's Alpha	No. of Items	Cronbach's Alpha	No. of Items
0.965	23	0.953	15

2.11.1 Statistical Tools Used

The following are the various statistical tools used in the study for analyzing the collected data.

a. Mean
b. Standard deviation
c. Range
d. Chi-square test
e. Exploratory factor analysis.

2.11.1.1 Hypotheses

H_0: There is no significant association between the attributes of online teaching and superior factors of online teaching over face-to-face teaching.

H_1: There is a significant association between the attributes of online teaching and superior factors of online teaching over face-to-face teaching.

2.11.2 Analysis and Interpretations

The data was collected using the online questionnaire and was examined using various statistical tools that will help to achieve the objectives of the study. The demographics and attributes of online learning play a significant role in examining the effectiveness of the delivery of the course content. The various demographic profiles and attributes of online learning are explained with the help of the following content.

2.11.2.1 Demographics and Attributes of Online Learning

The demographic profile of the students was analyzed to identify its role in the facilitation of online learning. The variables involved in the demographic profile were Age, Number of Siblings, Monthly Expenditure for the Internet, Education, and Parent's Occupation. The attributes of online learning were measured with the following variables: Number of Electronic Gadgets, Time Spent Online, Time Spent

TABLE 2.1
Demographic Profile and Attributes of Online Learning

	Demographic Profile and Attributes of Online Learning					
Statistics	Age	No. of Siblings	Monthly Expenditure (Rs.)	No. of Electronic Gadgets	Time Spent Online per Week	Time Spent for Online Education
Mean	22.59	1.74	688.79	1.96	6.75	4.48
Std. Deviation	3.92	1.02	535.43	0.78	3.06	2.05
Range	27	4	2900	4	12	9
Minimum	18	0	100	1	1	1
Maximum	45	4	3000	5	13	10

Source: Primary Data.

Online for Education, Methods Adopted in Online Teaching, Applications Used for Online Teaching, and User-Friendly Application for Effective Learning. This variable helps to bring out the effect of the demographic profile and attributes on the effectiveness of online learning (see Table 2.1).

The average age of the students who participated in the comparative assessment of online teaching and face-to-face teaching was 23 years. The standard deviation of the age groups reveals that it is stable without much difference, and it stands at 3.92 years. The age factor of the students has a range of 27 years, with a maximum age of 45 years and a minimum age of 18 years. Online teaching is mostly inculcated to the younger generation, and the sample involved in the study substantiates this fact.

The number of siblings in the family may create conflict in terms of the usage of electronic gadgets, and this variable has also been used in the study. The number of siblings in the family among the participants of the study was 1.74, which can be rounded off to 2 siblings, which gives a standard deviation of 1 sibling. The range is 4 siblings, as the maximum number is 4 siblings and the minimum is 0. The majority of the sample respondents had at least one sibling in their family.

The monthly expenditure that is spent on Internet access can add to the financial burden of the family, which gives the scope for studying the variable. The average monthly expenditure of the sample group stands at Rs. 689 with a standard deviation of Rs. 535, which implies that there are vast differences in the expenditure pattern made by each member involved in the study. The maximum expenditure of a single person amounts to Rs. 3,000 and the minimum amounts to Rs. 100.

The availability of electronic gadgets is vital for attending online classes, and their numbers have a tale to tell. The number of electronic gadgets in the family is averaged to be 1.96, which can be rounded off to 2 gadgets. The standard deviation stands at 0.78, where the maximum number of gadgets is 5 and the minimum number of gadgets is 1.

The time spent online gives the scope for learning from online sources, which is very much needed in the time of crisis. The mean hours spent by the sample group online is accounted to be 6.75 hours, while the standard deviation is 3.06 hours. The maximum number of hours spent online is 13, and the minimum number of hours is 1. The number of hours spent online gives an idea of to what extent the youngsters are spending their time on the Internet.

The time spent online for education is very much relevant to the study as it gives an idea about the age group of people judging the effectiveness of online learning in the study area. The average hours spent on online education seemed to be 4.48 hours, with a deviation of 2 hours. The group has a significant understanding of the working of online education.

Table 2.2 explains the second part of the demographic profile and attributes of online learning, which has a nominal scale.

The majority of the students involved in the study were undergraduates, and they had been undergoing online education in the current situation. The education qualification gives some clarity about the usage and their effectiveness in a particular group.

TABLE 2.2

Second Part of the Demographic Profile and Attributes of Online Learning

Demographics	Category	Frequency	Percentage
Education	Undergraduation	248	79.20
	Post-graduation	53	16.90
	Research	12	3.80
	Total	313	100.00
Parent's Occupation	Govt. Employee	110	35.10
	Private Employee	84	26.90
	Self-Employed	119	38.00
	Total	313	100.00
Methods Adopted in Online Teaching	Lecture	236	75.40
	Workshop	6	1.90
	Group Discussion	24	7.700
	Others	47	15.00
	Total	313	100.00
Applications Used for Online Teaching	Zoom	119	38.00
	Google Meet	138	44.10
	Cisco Webex	2	0.60
	Others	54	17.30
	Total	313	100.00
User-Friendly Applications for Effective Online Learning	Zoom	90	28.80
	Google Meet	148	47.30
	Cisco Webex	16	5.10
	Skype	3	1.00
	Others	56	17.90
	Total	313	100.00

Source: Primary Data.

The Parent's Occupation reveals that self-employed people are having the burden of preparing all the facilities for their children to attend online classes. The majority belong to this group, and it adds to the emotional responses of the parents in the long term. The methods adopted in online teaching seem to be the lecture method, which is the same-old technique used in face-to-face learning, and it continues in online teaching as revealed by the opinions of samples involved in the study.

There are numerous applications for the delivery of online teaching, and the current situation has given rise to the number of applications. The students mention that the majority of the classes are handled with the Google Meet application, which is clear from the responses, and equally strong is the application named Zoom.

The most User-Friendly Application for Effective Online Learning, from the perspective of the students, also seems to Google Meet, followed by Zoom. The students feel that the tools and techniques that can be used in Google Meet are sophisticated to facilitate effective online learning.

2.11.3 Factors That Reveal the Superiority of Online Learning over Face-to-Face Learning

The comparative assessment of online learning against face-to-face learning is assessed with the help of 23 variables exclusively identified for the study based on the reviews. The following analysis gives the comparative factors that give the edge for online learning over the face-to-face learning in the Indian scenario of online teaching.

The KMO and Barlett's test explains the normality of the variables involved in factor analysis (see Table 2.3). The p-value is significant and explains the validity of factors formed based on factor analysis used for assessing the factors of online teaching. Table 2.4 explains the variance involved in the formation of the factors and also the variance that is existent among each variable of the factor analysis.

The variance table explains the number of variables formed with the help of eigenvalues based on the factor analysis. There are three factors formed, which is evident from the fact that the three factors have an eigenvalue of 1. The three factors can have an effect of 69% on the total impact among the 23 variables, and thus, the factors formed are reliable and effective.

TABLE 2.3
KMO and Barlett's Test for Online Learning

"KMO Measure of Sampling Adequacy"		**0.920**
Barlett's Test of Sphericity	Chi-Square	6,496.219
	Df	253
	Sig.	**<0.001****

Note: ** indicates significance @ 1% level and * indicates significance @ 5% level.

TABLE 2.4
Total Variance Explained for Online Teaching

	Initial Eigenvalues			Extraction Sum of Squared Loadings			Rotation Sum of Squared Loadings		
Component	Total	% of Variance	Cumulative %	Total	% of Variance	Cumulative %	Total	% of Variance	Cumulative %
1	13.10	56.96	56.96	13.10	56.96	56.96	6.56	28.52	28.52
2	1.77	7.68	64.64	1.77	7.68	64.64	5.76	25.05	53.57
3	1.17	5.09	69.73	1.17	5.09	69.73	3.72	16.16	69.73
4	0.82	3.58	73.31						
5	0.69	2.98	76.29						
6	0.58	2.52	78.81						
7	0.56	2.42	81.23						
8	0.51	2.23	83.47						
9	0.44	1.91	85.38						
10	0.39	1.70	87.07						
11	0.38	1.67	88.74						
12	0.36	1.57	90.31						
13	0.33	1.45	91.76						
14	0.32	1.37	93.13						
15	0.27	1.19	94.32						
16	0.24	1.02	95.34						
17	0.23	0.98	96.32						
18	0.21	0.90	97.22						
19	0.17	0.74	97.96						
20	0.15	0.66	98.62						
21	0.13	0.57	99.19						
22	0.12	0.53	99.72						
23	0.06	0.28	100.00						

The rotated component matrix helps to examine the number of factors formed and the variables that influence the formation of the factors (see Table 2.5). There are three factors formed that explain the superiority of online teaching over face-to-face learning. The three factors formed are as follows.

2.11.4 Parameter Freedom to Communicate and Interact

The first factor is formed with the variables "Enables to understand easily due to the usage of electronic components" (0.795), "Gives freedom to communicate and interact freely with instructors" (0.780), "Interaction in various modes of chats, questions, and description is easier" (0.745), "Flow of information from the syllabus can be more informative and technical" (0.726), "The coverage of the syllabus is complete and quicker" (0.718), and Outcome-based education can be implemented effectively" (0.707). It can be deduced that the students can communicate and interact freely

TABLE 2.5
Rotated Component Matrix for Online Learning

Parameters	Component 1	Component 2	Component 3
Enables to understand easily due to usage of electronic components	0.795		
Gives freedom to communicate and interact freely with instructors	0.780		
Interaction in various modes of chats, questions, and description is easier	0.745		
The flow of information from the syllabus can be more informative and technical	0.726		
The coverage of the syllabus is complete and quicker	0.718		
Outcome-based education can be implemented effectively	0.707		
Online courses help to overcome the hesitation to clear the doubts that normally exists in the face-to-face learning			
The tools of imparting lessons are attractive and make students attentive			
Feedback regarding the classes can be unbiased			
Submission of internal assignment documents is easy		0.861	
Round-the-clock learning is made possible		0.759	
Examinations are easy to attend without much pressure of traveling during exam time		0.741	
Ability to provide information or resource persons for specialized knowledge from every corner of the world		0.720	
Teachers' ability to use online tools can be tested		0.702	
Access to course materials is universalized and easy			
Mode of evaluation is easier and can be assessed by the students			
It helps to improve the self-efficacy of the students			
One-to-one learning is effective			
Flexibility in the timings of the classes			0.737
The need for the movement from one class to another class is avoided			0.709
Adoption of online tools can have an attitudinal change in the students toward teachers			
Recording and reusing materials help to revise the syllabus			
Motivation to attend the class based on the classroom environment changes			

Note: (a) Extraction Method: Principal Component Analysis; (b) Rotation Method: Varimax with Kaiser Normalization."

during online learning when compared to face-to-face learning, which is evident from the variables involved in the formation of the factor.

2.11.4.1 Submission and Seeking Clarity at Ease

The variables that are influential in the formation of the factor are "Submission of internal assignment documents is easy" (0.861), "Round-the-clock learning is made possible" (0.759), "Examinations are easy to attend without much pressure of traveling during exam time" (0.741), "Ability to provide information or resource persons for specialized knowledge from every corner of the world" (0.720), and "Teachers' ability to use online tools can be tested" (0.702). The submission of assignments and seeking clarity of information from the teachers are made easy in online learning.

2.11.4.2 Flexibility Factor in Movement and Timings

The learning schedule becomes flexible, and the recording features available in the applications also make it convenient to learn. Online education reduces the need for movement for different classes for the lecture. The factor is formed with the variables of "Flexibility in the timings of the classes (0.737) and "The need for the movement from one class to another class is avoided" (0.709).

The three factors that determine the superiority of online teaching over face-to-face teaching can be identified as ***(I) Freedom to communicate and interact, (II) Submission and seeking clarity at ease, and (III) Flexibility factor in movement and timings.***

2.11.5 Dominant Problems of Online Teaching

Online teaching has some structural problems in delivering content to the students. The problems faced by the students were identified through a survey of literature, and 15 prominent problems were selected for the study. The opinions given were used to extract the following results with the usage of factor analysis.

The p-value of the KMO and Barlett's test (see Table 2.6) explains the significance of the normality of variables involved in the study, and it satisfies the condition necessary for identifying the reliable factors from a normal distribution of variables (see Table 2.7).

The variance table reveals that two factors have been formed with the help of 15 variables, and those factors capture the effect of 66% of the total problems investigated in the study.

The rotated component matrix has laid out that two dominant problems are identified with the usage of factor analysis, and they are explained in Table 2.8.

2.11.6 Need for Technical Upgradation of Hardware and Software

The first dominant problem of Need for Technical Upgradation of Hardware and Software is formed with the variables "Lack of updated software and browsers will force to miss classes" (0.845), "Internet usage is always a worry for learners" (0.840), The cost of Internet usage apart from the fees adds to financial expenditure" (0.802), "Speed of the

TABLE 2.6
KMO and Barlett's Test for Problems in Online Teaching

KMO Measure of Sampling Adequacy		**0.925**
Barlett's Test of Sphericity	Chi-Square	3,818.19
	Df	105
	Sig.	**<0.001****

Note: ** Indicates significance @ 1% level.

TABLE 2.7
Total Variance Explained for Problems in Online Teaching

	Initial Eigenvalues			Extraction Sum of Squared Loadings			Rotation Sum of Squared Loadings		
Component	Total	% of Variance	Cumulative %	Total	% of Variance	Cumulative %	Total	% of Variance	Cumulative %
1	9.08	60.50	60.50	9.08	60.50	60.50	5.98	39.87	39.87
2	0.96	6.37	66.88	0.96	6.37	66.88	4.05	27.01	66.88
3	0.87	5.78	72.65						
4	0.72	4.80	77.45						
5	0.55	3.64	81.09						
6	0.49	3.30	84.39						
7	0.47	3.15	87.53						
8	0.37	2.48	90.02						
9	0.34	2.26	92.27						
10	0.27	1.77	94.04						
11	0.23	1.55	95.59						
12	0.21	1.37	96.96						
13	0.16	1.08	98.04						
14	0.16	1.05	99.09						
15	0.14	0.91	100.00						

Internet tends to make the learners miss the classes" (0.787), "Lack of updated gadgets to adapt to online learning" (0.735), and "Students' understanding level cannot be assessed until the examination" (0.729). The problems primarily focus on one area, namely technical upgradation, which is affecting the efficiency of online teaching.

2.11.7 Increase in Assessments and Lack of Differentiation in Teaching Strategies

The Increase in Assessments and Lack of Differentiation problem is formed with the variables "Students have to be familiar with the learning platform" (0.793), "Lack of differentiation in teaching strategies can make listening to classes monotonous for students" (0.778), and "The assignment and testing frequency is increased" (0.767).

The dominant problems that affect the efficiency of online learning are Technical Upgradation of Hardware and Software, Increase in Assessment, and Lack of Differentiation in Teaching Strategies.

2.11.8 Association Between Attributes of Online Teaching and Factor Scores Denoting Superiority of Online Learning over Face-to-Face Learning

The attributes of online teaching have a significant impact on the usage of online learning. The facilities have to be there at the disposal of the students to make use of the online classes. The attributes are tested for association with the regression scores

TABLE 2.8
Rotated Component Matrix in Online Teaching

Parameters	Components 1	Components 2
Lack of updated software and browsers will force to miss classes	0.845	
Internet usage is always a worry for learners	0.840	
The cost of Internet usage apart from the fees adds to financial expenditure	0.802	
Speed of the Internet tends to make the learners miss the classes	0.787	
Lack of updated gadgets to adapt to online learning	0.735	
Students' understanding level cannot be assessed until the examination	0.729	
Infrastructure creation from the institution's side can increase the fees for students		
Discussion among the groups and peers cannot be effective		
Critical thinking cannot be imparted		
Lack of teachers' expertise in handling online tools is a cause for concern		
Students have to be familiar with the learning platform		0.793
Lack of differentiation in teaching strategies can make listening to classes monotonous for students		0.778
The assignment and test frequency is increased		0.767
Technical errors can cause a postponement of classes at regular intervals		
Lack of proper evaluation tools and technical errors can cause reappearing for exams		

Note: (a) Extraction Method: Principal Component Analysis; (b) Rotation Method: Varimax with Kaiser Normalization."

TABLE 2.9
Chi-Square Test – Attributes of Online Teaching/Factor Scores of Online Teaching

Attributes of Online Teaching/Factor Scores of Online Teaching	Freedom to Communicate and Interact	Submission and Seeking Clarity at Ease	Flexibility Factor in Movement and Timings
Monthly expenditure	<0.001**	<0.001**	<0.001**
Electronic gadgets	<0.001**	<0.001**	<0.001**
Time spent for online education	<0.001**	<0.001**	<0.001**
Methods adopted in online teaching	<0.001**	<0.001**	<0.001**
Apps used for online teaching	<0.001**	<0.001**	<0.001**
User friendly apps in terms of effective learning	<0.001**	<0.001**	<0.001**

Note: ** Indicates significance @ 1% level.

of factors formed in the factor analysis of superiority of online learning over the face-to-face learning (see Table 2.9).

The demographic variables of monthly expenditure, electronic gadgets, time spent for online education, methods adopted in online teaching, the application used for online teaching, and user-friendly apps in terms of effective learning have a significant association with the factor scores denoting the superiority of online teaching. The presence of attributes of online teaching has a significant role in the effectiveness of online learning for students.

2.12 LIMITATIONS OF THE STUDY

a. "The survey, focus groups, and interviews addressed in this study were opportunities to answer the particular question of whether and how online learning affected teachers' face-to-face practice."
b. Lack of immediate input, issues with technological infrastructure, lack of computer familiarity, lack of non-verbal contact signals, lack of social presence, the amount of time needed to prepare online materials, and academic concerns about the maintenance and viability of online programs or emerging technologies are some of the limitations.
c. Without unpacking what a system does and how individuals communicate with the technology to create practice, online instruments are collapsed into blanket categories.
d. As the unit of study, a methodological emphasis on persons marginalizes the substantial effect of experiences at the organizational level and systemic context on blended instructional practice.

2.13 SCOPE FOR FURTHER RESEARCH

Future research can aim to examine the impact created by online education on the overall educational scenario that exists in the country. The inculcating effects of online education can be tested in comparison with face-to-face learning to give the best results to adjudicate the effectiveness of online education. Outcome-based testing based on the comparative analysis will be an addition to the literature on online learning. The technological adoption in the education field and the overall infrastructure facilities at the learners' end are proving to be significant problems that have to be assessed from the learner's point of view. The adoption of technological developments and materialistic requirements for the learning environment will prove to be a bottleneck for the growth of online education in the country. Research on these areas will prove to be crucial for the growth of online education.

2.14 RESULTS AND DISCUSSION

The study report indicates that demographic profiles and attributes of online education have an important part to play in making online teaching more successful. Online learning has offered the students a chance to have a free conversation without the fear of the classroom atmosphere. The socio-economic background of the learners has to be developed in order to have access to online learning. The role of the government in building the infrastructure for imparting education by online teaching cannot be omitted. There is a need to assess the factors that make online learning superior to comparison to face-to-face learning (Shea and Pickett 2014). The students feel that online teaching is superior when compared to face-to-face learning in the areas of (I) freedom to communicate and interact, (II) submission and seeking clarity at ease, and (III) flexibility factor in movement and timings. The results of first objective reveal that online teaching is dominant in specified areas, which compels to shift toward online learning from the viewpoint of the students.

There is a need for knowing the various areas that are detrimental to online education (Song et al. 2004). The following are the areas based on which the effectiveness of online teaching is affected by the problems: Need for Technical Upgradation of Hardware and Software, Increase in Assessment, and Lack of Differentiation in Teaching Strategies. The focus on making online teaching an effective platform for teaching completely depends on the availability of state-of-the-art technological features of systems and software that can be operated by both the students and teachers.

The study gives a new insight that online teaching has become the need of the hour, with students having a predominant opinion that online teaching provides them with numerous benefits. Online teaching has to be blended with face-to-face teaching, which makes the students and teachers be updated with the rigorous need for both types of teaching. The teaching format has to be reformed with blended teaching, which involves both online teaching and face-to-face learning at the right composition to bring out better learning opportunities for the students. There is future scope of research on higher education learning frameworks like deep and meaningful learning, learning higher-order thinking, contextual learning, pre-class and post-class surveys, and e-portfolio.

REFERENCES

Allen, I. Elaine, and Jeff Seaman. 2017. *"Online report card: Tracking online education in the United States"* Babson Survey Research Group. Babson College, 231 Forest Street, Babson Park, MA 02457, 2016.

Ben-David, Shai, Eyal Kushilevitz, and Yishay Mansour. 1997. "Online Learning versus Offline Learning." *Machine Learning* 29 (1): 45–63. https://doi.org/10.1023/A:1007465907571.

Bestiantono, Della Shinta, Putri Zulaiha Ria Agustina, and Tsung-Hui Cheng. 2020. "How Students' Perspectives about Online Learning Amid the COVID-19 Pandemic?" *Studies in Learning and Teaching* 1 (3): 133–39. https://doi.org/10.46627/silet.v1i3.46.

Brown, Michael Geoffrey. 2016. "Blended Instructional Practice: A Review of the Empirical Literature on Instructors' Adoption and Use of Online Tools in Face-to-Face Teaching." *Internet and Higher Education* 31: 1–10. https://doi.org/10.1016/j.iheduc.2016.05.001.

Chigeza, Philemon and Kelsey Halbert. 2014. "Navigating E-Learning and Blended Learning for Pre-Service Teachers: Redesigning for Engagement, Access and Efficiency." *Australian Journal of Teacher Education* 39 (11): 133–46. https://doi.org/10.14221/ajte.2014v39n11.8.

Daniel, Emmanuel I., Christine Pasquire, and Graham Dickens. 2015. "Exploring the Implementation of the Last Planner® System through IGLC Community: Twenty One Years of Experience." Proceedings of IGLC 23 – 23rd Annual Conference of the International Group for Lean Construction: Global Knowledge – Global Solutions 2015-Janua (July): 153–62. https://doi.org/10.13140/RG.2.1.4777.2000.

Daniel, Mayra C., Gail Schumacher, Nicole Stelter, and Carolyn Riley. 2016. "Student Perception of Online Learning in ESL Bilingual Teacher Preparation." *Universal Journal of Educational Research* 4 (3): 561–69. https://doi.org/10.13189/ujer.2016.040313.

Delgado Kloos, Carlos, Pedro J. Muñoz-Merino, Carlos Alario-Hoyos, Iria Estévez Ayres, and Carmen Fernández-Panadero. 2015. "Mixing and Blending MOOC Technologies with Face-to-Face Pedagogies." IEEE Global Engineering Education Conference, EDUCON 2015-April (March): 967–71. https://doi.org/10.1109/EDUCON.2015.7096090.

Dikkers, Amy Garrett. 2015. "The Intersection of Online and Face-to-Face Teaching: Implications for Virtual School Teacher Practice and Professional Development." *Journal of Research on Technology in Education* 47 (3): 139–56. https://doi.org/10.1080/02773813.2015.1038439.

Heggart, Keith R. and Joanne Yoo. 2018. "Getting the Most from Google Classroom: A Pedagogical Framework for Tertiary Educators." *Australian Journal of Teacher Education* 43 (3): 140–53. https://doi.org/10.14221/ajte.2018v43n3.9.

Hodges, Charles, Stephanie Moore, Barb Lockee, Torrey Trust, and Aaron Bond. 2020. "Remote Teaching and Online Learning." *Educause Review*, 1–15.

Jaggars, Shanna Smith. 2014. "Choosing Between Online and Face-to-Face Courses: Community College Student Voices." *American Journal of Distance Education* 28 (1): 27–38. https://doi.org/10.1080/08923647.2014.867697.

Johnson, David and Chris C. Palmer. 2015. "Comparing Student Assessments and Perceptions of Online and Face-to-Face Versions of an Introductory Linguistics Course." *Journal of Asynchronous Learning Network* 19 (2). https://doi.org/10.24059/olj.v19i2.449.

Knott, Jessica Lucille and Jessica Lucille Knott. 2015. "Online Teaching and Faculty Learning: The Role of Hypermedia in Online Course Design."

Lewis, Laurie and Basmat Parsad. 2009. "Distance Education at Degree-Granting Postsecondary Distance Education at Degree-Granting Postsecondary Institutions : 2006–07." World Wide Web Internet And Web Information Systems, 2000–2001. http://nces.ed.gov/pubs2009/2009044.pdf.

Lim, Cheolil and Hyeongjong Han. 2020. "Development of Instructional Design Strategies for Integrating an Online Support System for Creative Problem Solving into a University Course." *Asia Pacific Education Review* 21 (4): 539–52. https://doi.org/10.1007/s12564-020-09638-w.

Mohd Khalid, M. N. and Don Quick. 2016. "Teaching Presence Influencing Online Students' Course Satisfaction at an Institution of Higher Education." *International Education Studies* 9 (3): 62. https://doi.org/10.5539/ies.v9n3p62.

Mccutcheon, Karen, Maria Lohan, Marian Traynor, and Daphne Martin. 2015. "A Systematic Review Evaluating the Impact of Online or Blended Learning vs. Face-to-Face Learning of Clinical Skills in Undergraduate Nurse Education." *Journal of Advanced Nursing* 71 (2): 255–70. https://doi.org/10.1111/jan.12509.

McFarland, Joel, Bill Hussar, Cristobal de Brey, Tom Snyder, Xiaolei Wang, Sidney Wilkinson-Flicker, Semhar Gebrekristos, et al. 2017. "The Condition of Education 2017." *National Center for Educational Statistics* NCES 2017-: 1–133. https://nces.ed.gov/pubs2017/2017144.pdf.

Means, B., Toyama, Y., Murphy, R., & Baki, M. 2013. "The effectiveness of online and blended learning: A meta-analysis of the empirical literature" *Teachers college record*, *115*(3), 1–47.

Monkhouse, W. S. 1992. "M.J.T. FitzGerald: Undergraduate Medical Anatomy Teaching: Journal of Anatomy (1992) 180, 203-209 [1]." *Journal of Anatomy* 181: 177

Morse, Ken. 2007. "'Learning on Demand.'" *The Challenges of Educating People to Lead in a Challenging World*, 33–49. https://doi.org/10.1007/978-1-4020-5612-3_2.

Nguyen, Tuan. 2015. "The Effectiveness of Online Learning: Beyond No Significant Difference and Future Horizons." *MERLOT Journal of Online Learning and Teaching* 11 (2): 309–19.

Nortvig, Anne Mette, Anne Kristine Petersen, and Søren Hattesen Balle. 2018. "A Literature Review of the Factors Influencing E-Learning and Blended Learning in Relation to Learning Outcome, Student Satisfaction and Engagement." *Electronic Journal of E-Learning* 16 (1): 45–55.

Petty, Teresa M., Amy J. Good, and Tina L. Heafner. 2019. "A Retrospective View of the National Board Certification Process." *Educational Forum* 83 (2): 215–30. https://doi.org/10.1080/00131725.2019.1576245.

Raymond, Anita, Elisabeth Jacob, Darren Jacob, and Judith Lyons. 2016. "Peer Learning a Pedagogical Approach to Enhance Online Learning: A Qualitative Exploration." *Nurse Education Today* 44: 165–69. https://doi.org/10.1016/j.nedt.2016.05.016.

Rockinson-Szapkiw, Amanda J., Jillian Wendt, Mervyn Wighting, and Deanna Nisbet. 2016. "The Predictive Relationship among the Community of Inquiry Framework, Perceived Learning and Online, and Graduate Students' Course Grades in Online Synchronous and Asynchronous Courses." *International Review of Research in Open and Distance Learning* 17 (3): 18–35. https://doi.org/10.19173/irrodl.v17i3.2203.

Shea, Peter and Alexandra Pickett. 2014. "Faculty Bio-Sketches." *Clinical Biochemistry* 47 (9): 786–803. https://doi.org/10.1016/j.clinbiochem.2014.05.055.

Shi, Yang, Kaoru Yamada, Shane Antony Liddelow, Scott T. Smith, Lingzhi Zhao, Wenjie Luo, Richard M. Tsai, et al. 2017. "ApoE4 Markedly Exacerbates Tau-Mediated Neurodegeneration in a Mouse Model of Tauopathy." *Nature* 549 (7673): 523–27. https://doi.org/10.1038/nature24016.

Singh, H. 2003. "Building Effective Blended Learning Programs." *Educational Technology* 43 (6): 51–54.

Singh, Shweta, David H. Rylander, and Tina C. Mims. 2012. "Efficiency of Online vs. Offline Learning: A Comparison of Inputs and Outcomes." *International Journal of Business, Humanities and Technology* 2 (1): 93–98. http://ijbhtnet.com/journals/Vol_2_No_1_January_2012/12.pdf.

Song, Liyan, Ernise S. Singleton, Janette R. Hill, and Myung Hwa Koh. 2004. "Improving Online Learning: Student Perceptions of Useful and Challenging Characteristics." *Internet and Higher Education* 7 (1): 59–70. https://doi.org/10.1016/j.iheduc.2003.11.003.

Szeto, Elson. 2015. "Community of Inquiry as an Instructional Approach: What Effects of Teaching, Social and Cognitive Presences Are There in Blended Synchronous Learning and Teaching?" *Computers and Education* 81: 191–201. https://doi.org/10.1016/j.compedu.2014.10.015.

The White House Summit on Community Colleges. 2011. "The White House Summit on Community Colleges," 1–32.

Yang, Jie Chi, Benazir Quadir, Nian Shing Chen, and Qiang Miao. 2016. "Effects of Online Presence on Learning Performance in a Blog-Based Online Course." *Internet and Higher Education* 30: 11–20. https://doi.org/10.1016/j.iheduc.2016.04.002.

Zhang, Huaihao, Lijia Lin, Yi Zhan, and Youqun Ren. 2016. "The Impact of Teaching Presence on Online Engagement Behaviors." *Journal of Educational Computing Research* 54 (7): 887–900. https://doi.org/10.1177/0735633116648171.

3 Issues and Challenges Faced by College Students in Online Learning during the Pandemic Period

Twinkle Sanghavi

CONTENTS

DOI: 10.1201/9781003132097-3

3.1 INTRODUCTION

Humanity is experiencing its worst face, as numerous individuals are falling prey to this invisible disease. The past few months have been a real struggle for the entire world with regards to facing hardships and moving ahead in the "new-normal." Every sector opened up with a new phase, making technology the crucial element. Schools and colleges all around the world also sought to resume their operations with the help of the Internet (Corlatean 2020).

Education is going through a major transformation throughout the world due to the spread of the virulent disease. However, online education has always been an alternative to school education before. Due to the presence of innumerous loopholes, web-based education was never implemented as a full-time approach until now. There have been both favorable and unfavorable impacts of online education on society, which has jumped into the experiential bandwagon of online learning to follow administrative protocols and mitigate the cause of COVID-19 (Zhou et al., 2020). The moment of the crisis demands an open-source online education model that can be easily adopted by educators and learners for a smoother operation. Lastly, the integral multi-prolonged policies are mandatory to create a resilient learning process in the world that can assure enhancement, scope for employment, and learners' effectiveness (Kapasia et al., 2020).

Also, during the outbreak of COVID-19, different policy initiatives have been introduced by governments as well as tertiary institutions throughout the world for continuing the teaching activities for keeping students and teachers safe from the pandemic. However, a lot of disagreement, as well as ambiguity, has been witnessed regarding what needs to be taught, how it needs to be taught, the workload of the professors and the staff, the environment of learning, and the implications of education equity (Zhang et al., 2020).

A lot of efforts are being taken at the national level for utilizing technological advancements for supporting remote learning, online education, and remote learning during this global pandemic. These are quickly evolving and emerging now. There are certain deficiencies as well as weaknesses in the infrastructure of online education, inexperience in the methods of teaching, a gap in information, the complex environment at home, etc. (Murgatrotd, 2020). However, in spite of some restrictions and loopholes, the current scenario demands action for making the education system effective and smooth. For instance, China initiated the policy of suspending classes of students without stopping learning to ensure that learning is not compromised because of the pandemic or the lockdown. This is among the multiple policies that China has introduced to ensure that students' education is not affected because of lockdowns or suspension of classes. For handling this issue, researchers suggest that the education system and the government further promote the construction of the education system, consider training the students and the teachers with standard home-based learning and teaching system, conduct online training, and support academic research with online education (Huang et al., 2020).

The use of ICT or Information Communication and Technology has been inevitable in the education system, especially during this pandemic. Facilitators and teachers are being encouraged to integrate technology into instructional practices because ICT has the potential to revolutionize an outdated education system (Aczel et al., 2008).

Some researchers also consider that integrating ICT into the education system is important because technology has become the need, and it is no more just a "want" in the lives of people. ICT-induced pedagogy also favors the learners since they prefer discovering and creating unique solutions for the challenges related to learning. Thus, the learners do not consider the fact that a facilitator or a professor would have the answers to all the questions; rather, they consider them as resource people, model, and just a supporting system which promotes exploration.

3.2 LITERATURE REVIEW

3.2.1 Pandemic and Changing Mode of Learning

Online learning has been a topic of wide interest and choice for scholars around the world. During the times of COVID-19, exponential growth has been observed in a number of studies in this area. In this section of the paper, some important and representative studies have been reviewed to set up the foundation of the study. Zhong (2020) found that the sudden changes in the livelihood of people engaged in various sectors have affected them badly. It is crucial, for this reason, that academic rulers take instant steps to create and execute policies that can diminish the pedagogical impact of this disease. It is believed that the administration can aid education leaders in chalking out the desired e-learning response. The most initial and easiest form of collaboration is to exchange information about the present proceedings of educational institutions, communities, and nations during the cataclysm.

Ali (2020) stated that due to the growing concerns regarding the spread of the coronavirus and the invariable attempts to alleviate its effects, an increasing number of educational institutions have shut down, eliminating practical classes universally. The pandemic has exposed (Jena, 2020) the impending inconsistencies that exist in the universal educational system. The present situation has lucidly demonstrated the need for an independent and strong education system that can overcome a catastrophically worst future. The study shows that universities throughout the world are accepting online education as their key source of learning for an infinite time. The study also reveals that apart from the availability of technological resources, staff participation, and confidence, student's impetus has a major role to play. Educators should utilize technological gadgets to boost learning, particularly in such times.

The educational institutions and guardians were not prepared for the situation when schools closed due to COVID-19 (Montacute, 2020). According to parents, schools served as a balancing element in a child's life; getting the school home was another act of learning imbalance. Under this research, with the use of society's COVID-19 data, it was explored that kids who were economically weak, were fed by free school meals, and were brought up by single parents devoted less time to working on school assignments at home in comparison to others. The regular school's home assignments were given more importance; however, if there is a provision for routine assignments assessment, children will invest some quality time in home assignments, therefore eradicating differences (Bayrakdar and Guveli, 2020).

Frenette et al. (2020) reflect the various non-health precautionary measures that are adopted to mitigate the spread of the pandemic in Canada. The most vital step that worked immensely was the closure of schools. This abrupt step may help in social distancing and save students from the horror of the disease but will have a contrary effect on their academic careers. According to O'Doherty et al. (2018), the requirement to shift students' pedagogical activities to an entirely online environment of their home was a rigorous attempt at degrading the quality of education. Several other implications, especially the technological barrier, can affect the academic growth of students who have not attained enough maturity to make the best use of this technological invention.

3.2.2 ICT and Education

Similarly, there is substantial literature and material about the integration of ICT in classroom learning. In the global scenario, the developed and the developing economies consider the value of the integration of the tools of ICT for the growth and development of the economy. Developed countries such as the United States, for example, spend a huge amount of money annually on education technology in public schools. Likewise, Australia invests a lot of money in the activities related to ICT in schools (Albugarni and Ahmed, 2015). In a similar way, several developing nations such as Uganda and India are adopting programs that aim at the implementation of ICT-integrated activities for reinforcing the process of learning and teaching (Ssewanyana and Busler, 2007). The reason why these countries have immensely invested and are still investing in ICT is that the tools of ICT act as the drivers for boosting the education system of a country, which leads to economic growth and development.

While considering the factors which affect the implementation of ICT in the education system, they can be categorized into endogenous and exogenous factors, and these categories may be divided further into teacher-level and school-level factors (Drent and Meelissen, 2008).

Under this framework, the perception of teachers about the use and implementation of ICT for a student-oriented education system has been determined by a host of interacting factors that have been segregated into endogenous and exogenous conditions at the school level or teacher level (Cuban et al., 2001). Exogenous conditions are mainly related to non-manipulative conditions, and endogenous conditions refer to manipulative conditions, which implies that they may be changed easily. That is why the factors are divided into four categories: teacher-level endogenous, teacher-level exogenous, school-level endogenous, and school-level exogenous. There are two main benefits of using the categories. First, categorization is important while discussing the results of the analysis of the factors since it helps in examining the factors related to the properties, and it also helps the stakeholder's analysis at the level of the teacher as well as students. Second, it is important for the policy makers to consider the areas which need to be focused upon for bringing about a useful change. Among the categories of the factors that have been identified by the authors which affect the educational implementation of ICT, the endogenous-level factors related to teachers are time, quality, and the educational beliefs of teachers regarding the effectiveness of the use of ICT in the education system (Dawes, 2001; Larner and Timberlake, 1995). At the exogenous level, demographic factors are also considered. Endogenous-level school factors comprise technical assistance, easy access to the ICT resources, school management, leadership, etc. Lastly, the location of the school comes under the category of exogenous-level school factors.

The barriers perceived by students regarding education have also been documented (Muilenburg and Berge, 2005). Researchers also consider that academic skills, issues related to administration, technical skills, and time, as well as support toward studies, social interaction, cost, access to the Internet, and other technical glitches are a few challenges related to online learning.

The past two decades have witnessed the growth of research and related discussions on the use of ICT by teachers in the education system. We have also witnessed the evolution of the meaning of the use of ICT by teachers in the education system. During the 1990s, OTA or the Office of Technology Assessment of the US Congress stated that ICT has the potential to empower three major parts of the job of a teacher: enhancing instructions, simplification of routine tasks, and fostering professional activities. Firstly, in order to enhance the instruction-giving system, it was stated that ICT does not influence teaching and learning programs directly. Instead, it is effectively integrated with the contents as well as the pedagogies of teaching.

Another study suggests the use of ICT to improve productivity for conducting day-to-day tasks lies in maintaining records of the students, preparing lessons, communicating with the parents and colleagues, etc. (Chen, 2010). Another benefit that ICT offers is that it opens the door to teachers for practicing renewal of their knowledge by themselves rather than getting the information passively in a predetermined manner. Studies also suggest that the training of teachers could also help in changing the beliefs of teachers regarding the value of some particular technologies by providing them the opportunities for working on them (Russell et al., 2003). Recently, the use of ICT for improving instructions, especially its significant support toward student-centric education, has been discussed widely (Drent and Meelissen 2008).

The American Psychological Association has recommended that the teachers should be encouraged to consider appropriate practices of instructions and technology for facilitating student-centric learning. However, it has also been found that, in the case of primary schools, in one of the countries where all the teachers were well equipped with computers and were also well connected with the Internet, the teachers were not integrating ICT actively into education because of other factors including the poorly structured design of the curriculum and shortage of time. Meanwhile, in another report, it was suggested that the use of this technological advancement for managing information and for preparing presentations was received well in schools. Following such trends, the recent studies depict that the trend of ICT has been adopted in order to support a student-centric education system. For instance, in one of the research reports, the implementation of locally conceptualized videos was supported by PBL or Problem-Based Learning for fostering a proper education system regarding fire in the forest in Tanzania. The study suggests that ICT is definitely efficient for nurturing the ability of students to think critically and search for solutions to problems proactively. Recently, the case of employing virtual learning was also explored by researchers for supporting the PBL of students. As an effective tool of ICT, the virtual learning environment encouraged students to learn by participating in the problem-solving abilities on individual bases, and exchanging ideas with each other. The study also suggests that the students learning through the virtual learning environment performed better on the efficiency test.

3.2.3 Online Classrooms

The online classroom method has worked as a great way for students learning nursing and has enabled students to have a real-time conversation with other nursing students and teachers. The tutors can decide on, prepare, and exhibit hand hygiene reinforcement using a practical classroom approach. The particular three-step method can help in improving the hand hygiene of clinical nursing students (Oldenburg and Marsch, 2020). The virtual setup is only restricted to certain practical experiments that can be taught through virtual presence. There has been a surge in the tandem transportation mode having universal connectivity in the areas of calamity. The application of online classrooms is a great alternative in countries where classes cannot be physically summoned due to the growing number of positive COVID-19 cases (Ng and Peggy, 2020).

Adnan and Anwar (2020), in their exploration, elaborate on the attitudes of Pakistani youths in higher education toward obligatory online learning university courses in the wake of the universal epidemic of COVID-19. A survey conducted on undergraduate (UG) and post-graduate (PG) students portrays that e-learning is unable to produce likely results in an under-progressed nation like Pakistan. The inaccessibility to proper technological resources, poor economic conditions, and network issues are the main problems faced by the majority of the students. The absence of physical touch with the educator, inappropriate response times, lack of conventional classroom setting, and lack of communication with classmates were additional issues highlighted by the students. Crompton and Burke (2018) concluded that mobile learning protrudes education through the usage of mobile devices. It has numerous benefits; firstly, it is flexible and can occur anywhere and anytime, and physical presence is not required for this. Secondly, educators are free to customize instructions based on their students' academic flow (Corbeil and Valdes-Corbeil, 2007).

The pandemic has dismantled the scheduled learning habit of day-scholars. The stay-at-home concept has hampered the routine life of students who spent their first half of the day at school. The untimely classes and the lack of obligation to perform certain activities have taken a toll on their physical health. Online classes have also made them value education less (Stewart et al., 2018). Therefore, if there is an adequate online engagement strategy to teach and evaluate assignments, learners will be able to indulge themselves in effective learning and develop good skills without compromising their professional development. The abovementioned activities require the active participation of teachers, identification of appropriate digital learning platforms, planning educational schedules, and accurate preparation of activities to get the desired outcomes. The policies will assist an educational institution to effectually conquer the educational impediment that has happened during a moment of crisis when public health is at stake, and other sectors are also facing hullabaloo (Zayapragassarazan, 2020).

Since the disease has no certainty to leave any soon, there is a dire need to embrace the digital platform and make maximum use of it so that students do not feel the difference between an offline and online setup. The notion of "work from home" has much more significance in the present situation (Alexandria University,

2002). India, as a responsible nation of this planet, should build innovative strategies to ensure that every single learner has an appropriate way of learning that is sustainable. The policies should involve several people from various other backdrops, consisting of remote areas and underprivileged and minority sections of the society, for a fortified education system. Since online learning is becoming a way of life, it should be alternatively used post-lockdown along with practical classroom learning (Liu et al., 2010).

3.2.4 Issues and Challenges of Online Learning

Aboagye et al. (2020) found that in spite of the fact that online learning is the only alternative in certain situations such as a pandemic, students, teachers, and educational institutions are facing many issues. The difficulties faced by students are mainly related to adapting to the technology, positing scanned assignments, and logging into their respective classes with low Internet speeds. Since there is no physical interaction, students find online classes less interesting. "Poor network connection" and "Lack of adequate technical resources" came out to be the most important challenges for conducting the classes in a smooth manner. Along with this, there were other issues, too, such as "social issues," "lecturer issues," "academic issues," and "generic issues." Further, the students were not prepared to go to online classes. The authors recommend a blended approach of teaching that comprises both conventional and online teaching methods both so that the students enjoy their online classes.

Kebritchi et al. (2017) revealed that the issues pertaining to online learning are expectations, readiness, identity, and participation in online courses. The access to technology and the Internet in remote areas (semi-urban and rural) is not as good as it is in the metro cities. We cannot forget the problems of the students who are living in such areas and facing huge academic loss due to limited access to technology and the Internet. Financial and economic constraints also make this problem even bigger since the parents cannot afford the cost of the required digital devices and Internet plans, which are the key requirements of online learning.

Subedi et al. (2020) found that students suffered from disturbances that were created due to poor Internet connectivity and power supply during the online classes and hence were compelled to buy better data packs with faster speed and data volume. Continuous disconnection leads to the loss of interest in the online learning process. Similar to this, the study conducted by Arora and Srinivasan (2020) revealed some typical problems in online classes such as "network issues," "a lack of training," "a lack of awareness," "a lack of interest," "less attendance," "a lack of personal touch," and a "lack of interaction. Kaup et al. (2020) added to the same in their study and concluded that the "technology," "training," and "student's engagement" are the major issues in online learning. Ali (2020) revealed that along with the "resources," "staff readiness," and "confidence," the motivation and the accessibility of the students also play very important roles in ICT-integrated learning.

Khati and Bhatta (2020) found that though online learning provides flexibility, "Poor network connection," "the security of the Internet," and the possibility of

"Internet addiction" are the most prominent challenges. The teachers and the administrators can be motivated by considering that this online education is only for a short period, and this, in turn, can increase the effectiveness of online teaching and learning as well. Teachers and students find online teaching and learning flexible, but at the same time, there are so many challenges and issues that both of them are facing (Gillett-Swan 2017). Student's attention and their safety are some of the major challenges with respect to the online classes. The children that belong to low-income working-class families are not able to join the classes since they do have access to computers or other devices, and they do not even have reliable Internet services. There are a number of academic staff who think that the decision of shifting to online classes was not good when there is limited infrastructure and training.

The current pandemic has put a lot of stress on the education system, right from the primary level to the higher education system. During the crisis, both the higher education learners and the primary education learners are now moving from traditional classroom learning to online education, which can be pursued right from the comfort of their houses. These novel contexts, as well as setups, are diverse now, and they are prominently dissimilar from one another. This has completely changed the manner in which learners engage themselves and teach (Xie et al., 2019).

3.3 RESEARCH GAPS

There have been many studies carried out in the area of online education. However, most of the studies have explored the teachers' viewpoint when it comes to the issues and challenges of online learning. Further, the earlier studies do not establish any causal relationship of the identified issues and challenges with the overall experience of online learning. This study not only establishes such relationships but also develops a scale on problems related to online learning for higher education students.

3.4 RESEARCH QUESTIONS

1. What are the various issues and challenges faced by students in higher education while studying online?
2. Is there any impact of the various issues and challenges on the overall experience of online learning?

3.5 OBJECTIVES OF THE STUDY

1. To explore the various issues and challenges faced by students in higher education while studying online.
2. To find out the impact of the various issues and challenges on the overall experience of online learning.

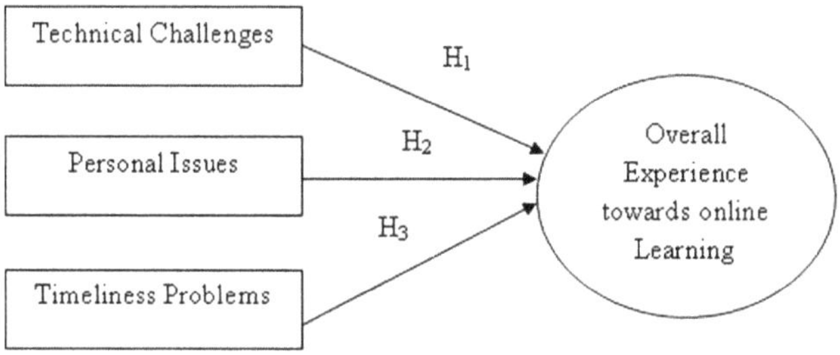

FIGURE 3.1 Theoretical framework.

3.6 THEORETICAL FRAMEWORK OF THE STUDY

Figure 3.1 shows the theoretical framework of the study. There are three major constructs: Technical Challenges, Personal Issues, and Timeliness Problems. The dependent variable is the "Overall experience of online learning." The research focuses on establishing the impact of the issues and challenges on the overall experience. The study is specific and focused on the pandemic situation. It is important to mention the scenario of the pandemic for such a study because the stakeholders of online learning did not get any chance to prepare for online classes or e-learning settings. The hypotheses derived from the theoretical framework of this study are given as below.

3.7 HYPOTHESES OF THE STUDY

H_1: Technical Challenges have a significantly negative influence on the overall experience of online learning.
H_2: Personal Issues have a significantly negative influence on the overall experience of online learning.
H_3: Timeliness Problems have a significantly negative influence on the overall experience of online learning.

3.8 SCOPE OF THE STUDY

The present study includes the experience of students in online learning in a pandemic situation. The theoretical scope of the study is confined to the study of how the issues and challenges of online learning affect the overall experience of online learning. Geographically, the study has taken into consideration B-school students pursuing UG and PG commerce, management, and arts and humanities courses at Delhi NCR. The students from other branches such as technical, engineering, and computer science were not taken into consideration because these courses require practical and lab sessions, and such a requirement could make the study biased.

3.9 RESEARCH METHODOLOGY

The present study is descriptive with hypothesis testing. In this study, the students have been chosen from the academic streams of commerce, management, and arts and humanities. The students from other disciplines have not been included in this study because they require "Laboratory Settings," and their inclusion could make the study biased.

The study has explored the various factors that negatively affect the online teaching–learning process. According to Green (1991), the minimum sample size in multiple regression should be $N \geq 104 + m$, where m = number of predictors. In the present study, there are three predictors (after précising by factor analysis), and hence, the minimum sample size required is 104 + 3 = 107. The data were collected from 205 B-school students pursuing UG and PG management courses at Delhi NCR colleges through structured questionnaires containing a 5-point Likert scale for measurement.

A two-step data analysis approach was applied. The details have been provided below:

a. For a precise understanding of the data, "Exploratory Factor Analysis (EFA)" was applied on all the ten items, measuring and taking into consideration issues and challenges in online learning, and subsequently, the reliability of the same was measured. Based on the item-to-total correlation criterion, one item that had an item-to-total correlation of below 0.3 was dropped (Nunnally, 1978).
b. To establish the causal effect, multiple regression was applied. The measurement of independent variables was based on the "Factor Scores" obtained from the EFA process. The dependent variable was the "**Overall experience of online learning.**" The following multiple regression model was formed:
   Yc (Overall experience of online learning) = a (intercept) + b_1*x_1 (Technical Challenges) + b_2*x_2 (Personal Issues) + b_3*x_3 (Timeliness Problems).
   b_1, b_2, b_3 = regression coefficients (B or understanders beta) for the respective independent variables.

3.10 DATA ANALYSIS AND INTERPRETATION

Table 3.1 shows that the percentages of participants from two different age groups, namely 17–21 and 21 and above, are almost equal (54.65% and 45.37%, respectively). In this study, 52.68% of the students were pursuing UG courses, and the remaining 47.32% were pursuing PG courses. With respect to the streams, it was found that 50.24% were pursuing the commerce or management stream, and the remaining 49.76% were pursuing the Arts or Humanities stream.

3.10.1 EXPLORATORY FACTOR ANALYSIS

EFA is a technique based on correlation, which puts together the highlight correlated variables (manifest) in the form of constructs (latent) and makes the data fit for establishing causal relationships. Table 3.2 shows the results of the KMO and Bartlett's test. The KMO value is more than the recommended value of 0.6 (Kim and Mueller,

TABLE 3.1
Demographic Profiling

	Categories	Respondents	Percentage
Age profile	17–21	112	54.63
	21 and above	93	45.37
	Total	**205**	**100**
Level	Undergraduate	108	52.68
	Post-graduate	97	47.32
Education stream	**Total**	**205**	**100**
	Commerce/Management	103	50.24
	Arts/Humanities	102	49.76
	Total	**205**	**100**

TABLE 3.2
KMO and Bartlett's Test

KMO		**0.836**
Bartlett's Test of Sphericity	Approx. Chi-Square	826.285
	Df	36
	Sig.	0.000

1978), which determines that the sample is adequate for the factor analysis. The significance value is 0.000, which shows that the correlation matrix is not an identity matrix. Hence, the data fulfills the initial diagnostics of the EFA.

Table 3.3 presents the number of factors derived with the corresponding variation. Three factors were extracted that explain the total variance of 74.468%.

TABLE 3.3
Total Variance Explained

	Initial Eigenvalues			Rotation Sums of Squared Loadings		
Variable	**Total**	**% of Variance**	**Cumulative %**	**Total**	**% of Variance**	**Cumulative %**
1	4.330	48.113	48.113	2.726	30.289	30.289
2	1.280	14.227	62.339	2.295	25.501	55.789
3	1.092	12.129	74.468	1.681	18.679	74.468
4	0.527	5.859	80.327			
5	0.422	4.689	85.016			
6	0.405	4.501	89.517			
7	0.374	4.159	93.676			
8	0.330	3.667	97.343			
9	0.239	2.657	100.000			

Note: The variance explained has been shown only for the rotated solution.

TABLE 3.4
Factors, Factor Loading, and Co-Efficient Alpha

Serial No.	Provisions for Different Skills	Factor Loading	Co-Efficient Alpha
Factor 1	**Technical Challenges**		**0.861**
1.	Poor network connection	0.832	
2.	Poor technical knowledge	0.824	
3.	Lack of adequate technical resources	0.759	
4.	Delayed feedback	0.746	
Factor 2	**Personal Issues**		**0.818**
1.	Lack of experience	0.861	
2.	Lack of personal touch	0.804	
3.	Lack of self-motivation	0.784	
Factor 3	**Timeliness Problems**		**0.756**
1.	Issues of time management	0.880	
2.	Unrealistic deadlines for assignments	0.847	

Source: SPSS output compiled by the author.

Notes: Factor loadings should be greater than 0.5 (Hair et al., 1995). The alpha value to check the reliability should be 0.7 or higher (Nunnally, 1978). The results obtained from the analysis in the aforementioned case satisfy all these criteria.

The obtained factors were labeled as Technical Challenges, Personal Issues, and Timeliness Problems, as depicted in Table 3.4.

Factor 1 – Technical Challenges: This factor is constituted by the following variables: Poor network connection, Poor technical knowledge, Lack of adequate technical resources, and Delayed feedback. It reveals the maximum variance among all three factors (30.289%).

Factor 2 – Personal Issues: The constituents of this factor are as follows: Lack of experience, Lack of personal touch, and Lack of self-motivation. This factor explains 25.501% of the variance.

Factor 3 – Timeliness Problems: This factor explains 18.679% of the variance. There are two items under this construct, namely Issues of time management and Unrealistic deadlines for assignments.

3.10.1.1 Reliability Analysis

To establish the reliability of the constructs, the minimum cut-off value is 0.7 or more (Nunnally and Bernstein, 1978). This criterion has been fulfilled in this study by all the constructs and also by all the variables put together as the scale (Table 3.5).

3.10.1.2 Regression Analysis

In this study, to measure the impact of all three constructs, namely Technical Challenges, Personal Issues, and Timeliness Problems, on the "Overall experience of online learning," multiple regression was applied. The model explains more than 55% of the variance (R square = 0.553), and hence, the model can be termed as a good fit.

TABLE 3.5
Reliability Statistics

Cronbach's Alpha	Number of Items
0.858	9

Source: SPSS output.

TABLE 3.6
ANOVA[a]

Model		Sum of Squares	Df	Mean Square	F	Sig.
1	Regression	119.662	3	39.887	82.801	0.000[b]
	Residual	96.826	201	0.482		
	Total	216.488	204			

[a] Dependent Variable (DV): **Overall experience of online learning**.
[b] Predictors (IDVs): (Constant), and independent variables (Technical Challenges, Personal Issues, and Timeliness Problems).

Table 3.6 (ANOVA) shows whether the IDVs have a significant impact on the DVs. The significance value is less than 0.05 (0.000), which reflects that one or more of the IDVs significantly influence the DV.

Table 3.7 shows that all three variables, namely *Technical Challenges, Personal Issues, and Timeliness Problems*, have a significantly negative impact on online learning. All the hypotheses proposed in the study were supported.

The summary of the results of the hypotheses testing has been presented in Table 3.8. It has been found that all the relationships are significant; the independent

TABLE 3.7
Coefficients[a]

		Unstandardized Coefficients		Standardized Coefficients		
Model		B	Std. Error	Beta	T	Sig.
1	(Constant)	2.439	0.048		50.315	0.000
	Technical Challenges	–0.474	0.049	–0.461	–9.763	0.000
	Personal Issues	–0.305	0.049	–0.296	–6.285	0.000
	Timeliness Problems	–0.518	0.049	–0.503	–10.658	0.000

[a] Dependent Variable: **Overall experience of online learning.**

TABLE 3.8
Summary of Results of Hypotheses Testing

	Hypothesis	Results	Direction of Results
Ha1	Technical Challenges influence the Overall experience of online learning	Supported	Negative
Ha2	Personal Issues influence the Overall experience of online learning	Supported	Negative
Ha3	Timeliness Problems influence the Overall experience of online learning	Supported	Negative

variables significantly influence the dependent variable negatively. To conclude, Technical Challenges, Personal Issues, and Timeliness Problems influence the Overall experience of online learning.

3.11 DISCUSSION

In this study, it has been found that there are some common issues that affect the overall experience of online learning. This study has been carried out from the perspective of college students who have been pursuing a graduation or post-graduation degree in the management or commerce stream. Online and offline learning cannot be compared until the adequate time is given to the students and instructors for the preparation for both. Since online learning, as a regular phenomenon, was a new concept and participants were not prepared, a lot of important issues were highlighted by the stakeholders from time to time. The lack of resources and knowledge regarding the technological aspects makes students uncomfortable while attending the classes. Every interruption during online classes creates an interruption in the overall learning and understanding. Most of the students are not comfortable with fixing the technical issues; hence, they have to sacrifice the learning whenever there are any such kinds of issues. The availability of compatible hardware and expertise in using them is also a big challenge for the stakeholders.

Since the students do not have any experience in attending online classes and dealing with the challenges thereof, their self-motivation is also down, which further downsizes their performance in class. The lack of personal touch makes them feel as though they are a machine sitting in front of a machine as freewheel learning and discussions are restricted.

The timeliness problems highlighted in the present study contain two major aspects: issues in time management and unrealistic deadlines for the assignments. However, students get more flexibility while attending classes in an online format, but being in their own homes, they get involved in other things along with the live online classes, and this multitasking spoils their dedication and time management. The physical classroom setting proposes an active and more engaging experience for students as compared with the online classes. Though, instead of the various challenges of online classes, in the era of rapidly changing technology innovations, one cannot deny the importance of e-learning. The challenges can be taken care of, and the issues can be resolved. E-learning is a good alternative arrangement but cannot replace physical classroom learning. For a better learning experience and better educational environment, both have to co-exist.

In the case of e-learning, the challenges and problems are not limited to technical and non-technical issues. Experts have started raising the questions of the health and well-being of the students learning through online classes. Online learning, over a period of time, causes stress, anxiety, irritation, and even depression. Too much screen time is harmful to the eyes, and it also causes certain other physical and psychological issues. However, the present study does not cover this aspect, but it could be an important area for potential research.

3.12 CONCLUSION

Online learning is a good alternative to face-to-face learning. However, some of its serious limitations restrict it in such a way that it cannot replace face-to-face learning. This study has found that technical challenges are prevalent during the course of online learning, along with personal issues and timeliness problems. The overall learning experience in the online space is negatively affected by the technological barrier, unavailability of technical resources, poor economic status, lack of knowledge about the platform, and many others. The main motive of educators and educational institutions is to deliver a close-to-class learning experience to students through online learning (Daniel, 2020). The results of this study are consistent with the earlier studies carried out by Crompton and Burke (2018), Adnan and Anwar (2020), Subedi et al. (2020), Arora and Srinivasan (2020), and many others. The issues and challenges stated above cannot be avoided. To restructure the e-learning model, educational institutions should take the help from such digital setups that provide a more interactive experience and facilitate the participation of students. The biggest problem in the way of advanced digital setups for better online learning is the cost of such setups for the students and Internet connectivity. Here, a blended setup may be recommended, which gives multiple opinions on online learning by providing the recording of all lectures and an online resource center that provides notes, PowerPoint presentations, and explanations of the technical concepts. The assignments should have flexibility in form and timeliness of submission. Overall, the truth of the present time is that neither is online learning an effective replacement for offline learning nor can it be fully eliminated. In the future, when the pandemic is over, online learning will keep contributing in contexts where face-to-face learning has no limitations.

3.13 IMPLICATIONS OF THE STUDY

The findings of the study are enough to raise the eyebrows of academic institutions, which have adopted e-learning in haste and hurry. The execution of effective classes through online platforms requires adequate training for the students, continuous counseling and mentorship, uninterrupted Internet with good speed, and basic knowledge to operate the gadgets and apps facilitating the learning. There are certain important implications of the study for policy makers and heads of institutions in the field. The stakeholders who are deciding upon online learning should think and re-think about the curriculum, implementation of the classes, constraints that students are facing, and so on. The students should be motivated to learn the new technology, and such learning sessions should be repetitively provided by the institutions to the students at their convenient timings.

The issues and challenges highlighted in this study are those which must be addressed at the initial state. Online learning is the need of the hour. The most important differences between online learning and physical classroom learning are eye-to-eye contact between the teacher and students, which engages the students in the class, and the fulfillment of social needs. Since there are no alternatives to these two, both physical classroom learning and online learning will exist in the future. However, to make e-learning effective, policy makers have to address the issues and challenges figured out in this study. A robust mechanism toward this effort will prepare the future learners for any other situation similar to the current pandemic, such as the closure of schools due to sudden strikes, lockdowns due to court verdicts on controversial and communal cases, political instability, and protests. Overall, the future of e-learning is bright, only if it is handled effectively and efficiently by the stakeholders in the field. It is a good alternative in situations of crises and does not let the loss of studies happen by providing students with the flexibility to learn anytime and from anywhere.

3.14 LIMITATIONS OF THE STUDY

This study includes only primary data collected through questionnaires; hence, minor response biases cannot be avoided. The study does not take into consideration the responses of the students who are studying in courses that require laboratory-based settings; hence, the results of this study cannot be generalized for those students. Another limitation of the study is that it covers the students of Delhi NCR; hence, the findings may be generalized for urban India and metro cities only because the conditions for online learning setups in Delhi NCR are better than those in the rural and semi-urban regions of India.

REFERENCES

Aboagye, E., Yawson, J. A., & Appiah, K. N. 2020. COVID-19 and e-learning: The challenges of students in tertiary institutions. *Social Education Research*, 2:1–8.

Aczel, J. C., Peake, S. R., & Hardy, P. 2008. Designing capacity-building in e-learning expertise: Challenges and strategies. *Computers & Education*, 50(2): 499–510.

Adnan, M. & Anwar, K. 2020. Online learning amid the COVID-19 pandemic: Students' perspectives. *Journal of Pedagogical Sociology and Psychology*, 2(1): 45–51.

Albugarni, S. & Ahmed, V. 2015. Success factors for ICT implementation in Saudi secondary schools: From the perspective of ICT directors, head teachers, teachers and students. *International Journal of Education and Development using Information and Communication Technology*, 11(1): 36–54.

Ali, W. 2020a. Online and remote learning in higher education institutes: A necessity in light of COVID-19 pandemic. *Higher Education Studies*, 10(3):16–25.

Arora, A. K. & Srinivasan, R. 2020. Impact of pandemic COVID-19 on the teaching- learning process: A study of higher education teachers. *Prabandhan: Indian Journal of Management*, 13: 43–56.

Bayrakdar, S. & Guveli, A. 2020. Inequalities in home learning and schools' provision of distance teaching during school closure of COVID-19 lockdown in the UK. ISER Working Paper Series 2020-09, Institute for Social and Economic Research. https://www.iser.essex.ac.uk/research/publications/working-papers/iser/2020-09.pdf

Chen, R. J. 2010. Investigating models for pre-service teachers' use of technology to support student-centred learning. *Computers & Education*, 55(1): 32–42.
Corbeil, J. R. & Valdes-Corbeil, M. E. J. E. Q. 2007. Are you ready for mobile learning? *Educause*, 30(2): 51.
Corlatean, T. 2020. Risks, discrimination and opportunities for education during the times of COVID-19 pandemic. RAIS Conference Proceedings. https://ideas.repec.org/p/smo/spaper/004tc.html
Crompton, H. & Burke, D. J. C. 2018. The use of mobile learning in higher education: A systematic review. *Computers & Education*, 123: 53–64.
Cuban, L., Kirkpatrick, H., & Peck, C. 2001, *High access and low use of technologies in high school classrooms: Explaining an apparent paradox*. American Educational.
Daniel, J. 2020. Education and the COVID-19 pandemic. *Prospects*, 49: 91–96.
Drent, M. & Meelissen, M. 2008. Which factors obstruct or stimulate teacher educators to use ICT innovatively? *Computers & Education*, 51(1): 187–199.
Frenette, M., Frank, K., & Deng, Z. 2020. School closures and the online preparedness of children during the COVID-19 pandemic. *Economic Insights*. https://www150.statcan.gc.ca/n1/pub/45-28-0001/2020001/article/00001-eng.htm
Gillett-Swan, J. 2017. The challenges of online learning: Supporting and engaging the isolated learner. *Journal of Learning Design*, 10: 20–30.
Green, S. B. (1991). How many subjects does it take to do a regression analysis? *Multivariate Behavioral Research*, 26: 499–510.
Hair, J. F. Jr., Anderson, R. E., Tatham, R. L., & Black, W. C. 1995. *Multivariate data analysis* (4th ed.). Englewood Cliffs, N.J.: Prentice Hall.
Hair, J. J., Black, W. C., Babin, B. J., Anderson, R. T., & Tatham, R. L. 2006. *Multivariate data analysis* (6th ed.), Upper Saddle River: Prentice Hall.
Huang, R. H., Liu, D. J., Tlili, A., Yang, J. F., & Wang, H. 2020. Handbook on facilitating flexible learning during educational disruption: The Chinese experience in maintaining undisrupted learning in COVID-19 outbreak.
Jena, P. K. 2020. Impact of pandemic Covid-19 on education in India. *International Journal of Current Research*, 12: 12582–12586.
Kapasia, N., Paul, P., Roy, A., et al. 2020. Impact of lockdown on learning status of undergraduate and postgraduate students during COVID-19 pandemic in West Bengal, India. Children and Youth Services Review. https://www.ncbi.nlm.nih.gov/pmc/articles/PMC7308748/
Kaup, S., Jain, R., Shivalli, S., Pandey, S., & Kaup, S. 2020. Sustaining academics during COVID-19 pandemic: the role of remote teaching learning. *Indian Journal of Ophthalmology*, 68: 12–20.
Kebritchi, M., Lipschuetz, A., & Santiague, L. 2017. Issues and challenges for teaching successful online courses in higher education. *Journal of Educational Technology Systems*, 46: 4–29.
Khati, K. & Bhatta, K. R. 2020. Challenges of online education during COVID-19 pandemic in Nepal. *International Journal of Entrepreneurship and Economic*, 4: 45–49.
Kim, J. O. & Mueller, C. W. 1978. *Factor analysis: Statistical methods and practical issues*. Beverly Hills, CA: SAGE Publications.
Larner, D. & Timberlake, L. 1995. Teachers with limited computer knowledge: Variables affecting use and hints to increase use. Curry School of Education, University of Virginia. http://eric.ed.gov/?id=ED384595
Liu, F., Black, E., Algina, J., Cavanaugh, C., & Dawson, K. 2010. The validation of one parental involvement measurement in virtual schooling. *Journal of Interactive Online Learning*, 9: 105–132.
Montacute, R. 2020. Social mobility and COVID-19. https://www.suttontrust.com/our-research/social-mobility-and-covid-19/

Muilenburg, Lin & Berge, Zane. 2005. Student barriers to online learning: A factor analytic study. *Distance Education*, 26: 29–48.

Murgatrotd, S. 2020. COVID-19 and online learning, mediating educational challenges amidst COVID-19 pandemic. 6(2). doi:10.13140/RG.2.2.31132.85120

Ng, Y. M. & Peggy Or, P. L. 2020. Coronavirus disease (COVID-19) prevention: Virtual classroom education for hand hygiene. *Nurse Education in Practice*, 45. https://www.sciencedirect.com/science/article/abs/pii/S1471595320302730

Nunnally, J. & Bernstein, I. 1978. *Psychometric theory* (1st ed.). New-York: MacGraw-Hill.

O'Doherty, D., Dromey, M., Lougheed, J., Hannigan, A., Last J, and McGrath, D. 2018. Barriers and solutions to online learning in medical education. *BMC Medical Education*, 18:130.

Oldenburg R. & Marsch A. 2020. Optimizing teledermatology visits for dermatology resident education during the COVID-19 pandemic. *Journal of American Academy of Dermatology*, 82: 229.

Russell, M., Bebell, D., O'Dwyer, L., & O'Connor, K. 2003. Examining teacher technology use - Implications for pre-service and in-service teacher preparation. *Journal of Teacher Education*, 54(4): 297–310.

Shengru, Li, Yamaguchi, Shinobu, & Takada, Jun-ichi. 2018. Understanding factors affecting primary school teachers' use of ICT for student-cantered education in Mongolia. *International Journal of Education and Development Using Information and Communication Technology (IJEDICT)*, 14(1): 103–117.

Ssewanyana, J. & Busler, M. 2007. Adoption and usage of ICT in developing countries: Case of Ugandan firms. *International Journal of Education and Development Using Information and Communication Technology*, 3(3): 49–59.

Stewart, H., Watson, N., & Campbell, M. 2018. The cost of school holidays for children from low income families. *Childhood*, 25: 516–529.

Subedi, S., Nayaju, S., Subedi, S., Shah, S. K., & Shah, J. M. 2020. Impact of E-learning during COVID-19 pandemic among nursing students and teachers of Nepal. *International Journal of Science and Healthcare Research*, 5: 68–76.

Xie, K., Heddy, B., & Vongkulluksn, V. 2019. Examining engagement in context using experience-sampling method with mobile technology. *Contemporary Educational Psychology*, 59.

Zayapragassarazan, Z. 2020. COVID-19: Strategies for online engagement of remote learners. *F1000, Research*, 9: 1–11.

Zhang, W., Wang, Y., Yang, L., & Wang, C. 2020, Suspending classes without stopping learning: China's education emergency management policy in the COVID-19 outbreak. *Journal of Risk and Financial Management*, 13(55): 1–6.

Zhong, R. 2020. The coronavirus exposes education's digital divide. *The New York Times.* https://www.nytimes.com/2020/03/17/technology/china-schools-coronavirus.html

Zhou, L., Li, F., Wu, S., & Zhou, M. 2020. School's out, but class's on, the largest online education in the world today: Taking china's practical exploration during the COVID-19 epidemic prevention and control as an example. *Best Evidence in Chinese Education*, 4: 501– 519.

4 Teacher's Perception toward Online Teaching in Higher Education during COVID-19

Pooja Kansra and Rajni Kansra

CONTENTS

4.1 INTRODUCTION

Coronavirus illness (COVID-19) is an infectious disease and designated as a "*Public Health Emergency*" (World Health Organization 2020). According to the World Health Organization (WHO), a mix of social distancing, contact tracing, testing, and isolation is essential to curtail the impact of the coronavirus (Kumar 2020). Social distancing has been known as the utmost preventive measure suggested by healthcare providers, advisories, and regulatory bodies worldwide (Wilder and Freedman 2020). One of the most detectable changes that happened due to the current pandemic was the shift of people from work in the office to online work. The residents in numerous nations, including India, are demanded to work online to lessen social contacts to a base in the flare-up of the pandemic COVID-19 (Ling and Ho 2020).

The current coronavirus crisis has brought a lot of disturbance in the education sector in history and has targeted around 1.6 billion learners in 190 or more countries. The shutdown of schools and colleges affected approximately 94 percent of the world's student population (Nuere and Laura 2020). The complete nationwide lockdown made educational institutions put a complete full stop to the physical delivery

DOI: 10.1201/9781003132097-4

of classes, conducting examinations, and internship programs and made them move toward online delivery of classes (George 2020; Jandric et al 2020).

The complete shutdown of the education sector due to COVID-19 has led to an unrivaled effect as the teachers were directed to teach and deliver through online platforms (Rapanta et al 2020). Due to lockdown, chances of adoption of innovative teaching were increased for the continual delivery of classes. COVID-19 has pushed the digital revolution in higher education due to the adoption of online lectures, teleconferencing, online examinations, and virtual interactions among teachers and students. Teachers can customize their procedures and processes with the help of online tools more effectively, such as audio, videos, and text (Li and Lalani 2020). Online teaching has positive implications for critical thinking and learning (Bao 2020; Dhawan 2020; Keegan 1993).

However, the online teaching–learning process is biased for the poor and marginalized students. It was observed that students with hearing-impaired difficulty were the sufferers during online learning (Kapasia et al 2020; Putri et al 2020). Understanding the teaching–learning process through an online mode was a difficult task for them, as well as for regular students, as it was creating chaos for them to concentrate (Konig et al 2020). Thus, lockdowns demolished the timetable of students, teachers, and parents (Joshi et al 2020).

Online teaching is an unavoidable alternative that was opted during lockdown when social and physical distancing is the only way out amid the COVID-19 pandemic (Mishra et al 2020). A study by Loeb (2020) highlighted that offline teaching is more effective than online teaching; however, online teaching is better than no classes. It has facilitated the uninterrupted process of teaching and learning. At the same time, it has been observed from the literature that education through traditional sources in comparison to online education contributes to the socioeconomic divide and has proven unsuccessful in making education affordable (Moralista and Ryan 2020; Nichols 2003). Lack of access to the Internet and technology has reduced the participation of the study during online learning (Loeb 2020).

First time in present-day history, teachers around the globe are compelled to work online through required restrictions forced by the public authority. The enforced working online impacts people who never had any desire to or were not permitted to due to organizational policies. As a consequence, there was a radical change in higher education, giving rise to online learning, whereby teaching is undertaken virtually on digital platforms. It was found that those who experienced online teaching have positive views about remote teaching compared to those who never taught online (National Communication Association 2019; Lee et al 2015). Understanding the views and perceptions of the teachers is required in order to effectively utilize the online learning platform to improve the learning outcomes (Farhan et al 2019).

Therefore, the present study explores the various challenges, perceptions, and willingness to do online teaching post-COVID-19. This is the first kind of comprehensive study where an endeavor was made to recognize the pandemic experience of teachers toward numerous challenges faced during online teaching. The study also demonstrated the experience, perception, and willingness of the teachers to conduct classes online in the future. To accomplish the objective, this chapter has been divided into four broad sections. Section 4.1 discussed the current scenario of online teaching during COVID-19. Section 4.2 deals with material and methods adopted to examine the various experiences, perceptions, and future willingness to work online. Section 4.3 describes the empirical findings of the study. Section 4.4 concludes the discussion along with policy implications.

4.2 MATERIALS AND METHODS

4.2.1 Research Design

In the present study, descriptive and cross-sectional designs has been applied and conducted in Punjab.

4.2.2 Sample Size and Sampling Technique

The study was based on a primary survey for the collection of data. The sample size in the study was 174 teachers working in public and private institutions in Punjab. The data has been collected as per purposive sampling.

4.2.3 Study Instrument

A structured questionnaire was designed for the collection of the data. The questionnaire consists of the demographic profile of the respondents, various challenges faced during online teaching, various perceptions toward online teaching, and their willingness to do online teaching post-COVID-19. The variables were measured on a 5-point Likert scale: strongly agree = 5, agree = 4, neutral = 3, disagree = 2, and strongly disagree = 1.

4.2.4 Statistical Analysis

The various challenges and benefits of teaching online were examined with the help of weighted average score (WAS). Logit binary regression was used to find out the willingness of the teachers to do online teaching post-COVID-19. The dependent variable in the Logit regression was assumed to be 1 if the teacher was willing to do online teaching in the future and 0 otherwise. The independent variable consists of various socioeconomic variables. The frequency and percentages were also calculated wherever deemed necessary.

4.3 RESULTS

Table 4.1 demonstrates that most of the respondents were female and married. It was observed that 38 percent, 33 percent, and 29 percent of the respondents were in the age group of 20–30 years, 30–40 years, and 40–50 years, respectively. The education-wise comparison revealed that most of the respondents were post-graduates followed by PhD scholars. The analysis of income has shown that 30 percent, 29 percent, 22 percent, and 20 percent of the respondents had a monthly income of up to ₹30000, ₹30000–₹40000, ₹40000–₹50000, and ₹50000 and above, respectively. The results of the study had revealed that 79 percent of the respondents were working in the public sector and 21 percent in private sector. It was observed that

TABLE 4.1
Demographic Profile of the Respondents

Variables	N (%)
Gender	
Male	79 (45)
Female	95 (55)
Marital Status	
Married	101 (58)
Single	73 (42)
Age	
20 to 30 years	66 (38)
30 to 40 years	58 (33)
40 to 50 years	50 (29)
Education	
Post-Graduation	110 (63)
PhD	64 (37)
Monthly Income	
Up to ₹30000	52 (30)
₹30000–₹40000	50 (29)
₹40000–₹50000	38 (22)
₹50000 and Above	34 (20)
Type of Institute	
Public	138 (79)
Private	36 (21)
Family Size	
Up to 3 members	77 (44)
4–6 members	41 (24)
6 and above	56 (32)
Teaching Experience	
Up to 5 years	135 (78)
5–10 years	9 (5)
10 years and above	30 (17)
Level Taught	
UG	53 (30)
PG	48 (28)
Both UG and PG	73 (42)
Which Platform Is Used for Online Teaching?	
Zoom	80 (46)
Google Classrooms	55 (55)
Google Meet	39 (39)

Source: Survey Results.

44 percent, 32 percent, and 24 percent of the respondents had a family size of up to 3 members, 4–6 members, 6 members and above respectively. It was exhibited that the majority of the respondents had up to 5 years of experience, followed by 10 years and above and 5–10 years. It was analyzed that majority of the respondents have taught both UG and PG students. It was seen that a greater part of the respondents had taken online classes through Google Classroom, followed by Zoom and Google Meet.

Table 4.2 depicted the numerous challenges faced during the COVID-19. It has been found that the very first challenge which the teachers have faced while doing online teaching was lack of proper communication with students (Mean = 3.943, SD = 0.851) followed by student assessment (Mean = 3.833, SD = 0.913), difficult to use of different digital tools (Mean = 3.793, SD = 1.038), time management (Mean = 3.879, SD = 2.986), high stress level, Internet issues (Mean = 4.546, SD = 5.014), lot of checks (Mean = 3.994, SD = 0.940), work life imbalance (Mean = 3.782, SD = 0.949), difficulty in child care (Mean = 3.603, SD = 1.182), lack of motivation (Mean = 2.707, SD = 1.263), social isolation (Mean = 2.707, SD = 1.263), more work pressure (Mean = 3.695, SD = 0.982) and unable to do research work effectively (Mean = 3.655, SD = 1.121).

Table 4.3 revealed that 53 percent of the respondents were not able to follow a good work routine while during online teaching. It was found that 59 percent of them do not have an appropriate workspace at home to conduct online classes. The

TABLE 4.2
Challenges of Online Teaching during COVID-19

S. No.	Variable	Mean	SD
1	Lack of proper communication with students	3.943	0.851
2	Student assessment	3.833	0.913
3	Difficult to use of different digital tools	3.793	1.038
4	Time management	3.879	2.986
5	High stress level	3.925	1.764
6	Internet issues	4.546	5.014
7	Lot of checks	3.994	0.940
8	Work life imbalance	3.782	0.949
9	Difficulty in child care	3.603	1.182
10	Lack of motivation	2.707	1.263
11	Social isolation	3.535	1.205
12	More work pressure	3.695	0.982
13	Unable to do research work effectively	3.655	1.121

Source: Survey Results.

TABLE 4.3
Experiences of Online Teaching during COVID-19

S. No.	Variables	N (%)
1	Is it possible to follow a good work routine during online teaching?	
	Yes	82 (47)
	No	92 (53)
3	Is it possible to find an appropriate workspace at home to conduct online classes?	
	Yes	71 (41)
	No	103 (59)
5	Have you spent money to upgrade technology such as laptops, printers, broadband, smartphones for conducting your online classes?	
	Yes	137 (79)
	No	37 (21)

Source: Survey Results.

study identified that 79 percent of the respondents had incurred additional expenses on laptops, printers, broadband, smartphones, etc. in order to conduct online classes.

The various perceptions related to online teaching have been described in Table 4.4. It can be observed that the first perception was "it is very difficult to get immediate student feedback about online classes" (Mean = 4.546, SD = 5.014), followed by "I possess the adequate IT skills to conduct my online classes" (Mean = 4.115, SD = 0.955), "there is no comparison between physical classroom teaching and online classroom teaching" (Mean = 3.994, SD = 0.940), "Student attendance was a bigger challenge while conducting online classes" (Mean = 3.966, SD = 0.859), "It is not possible to maintain direct contact with students while doing online classes" (Mean=, SD=), "Due to hesitation, students avoid asking questions in the online lectures" (Mean = 3.943, SD = 0.851), "Digital tools can be very helpful while conducting online classes" (Mean = 3.919, SD = 0.918), "There is a need to provide adequate training to teach before scheduling online classes" (Mean = 3.879, SD = 2.986), "There was not much difference between online classes and traditional classes" (Mean = 3.856, SD = 0.942), "It is not possible to teach all courses through online classes" (Mean = 3.833, SD = 0.913), "There should be flexible timing to conduct the online classes " (Mean = 3.828, SD = 0.843), "There was less student participation during online classes" (Mean = 3.793, SD = 1.038), "There is a need to give appropriate breaks in between the online classes" (Mean = 3.782, SD = 0.949), and "There were a lot of family distractions while conducting online lectures" (Mean = 3.603, SD = 1.182).

Table 4.5 demonstrates the various determinants of willingness to do online teaching in the future. Age ($P < 0.01$) was significantly associated with the willingness to pay. On the basis of the regression coefficient, it can be concluded that younger teachers were more likely to join online teaching than elders. Gender ($P < 0.01$) was

TABLE 4.4
Perceptions toward Online Teaching during COVID-19

S. No.	Perceptions Items	Mean	SD
1	It is very difficult to get immediate student feedback about online classes during COVID-19	4.546	5.014
2	I possess the adequate IT skills to conduct my online classes during COVID-19	4.115	0.955
3	There is no comparison between physical classroom teaching and online classroom teaching during COVID-19	3.994	0.940
4	Student attendance was a bigger challenge while conducting online classes during COVID-19	3.966	0.859
5	It is not possible to maintain direct contact with students while doing online classes during COVID-19	3.943	0.851
6	Due to hesitation, students avoid asking questions in the online lectures during COVID-19	3.925	1.764
7	Digital tools can be very helpful while conducting online classes during COVID-19	3.919	0.918
8	There is a need to provide adequate training to teach before scheduling online classes during COVID-19	3.879	2.986
9	There was not much difference between online classes and traditional classes	3.856	0.942
10	It is not possible to teach all courses through online classes during COVID-19	3.833	0.913
11	There should be flexible timing to conduct the online classes during COVID-19	3.828	0.843
12	There was less student participation during online classes during COVID-19	3.793	1.038
13	There is a need to give appropriate breaks in between the online classes during COVID-19	3.782	0.949
14	There were a lot of family distractions while conducting online lectures during COVID-19	3.603	1.182

Source: Survey Results.

significantly related to the willingness to do online teaching post-COVID-19. The coefficient of gender shows that male respondents were more likely to do online teaching in the future as compared to female teachers. The marital status ($P < 0.05$) of the teacher was also found to be significant, and it can be observed that married respondents were less likely to do online teaching as it may be due to more family distractions, etc. Family size was found to be significant ($P < 0.01$); thus, it can be observed that respondents with larger families have not shown their willingness to do online teaching in the future. However, income, education, experience, and type of institute were not found to be significant.

TABLE 4.5
Willingness for Conducting Online Teaching Post-COVID-19

Variable	Coefficient	Std. Error	z-Statistic	Prob.
C	9.884**	4.301	2.298	0.022
Gender	3.599***	1.137	3.164	0.002
Age	−2.015***	0.656	−3.071	0.002
Marital status	−2.576**	1.267	−2.033	0.042
Income	0.122	0.545	0.225	0.822
Family size	−2.018***	0.686	−2.943	0.003
Education	0.238	1.286	0.185	0.853
Experience	−1.265	0.859	−1.472	0.141
Type of institute	−0.769	1.121	−0.686	0.493
Model Summary				
McFadden R-squared	0.867	Mean dependent var		0.517
SD dependent var	0.501	SE of regression		0.157
Akaike info criterion	0.287	Sum squared resid.		4.047
Schwarz criterion	0.450	Log-likelihood		−15.974
Hannan–Quinn criterion	0.353	Deviance		31.948
Restr. deviance	241.008	Restr. log-likelihood		−120.504
LR statistic	209.061	Avg. log likelihood		−0.092

Source: Primary Survey.

***Significant at 1 percent, **Significant at 5 percent, *Significant at 10 percent.

4.4 DISCUSSION

The study demonstrated that young teachers were more likely to join online teaching than elders. The findings of the present study were similar to those of a study conducted in Saudi Arabia (Alenezi, 2012). It may be due to the fact that younger people have a stronger inclination toward e-learning. This may be due to the fact that they spend more time on Internet surfing and social networking. Gender was significant, and it was exhibited that female respondents found it more difficult to do online teaching than males. The findings of this study were consistent with the findings of that conducted by Crosbie and Moore (2004), as they identified that working women with young children found it difficult to do work from home. This may be due to the fact that the female has to play a dual role. They have to work for home and work from home for teaching online. Along with household chores, women have to take care of children, which made it difficult for them to teach online. The marital status ($P < 0.05$) of the teachers was also found to be significant and implies that married teachers are less likely to do online teaching, which may be due to more family distractions and responsibilities etc. Family size was found to be significant ($P < 0.01$), and thus, it can be observed that respondents with larger families have not shown their willingness to do online teaching in the future. Previous studies have reported similar findings, such as distraction from family members reducing the momentum of work and resulting in delays in various official tasks. Thus, it pushes the employees to move out for work instead of staying online (Troup and Rose, 2012). In a divergence from the present findings, some previous studies have reported that online work provides greater flexibility and helps to make a balance between work and family responsibilities (Ng and Khoo, 2000; Kossek, 2011; Subramaniam et al 2015; Srivastava et al 2015). The present study has clearly shown the various problems and challenges encountered by teachers while teaching online. Thus, it can be clinched that continuous support is required from the government, school authorities, parents, and family members to make online learning more effective during the pandemic. This present study has its own limitations. The small sample size adopted in the present study may influence the generalization of the findings of the study. Due to cultural and socioeconomic disparities, the result may vary with a different setup. The present study was cross-sectional in nature, and over a period of time, the perceptions of teachers may change.

4.5 POLICY IMPLICATIONS

The present study has numerous ramifications on widening online teaching in India. It can support the government, educationists, and policy-makers to plan and fathom the different issues experienced in online teaching. The coronavirus has brought a lot of changes in the education system around the globe within a short span of time. If the pandemic remains for a longer period, it may change the education system from a face-to-face method to an online one. Online teaching will decrease the cost of education and will reach outside the boundary of the country. Consequently, it is essential to adapt to the changes which took place in the education system due to COVID-19. The teachers should be capable in their role and gain the essential skills to guarantee that the online learning climate will be able to effectively facilitate student learning and positively impact student outcomes. The teachers expressed their need for training in order to conduct online classes. These training programs can bring about a noteworthy change in teacher behavior and student achievements. There is a need to create information and communication technology (ICT) capacities among teachers to viably use innovation in everyday classroom teaching. The findings of this research suggest that issues and difficulties related to online education must be tended to and online courses must be deliberately arranged and managed. The undesirable perceptions toward online teaching must be addressed on a prompt basis to improve the perspectives of faculty toward teaching online and maybe decrease confrontation against the adoption of online education. The various perceptions cited by the teachers revealed that they need training on promoting student engagement in an online environment. It is recommended that faculty be provided with robust resources and more technology infrastructure to support online education. The convenience of the use of the online teaching platform will assist in enhancing the teaching and learning climate.

4.6 CONCLUSION

It was observed that teachers confronted several challenges while conducting online classes during COVID-19. The various perceptions cited by teachers show that the teaching–learning process was slow during online teaching, and online teaching cannot replace the doctrines of physical classroom teaching. However, willingness to teach online post-COVID-19 depends on socio-demographic variables. The study suggested the need to prepare the teaching fraternity for the new normal with continuous support, training, and development so that the students can absorb the various lessons learned during online teaching. Teachers should also embrace the change and learn from the experiences brought by COVID-19. There is a critical requirement for higher institutions to put resources into the professional advancement of their facilities and update them with viable pedagogical tools with and without the use of online technologies. In the future, an attempt can be made to identify the differences in the satisfaction levels of teachers toward online teaching and traditional teaching.

REFERENCES

Alenezi, A.M. 2012. "Faculty Members' Perception of E-learning in Higher Education in the Kingdom of Saudi Arabia." https://ttuir.tdl.org/bitstream/handle/2346/45399/ALENEZIDISSERTATION

Bao, W. 2020. "COVID-19 and Online Teaching in Higher Education: A Case Study of Peking University." *Human Behavior and Emerging Technologies*, 2(2): 113–115.

Crosbie, T. and Moore, J. 2004. "Work-life Balance and Working from Home." *Social Policy and Society*, 3(3): 223.

Dhawan, S. 2020. "Online Learning: A Panacea in the Time of COVID-19 Crisis." *Journal of Educational Technology Systems*, 1(1): 1–18.

Farhan, W., Razmak, J., Demers, S., and Laflamme, S. 2019. "E-learning Systems versus Instructional Communication Tools: Developing and Testing a New E-learning User Interface from the Perspectives of Teachers and Students." *Technology in Society*, 59: 101192. https://doi.org/10.1016/j.techsoc.2019.101192

George, M.L. 2020. "Effective Teaching and Examination Strategies for Undergraduate Learning during COVID-19 School Restrictions." *Journal of Educational Technology Systems*, 49(1): 23–48.

Jandric, P., David, H., Ian, T., Paul, L., Peter, M., Thomas, R., Lilia, D. M. 2020. "Teaching in the Age of COVID-19." *Postdigital Science and Education*, 2(3): 1069–1230.

Joshi, A., Muddu, V., and Bhaskar, P. 2020. "Impact of Coronavirus Pandemic on the Indian Education Sector: Perspectives of Teachers on Online Teaching and Assessments." Interactive Technology and Smart Education. https://doi.org/10.1108/ITSE-06-2020-0087.

Kapasia, N., Pintu, P., Avijit, R., Jay, S., Ankita, Z., Rahul, M., Bikash, B., Prabir, D., and Pradip, C. 2020. "Impact of Lockdown on Learning Status of Undergraduate and Postgraduate Students during COVID-19 Pandemic in West Bengal, India." *Children and Youth Services Review*, 116: 105–194.

Keegan, D. 1993. *Theoretical Principles of Distance Education*. London: Routledge.

Konig, J., Daniela, J., Jager, B., and Nina, G. 2020. "Adapting to Online Teaching during COVID-19 School Closure: Teacher Education and Teacher Competence Effects among Early Career Teachers in Germany." *European Journal of Teacher Education*, 1: 1–15.

Kossek, E.E., Baltes, B.B., and Matthews, R.A. 2011. "How Work-Family Research Can Finally Have an Impact in Organizations." *Industrial and Organizational Psychology*, 4(3): 352–369.

Kumar, B. 2020. "Social Distancing in the Time of COVID-19: The Hidden Cost? Mental Health, Outlook." https://www.outlookindia.com/website/story/opinion-social-distancing-in-the-time-of-covid-19-the-hidden-cost-mental-health/354748

Lee, J., March, L., and Peters, R. 2015. "Faculty Training and Approach to Online Education: Is There a Connection? American University Center for Teaching, Research & Learning." https://edspace.american.edu/online/wpcontent/

Li, C. and Lalani, F. 2020. "The COVID-19 Pandemic Has Changed Education Forever. This is How." https://www.weforum.org/agenda/2020/04/coronavirus-education-global-covid19-online-digital-learning/

Ling, G.H.T. and Ho, C.M.C. 2020. "Effects of the Coronavirus (COVID-19) Pandemic on Social Behaviours: From a Social Dilemma Perspective." *Technium Social Sciences Journal*, 7(1): 312–320.

Loeb, S. 2020. "How Effective is Online Learning? What the Research Does and Doesn't Tell us." https://www.edweek.org/ew/articles/2020/03/23/how-effective-is-online-learning-what-the.html

Mishra, L., Tushar, G., and Abha, S. 2020. "Online Teaching-Learning in Higher Education during Lockdown Period of COVID-19 Pandemic." *International Journal of Educational Research Open*, 1: 100012.

Moralista, R. and Ryan, M.O. 2020. "Faculty Perception Toward Online Education in Higher Education during the Coronavirus Disease 19 (COVID-19) Pandemic." Available at SSRN 3636438 (2020).

National Communication Association. 2019. "Faculty Attitudes on Technology". Retrieved from https://www.natcom.org/sites/default/files/publications/NCA_CBrief_Vol9_1.pdf.

Ng, C. and Khoo, K.J. 2000. "Teleworking in Malaysia: Issues and Prospects." Economic and Political Weekly, 2308–2313.

Nichols, M. 2003. "A Theory for E-learning." *Journal of Educational Technology & Society*, 6 (2): 1–10.

Nuere, S. and Laura D.M. 2020. "The Digital/Technological Connection with COVID-19: An Unprecedented Challenge in University Teaching." *Technology, Knowledge and Learning*, 1: 1–13.

Putri, R.S., Agus, P., Rudy, P., Masduki, A., Laksmi, M.W., and Choi, C. 2020. "Impact of the COVID-19 Pandemic on Online Home Learning: An Explorative Study of Primary Schools in Indonesia." *International Journal of Advanced Science and Technology*, 1: 4809–4818.

Rapanta, C., Luca, B., Peter, G., Lourdes, G., and Marguerite, K. 2020. "Online University Teaching during and after the COVID-19 Crisis: Refocusing Teacher Presence and Learning Activity." *Postdigital Science and Education*, 1: 1–23.

Sanders, D.W. and Morrison-Shetlar, A.I. 2001. "Student Attitudes toward Web-Enhanced Instruction in an Introductory Biology Course." *Journal of Research on Computing in Education*, 33(3), 251–262.

Srivastava, K., Sethumadhavan, A., Raghupathy, H., Agarwal, S., and Rawat, S.R. 2015. "To Study the Indian Perspective on the Concept of Work from Home." *Indian Journal of Science and Technology*, 8(S4), 212–220.

Subramaniam, G., Overton, J., and Maniam, B. (2015). "Flexible Working Arrangements, Work Life Balance and Women in Malaysia." *International Journal of Social Science and Humanity*, 5(1), 35–38.

Troup, C. and Rose, J. 2012. "Working from Home: Do Formal or Informal Telework Arrangements Provide Better Work-Family Outcomes?" *Community, Work & Family*, 15(4): 471–486.

Wilder, S.A. and Freedman, D.O. 2020. "Isolation, Quarantine, Social Distancing and Community Containment: Pivotal Role for Old-Style Public Health Measures in the Novel Coronavirus (COVID-19) Outbreak." *Journal of Travel Medicine*, 27(2): 1–20.

World Health Organization. 2020. "Coronavirus." https://www.who.int/health-topics/coronavirus#tab=tab_1.

5 Challenges Faced by Faculty and Students in Online Teaching and Learning

A Study of Higher Education Institutions in Oman

Kavita Chavali and Shouvik Sanyal

CONTENTS

5.1 INTRODUCTION

This year, the world has seen the emergence of a deadly global pandemic caused by the SARS-CoV-2 coronavirus, commonly known as COVID-19 (Buheji et al. 2020). The rapid spread of the virus across the world has led to more than 203 million infections and more than 4.3 million deaths so far. The pandemic has brought businesses and institutions to a grinding halt in several nations, as lockdowns and closures were enforced by governments in their effort to contain the spread of the virus. Although businesses and supply chains have been the worst hit, the education sector too has been affected. Schools, colleges, and universities have been shut down, and in-person classes have been suspended, and there has been a transition to online learning. Several educational institutions that have traditionally relied on classroom

DOI: 10.1201/9781003132097-5

teaching have been struggling to cope with this challenging situation. The sudden transformation from traditional to online teaching and the learning system due to the pandemic brought in challenges such as lack of preparation time for faculty to adjust to the new online learning system. Both faculty and students felt that they were left alone during the teaching and learning processes and a need to bring in effective teaching pedagogy to keep students motivated and engaged because the dropout rates of online learning are generally higher than that of in-campus-based learning. The lockdown and online teaching have tested the resources and abilities of students as well as teachers and educational institutions, especially in developing countries where Internet connectivity and availability and affordability of online study tools, learning systems, and gadgets have caused serious challenges to the education sector (Maggio et al. 2018). With no quick solution to getting insights into the pandemic, educational institutions and teachers have to come up with innovative solutions to engage students in class and achieve learning outcomes and goals in the best manner possible. Most institutions have utilized freely available tools and software like Zoom, Google Meet, and Google Classrooms, while some have purchased Learning Management Systems (LMSs) like Moodle and Webex.

Several arguments have been made regarding the benefits of e-learning. Accessibility, affordability, flexibility, and learning pedagogy are some of the major advantages of online learning and teaching (Abney et al. 2018; Chawinga 2017; Faizi, Afia and Chiheb 2013).

The accessibility and reach of online learning to rural and remote areas is greater and relatively cheaper than traditional institution-based learning. However, in many countries, students do not have the financial resources to purchase laptops and computers, especially in rural areas with low-income households. Erratic Internet connectivity also poses several challenges for effective online education.

The Sultanate of Oman has a rapidly expanding education sector in light of the growing rates of enrolments in schools and institutions of higher learning. The Higher Education Institutions (HEIs) in Oman also had to suspend classroom teaching in light of the increasing number of COVID-19 infections in the country and in line with the decision of the concerned higher education authorities. Several institutions transitioned to the online mode of learning in a short period to adhere to the academic calendar. In this process of transition, the students enrolled in the HEIs, and the instructors had to encounter several severe challenges. This research attempts to identify some of the crucial issues and challenges that are faced by the students and faculty members in HEIs in Oman and suggest possible solutions and strategies to cope with them. The study was carried out in three HEIs in the Dhofar Governorate in Oman.

5.2 REVIEW OF LITERATURE

Online education has become popular all around the globe, especially in HEIs (Allen and Seaman 2014). The reason for online education becoming increasingly popular is because of the opportunities it provides in terms of flexibility and accessibility (Li and Irby 2008; Luyt 2013). Online education is possible only because of access to the Internet, which has become a basic social requirement in the recent past. Infrastructure and online access are a prerequisite for shifting to online teaching and learning. There is a total transformation and a 360-degree turn as far as teaching and learning are concerned. Technology offers enhanced classroom teaching and learning experience (Ghavifekr et al. 2014). Online education is superior in providing timely feedback and reducing the geographical limitations to education compared to face-to-face learning (Chen and Yang 2006).

5.2.1 Challenges Faced by Faculty

There is an abundance of research available on the challenges faced by faculty in online teaching and challenges faced by students in learning online. Some of the challenges faced by faculty are identified as behavioral issues, adaptability from traditional to online teaching, attitudes of instructors (Brooks 2003), and not being comfortable and well versed with using technology (Arbaugh 2005). Adaptation of the faculty's teaching style to suit online learning and to keep the students engaged and the achievement of learning outcomes are major challenges in online teaching (Jacobs 2014). The absence of support from the universities to faculty and students is also a challenge in online teaching and learning (Yueng 2001). Faculty in traditional face-to-face learning environments get instant feedback in the form of cues from facial expressions, body language, and questions asked by the students who make known their level of understanding and engagement, which are found missing in online learning environments (Kenyon 2007). The paradigm shift in the role of a faculty in an online environment to facilitate students with the right method of managing ample information may also be a challenge to online teaching (Hemschik 2008; Huang 2018).

5.2.2 Challenges Faced by Students

There is also an abundance of research done in the past on the challenges of online learning. The transition of students from face-to-face to online, student's expectations, and change in their mindset is a major constraint in online learning (Li and Irby 2008). Adopting learning styles and skills required in online learning where mostly the students need to be self-motivated, highly disciplined, and self-driven can be a challenge for students in online learning (Luyt 2013; Mayes et al. 2011; Santrock and Halonen 2010). Research ascertains that there have been cases of high dropout rates and performance problems in online courses (Morris, Xu and Finnegan 2005). The adaptability of students from face-to-face to online learning has been a major challenge (Nambiar 2020). The student feels left out in online learning and gets distracted and disconnected because of the lack of peer group learning that was present in traditional learning (Koole 2014). According to the study of (Singh, Rylander and Mims 2012), online learning is taken casually by the students compared to traditional face-to-face classroom learning.

The literature review enables identifying the following research questions:

- What are the key challenges being faced by the students and faculty in HEIs in Oman regarding using online teaching and learning?
- What steps can be taken by HEIs, faculty, and other governmental institutions to facilitate online teaching and learning?

5.3 METHODOLOGY

5.3.1 Sample and Data Collection

The present study attempts to analyze the primary challenges faced by the students and faculty members in three HEIs in the Sultanate of Oman. The sample size consists of 100 faculty members and 150 students in different departments of these three HEIs. Data was collected using simple random sampling with the help of two structured questionnaires, one each for the students and faculty members. The questionnaires had ten items each. Likert scale ratings (1 – strongly disagree to 5 – strongly agree) have been used to record the responses of the students and faculty members to the questions. The current study distributed a total of 310 questionnaires, and 250 responses were valid for statistical analysis, thus giving a response rate of 80 percent, which is considered a good rate for management and behavioral sciences (Babbie 1995). The questionnaire was checked by experts before distribution. The questionnaire was developed by the researchers based on an extensive review of the literature. All the measures in the study were found to be valid and reliable. The respondents among faculty represent the gamut of teaching positions, from lecturer to professor, while for students, the sample included respondents from different departments to improve the scope of the survey.

5.4 ANALYSIS AND FINDINGS

The study was conducted in the Dhofar region of the Sultanate of Oman. A questionnaire is administered to understand the challenges of students and faculty in online teaching and learning of HEIs in Oman. The questionnaire was administered online. A sample of 150 students and 100 faculty filled the questionnaire. The data were analyzed using the Friedman Rank Test and principal component analysis (PCA) to understand the challenges in online teaching and learning in Oman.

The students were asked to rank these challenges on a scale of 1–10. According to Table 5.1, the Friedman Rank test signifies that the primary challenge in online learning is the adaptability from traditional to online classes. The students felt distracted and isolated, unlike in the traditional classroom system. The attention span on a digital device has been proved to be less compared to face-to-face interaction in traditional classrooms. The second most ranked challenge was the lack of interaction with peers, which affected their learning process. Peer group learning is a vital component and has a significant role to play in the learning process of students in HEI's. The HEIs in Oman, to sort this issue to some extent, have appointed student mentors to encourage, support, and guide the student in their online experience. The faculty plays the role of a facilitator in peer group learning and teamwork between students in smaller groups using technology like breakup rooms in online platforms used for delivery like Moodle and Zoom. The third most ranked challenge was the difficulty in understanding lectures online, probably because of the adaptability of students or the students' expectations from sessions in an online mode not matching with the reality. The fourth-ranked challenge was poor Internet connectivity in remote places and mountains. To handle the connectivity issues, the telecommunications sector in Oman has rendered support in the form of increasing the connectivity and providing free Internet for students to

TABLE 5.1

Challenges for Students in Online Learning – Friedman Rank Test

S. No.	Challenges for Students in Online Learning	Friedman Rank Test
1	I have poor Internet connectivity as I stay in a remote location or the mountains	7.20
2	I feel I am not ready for online learning as I am not equipped to operate the computer	6.82
3	I am missing the interaction with my peers/friends and it affects my learning process in the online system	8.93
4	I find it very difficult to adapt to online learning	5.30
5	I am unable to get one-to-one help from faculty like in the classroom	4.89
6	In the online system, I am postponing things I have to do because of the self-paced learning	6.00
7	I am unable to focus and get distracted in an online system	9.65
8	I do not have access to proper digital devices, which is a problem sometimes	5.46
9	I do not trust technology and am worried about data privacy	3.88
10	I face difficulty in understanding lectures delivered online	7.80

Source: Primary data, SPSS.

use LMSs like Moodle and for HEI websites. The fifth-ranked challenge was the student's readiness and lack of skills related to technology. To address this issue, HEIs have manuals prepared to educate the student on how to use technology and created helplines available 24/7 on any technology-related issue. The greatest advantage of online learning is that it is self-paced, but it can also be a disadvantage to students in that if they are not self-motivated, they would end up postponing things and would not be able to cope. It has been proven, in past research, to be the major reason for the dropout rates being high in online education (Tyler-Smith 2006, 77). Timelines should be set for every activity given, and constant feedback from faculty to some extent can reduce the impact of this. The next biggest challenge is not having access to proper digital devices like a laptop. During these pandemic times, as most activities are online, there is a huge demand for devices at home. With the family sizes being large at times, each one needing to use a digital device could be demanding on the family expenses at times. The last challenge is the inability to get one-to-one help from faculty, which is present in traditional learning. To solve this, the channels of communication should be very clear between the faculty and student, and instant feedback using some predefined channels would help to some extent.

The faculty were asked to rank these challenges on a scale of 1–10. According to Table 5.2, the Friedman Rank test signifies that the major challenge that faculty face in online teaching is excessive work and stress. This is because of the sudden transition into online teaching all over the globe due to the pandemic. The faculty

TABLE 5.2
Challenges for Faculty in Online Teaching – Friedman Rank Test

S. No.	Challenges for Faculty in Online Teaching	Friedman Rank Test
1	I feel handicapped as I am not proficient in using computers	6.70
2	I find it difficult to know the level of understanding of students in online classes as I miss cues like facial expression and body language	8.82
3	There is excessive work in online teaching, and I am stressed	9.46
4	Keeping students engaged is a challenge in an online system	5.30
5	I do not have familiarity with online teaching platforms, and I feel I am not as effective as in traditional face-to-face teaching	4.89
6	I have to change my pedagogy in online teaching for effective learning	6.88
7	I need to change my evaluation components in online teaching to achieve the learning objectives	7.65
8	I find it difficult to adapt to the role of a facilitator in online teaching	4.30
9	I have Internet connectivity problems	6.20
10	I need to adapt my teaching style to suit the online system	4.46

Source: Primary data, SPSS.

need time to prepare the course content suitable for online teaching and prepare themselves for a diverse and challenging scenario. The second biggest challenge is to know the level of understanding of students in online classes as the faculty are unable to get cues like facial expression and body language of the student, which exists in face-to-face teaching. To some extent, this can be solved with the faculty asking questions while taking online classes, motivating the students to participate in class by allocating some marks in the total evaluation components, and encouraging students to ask questions either in the chat box in the LMS or in private. There should be greater interaction and involvement in online classes between the faculty and students. The third biggest challenge is the need to change the evaluation components to achieve the learning objectives in online teaching. There is a large transformation in the evaluation components in traditional and online and also in the expectations of the faculty from the student. In online teaching, it is an open book system, and the faculty needs to be novel and creative in designing the evaluation components in online teaching. The fourth-ranked challenge faculty face is the change required in the pedagogy in online teaching. There is a need for instructors to make necessary changes to the pedagogy and the course delivery for a rewarding learning experience for their students. The fifth-ranked challenge is that not every faculty is proficient with technology and finds it a handicap. Several faculty have limited knowledge of online teaching to be able to effectively use online teaching aids. Preparing and delivering lectures online requires a certain level of expertise with the software to prepare presentations and classes. To avoid a digital divide among faculty, discussion forums can be created for discussing the kind of innovation and pedagogy each one is using in their respective classes so that each one can learn from others in the process. The fifth challenge is Internet connectivity problems for the faculty, like for the students. Faculty have also had to invest out of their pockets to purchase laptops and high-speed Internet connections. Besides, not all institutions have the required resources to invest in servers and high bandwidth Internet connections that are needed to implement online learning, which is a challenge to the stakeholders involved. The sixth-ranked challenge faculty face is keeping the student engaged in online teaching. The students need to be engaged in activities beyond classes like case studies and exercises. Active engagement of students enhances the learning process. The next ranked challenge is the faculty not having familiarity with online teaching platforms as they did not have experience in handling courses online. Faculty felt that they are not as effective in online teaching as in face-to-face teaching because of this unfamiliarity. The next challenge is the adaptation of the teaching style in online teaching. For effective learning, there should be a tweak in the style of teaching online. The last ranked challenge is faculty playing the role of a facilitator in engaging students online. According to the research conducted by Frazer et al. (2017), effective online teaching is when a faculty plays the role of a facilitator, the student feels connected with faculty in online classes, is approachable by establishing mutual comfort, and is responsive to students' needs.

Table 5.3 Kaiser–Meyer–Olkin (KMO) measures explain the sample adequacy. It indicates the proportion of variance in the variables that may be caused by underlying factors. Any value above 0.6 is considered adequate. It is 0.721, and the sample size is adequate to make conclusions.

TABLE 5.3
KMO and Bartlett's Test

KMO and Bartlett's Test		
Kaiser–Meyer–Olkin Measure of Sampling Adequacy		0.721
Bartlett's Test of Sphericity	Approx. Chi-square	518.255
	Df	105
	Sig.	0

Source: Primary data, SPSS.

PCA is a dimensionality-reduction method that is often used to reduce the dimensionality of large data sets by transforming a large set of variables into a smaller one (called principal components) that still contains most of the information in the large set. This study conducted a PCA on the challenges facing the students and faculty members in online teaching and learning. The PCA revealed that two components explained most of the variance in the challenges of online teaching and learning, which were identified as Adaptability Challenges (the first component) and Technological challenges (the second component). Although factor loadings above 0.50 are generally considered significant, the study included all factor loadings in these two dimensions to maintain the accuracy of the observations. The current study has used both the eigenvalue (EV > 1) and the cumulative percent of the variance to understand the significance of each item in explaining the total variance. Whereas there is a measure of consensus showing that EV must be 1, there is no fixed threshold regarding the cumulative percent of the variance. Any measure between 50 and 60 percent is considered significant enough (Williams, Onsman and Ted 2010). This study used orthogonal varimax rotation as this technique of rotation helps to provide more significant and interpretable factors with a smaller set of items.

Table 5.4 shows the result of the PCA on the challenges faced by students in online learning. As explained already, the first component is Adaptability Challenges (AC), which has an eigenvalue of 2,814 and explains as much as 47 percent of the variance in the challenges being faced by students in online learning. The item with the highest factor loading in this component (0.783) shows that the loss of physical interaction with peers and friends in the classroom environment is the biggest adaptability issue facing students. The item with the next highest factor loading (0.778) indicates that inability to focus and other distractions while studying is the biggest challenge that students face in online education. Difficulties in understanding lectures online and problems in adapting to online learning and lack of help and support are the other major challenges that students are facing in adapting to online learning.

The second principal component extracted is identified as Technological Challenges (TC) with an eigenvalue of 1.802 and it explains 28 of the variance in the challenges being faced by students in online learning. The item with the highest factor loading in this component (0.745) shows that the student is not confident enough to operate the computer for online learning, as they are not proficient or do not have any knowledge

TABLE 5.4
Principal Component Analysis

S. No.	Challenges for Students in Online Learning	1	2
1	I have poor Internet connectivity as I stay in a remote location/in the mountains	0.321	0.642
2	I feel I am not ready for online learning as I am not equipped to operate the computer	0.276	0.745
3	I am missing the interaction with my peers/friends and it affects my learning process in the online system	0.783	0.342
4	I find it very difficult to adapt to online learning	0.654	0.413
5	I am unable to get one to one help from faculty like in the classroom	0.621	0.230
6	In the online system, I am postponing things I have to do because of the self-paced learning	0.675	0.243
7	I am unable to focus and get distracted in an online system	0.778	0.395
8	I do not have access to proper digital devices, which is a problem sometimes	0.236	0.563
9	I do not trust technology and am worried about data privacy	0.112	0.534
10	I face difficulty in understanding lectures delivered online	0.643	0.256
Eigenvalues		2.814	1.802
Cumulative percentage of variance explained		47.14%	28.36%

Source: Primary data, SPSS.

of the software being used to deliver the lectures. This is a significant finding as all students in Oman go through a year-long foundation program where there are courses in information technology. The second most significant issue in TC is indicated by a factor loading of 0.642 which corresponds to the item of poor Internet connectivity in remote locations like mountains. The next two significant factors were no access to digital devices (0.563) and lack of trust in technology and data privacy issues.

Table 5.5 shows the result of the PCA on the challenges faced by faculty in online learning. The two main components which cumulatively explained as much as 66.57 percent of the variance were identified as Adaptability Challenges (AC) with an eigenvalue of 3.625 and Technological Challenges (TC) with an eigenvalue of 2.665. AC explains 44 percent of the variance in Online learning challenges for faculty. The item with the highest factor loading in this component (0.815) shows that there is an excessive workload for faculty members in teaching online because the faculty has to prepare lectures, PowerPoint presentations to teach and also have to deliver the online classes through an LMS which involves a lot of work leading to physical fatigue and stress. The second most significant factor with a loading of 0.789 indicates that faculty are having trouble in gauging the level of understanding of their students in the absence of cues like facial expression and body language which is present in a classroom environment. The third most significant factor in this component with a factor loading of 0.787 refers to the fact that faculty members are required to make suitable changes in their pedagogy and teaching styles to suit the online learning environment. Keeping the students engaged (0.722), the need to change the evaluation components to achieve the learning objectives (0.649), and difficulty in adapting to the role of a facilitator (0.603) are the other significant factors that lead to adaptability challenges for faculty.

The second principal component extracted is identified as Technological Challenges (TC) with an Eigenvalue of 2.665 and it explains 26.57 percent of the variance in the challenges being faced by faculty in online learning. The most significant items with the highest loadings include Internet connectivity issues (0.875) and a lack of proficiency in computers and online teaching (0.827), most probably because most faculty members have no prior experience in teaching online. These findings are in line with several other studies on challenges in online learning conducted in Europe and the United States (Davidson 2015; Graham and Misanchuk 2004).

5.5 DISCUSSION

The present study is an attempt to identify the various challenges being faced by faculty and students in Omani HEIs due to the transition to online learning, which has been accelerated due to the present COVID-19 pandemic. A few HEIs were already using online mode of delivery even before the pandemic, especially in courses meant for working executives and part-time courses. However, the onset of the pandemic necessitated the closing down of educational institutions and a shift to the online mode of teaching (Fardin 2020). Several HEIs were unprepared for this sudden transition as they lacked resources, and the faculty members, too, were not conversant with the online mode of teaching. However, they eventually coped and made the

TABLE 5.5
Principal Component Analysis

S. No.	Challenges for Faculty in Online Teaching	1	2
1	I feel handicapped as I am not proficient in using computers	0.325	0.827
2	I find it difficult to know the level of understanding of students in online classes as I miss cues like facial expression, body language	0.789	0.364
3	There is excessive work in online teaching and I am stressed	0.815	0.638
4	To keep the students engaged is a challenge in an online system	0.722	0.327
5	I do not have familiarity with online teaching platforms and I feel I am not as effective as in traditional face to face teaching	0.546	0.457
6	I have to change my pedagogy in online teaching for effective learning	0.787	0.415
7	I need to change my evaluation components in online teaching to achieve the learning objectives	0.649	0.536
8	I find it difficult to adapt to the role of a facilitator in online teaching	0.603	0.232
9	I have Internet connectivity problems	0.235	0.875
10	I need to adapt my teaching style to suit the online system	0.365	0.175
Eigenvalues		3.625	2.665
Cumulative Percentage of variance explained		44.00%	26.57%

Source: Primary data, SPSS.

transition to the online mode of learning. Some institutions used free conferencing tools like Zoom, Google Classroom, and Microsoft Teams, while others bought and installed paid LMSs like Moodle. Using separate structured questionnaires for faculty and students, data was collected from three Omani HEIs, wherein each questionnaire had ten items attempting to identify the key challenges facing them in implementing and following online teaching and learning. The collected data were analyzed using the Friedman Rank Test and PCA to identify the key challenges being faced by faculty and students in online learning. Loss of interaction with friends and peers and the lack of classroom environment were ranked the highest by students in the Friedman Rank Test, while faculty members gave excessive workload and stress the top ranking. This is in line with the findings of Anderson, Sandra and Standerford (2011) and Gillet-Swan (2017). Difficulty in understanding the students' level of understanding in the absence of visual cues like expressions and the need to change evaluation components in the online mode to achieve the learning objectives were the other key challenges identified by the faculty, while the absence of interaction with peers, difficulty in understanding lectures on the online mode, and poor Internet connectivity were rated the key challenges that were being faced by students. The PCA analysis was conducted to extract the factors that could best explain the variance in the dependent variable, i.e. challenges in online education. Two factors explained 75.5 percent and 70 percent of the total variance in the case of students and faculty members, respectively, and these were identified to be Adaptability challenges and Technological challenges. Konetes (2011), in his study, identified the same key factors. It was found that for the students as well as faculty members, adapting to the online mode of learning posed the greatest challenge. They faced several difficulties like the absence of interaction with peer groups and distractions at home during lectures for the students. Faculty members faced a lot of stress due to the increased workload of recording lectures, etc. They also found it difficult to gauge the level of understanding among students in the absence of visual cues like facial expressions and feedback. Most of these findings are corroborated by similar studies carried out by researchers in other countries.

5.6 CONCLUSIONS

The present study has attempted to analyze the challenges being faced by students and faculties in various HEIs in the Sultanate of Oman in the transition to an online teaching mode. Universities have had to raise resources in a very short time to be able to offer online teaching to students. Lack of funds, poor Internet connections, and reservations regarding the efficacy of online teaching have been key concerns that HEI managements have had to deal with in the transition process. However, most HEIs in Oman have now successfully made the transition and are delivering courses in the online mode. Some universities have also adopted the blended mode of teaching involving both online and on-campus classes. After the initial problems and hiccups, most issues have now been sorted out. However, students have faced difficulty in learning in an isolated environment and some have been facing issues with Internet connectivity as the bandwidth in Oman is limited. The faculty have had to cope with a heavy workload involving preparing lectures and assessments and then delivering them online. Several faculties had to invest out of their pockets for faster

Internet connections, cameras, and laptops to teach online. They have also faced the issue of judging the level of students' understanding in the absence of visual cues apart from the challenge of changing their pedagogy to suit the online teaching and keeping students engaged for the duration of the classes because of the multiple distractions that they face at home. It can be understood that HEIs need to support the faculty during these times by providing them with training and support in the use of computers and software to effectively deliver online teaching (Khati and Bhatta 2020). Certain training programs also need to be delivered to students to enhance their learning capabilities in the online environment. Faculty and students have to support each other and work closely to overcome the issues of engagement, loneliness, and other technical and psychological issues that have presented themselves in this scenario.

One of the strategies being commonly used by HEIs is the concept of blended learning. This is a mix of face-to-face and online learning and communication between instructors and students. Blended learning allows the faculty to personalize strategies and instructions for students. This unique format allows the students to study at their own pace and gives them flexibility in taking classes. Blended learning is a very useful technique, but it has its shortcomings. Managing motivation to study, which is a critical factor in blended learning as well as creating a customized curriculum for students, is a challenge for instructors. Technological literacy and the availability of necessary software and tools are other challenges. The technological resources employed in blended learning need to be reliable, easy to use, and unanimously accepted by all stakeholders of the learning process. Also, the technology infrastructure needed for blended learning can be costly and requires the HEI to commit a significant amount of money as an investment. Blended learning also leads to a significant increase in the workload of faculty. Lots of effort goes into finding the right balance between online and face-to-face learning, and not all are willing to put the effort in. With a great range of possibilities provided by the blended learning model, teachers may start overdoing educational activities and content. This may cause a cognitive load on the students, leading to ineffective outcomes. The availability of material and content on the Internet also increases the possibility of plagiarism in assignments. However, despite these constraints, the blended learning model has proved to be a very useful and effective method of teaching, especially in these troubled pandemic times.

5.6.1 Limitations of the Study

The primary limitation of the research is that the study has been done in the Dhofar Governorate of Oman, and the responses and findings produced from this study may not fully represent the total population and cannot be generalized to other countries. Also, there may be other challenges that are being faced by students and faculty that were not covered by this study.

5.7 PRACTICAL IMPLICATIONS

Online learning had become the preferred mode of education in countries across the world even before the present pandemic forced academic institutions to adopt this mode out of health concerns (Sun and Chen 2016). The reach, ease of access, cost-effectiveness, and other factors have all contributed to the popularity of this mode of learning. However, several challenges exist in the effective implementation of online learning. This research has identified several factors that pose challenges in the effective delivery of online education. The results of the study have several implications for the educators and administrators of HEIs in various countries that have transitioned either partially or fully to the online mode of teaching and learning. This study offers suggestions for HEIs involved in online education to enrich the role played by faculty to keep the students engaged. The blended learning environment and flipped classrooms, which combine face-to-face lectures and online teaching, have been proposed to be one of the strategies to help cope with the situation. This method of learning has many benefits and leads to improved learning among students. Students have flexibility in terms of time and location in online learning, thereby developing new skills in the process leading to life-long learning. The HEIs need to address the challenges in online education by providing professional development for instructors to learn how to operate LMSs effectively as well as provide training and technical support for students to enhance the effectiveness of learning and technical support for faculty to deliver online courses (Kebritchi, Lipschuetz and Santiague 2017). Efforts should be made to alleviate the stress that instructors are having to face due to the increased workloads in online teaching. To handle this sudden transformation into an online system during the pandemic, a team should exclusively be devoted to providing students and teachers with helplines on technology issues, monitor the situation daily and provide regular feedback, conduct training programs and online workshops on LMSs (e.g. Moodle), and provide manuals to help students and faculty comfortable with using technology. There is a need for HEIs in Oman to collaborate with other institutions and share learning from each other's experiences to increase the effectiveness of online teaching and learning.

For effective online learning, Graham, Craner and Duff (2001), in their research, come up with some principles for online teaching like encouraging interaction between faculty and students so that the student does not feel left out, giving tasks or assignments that encourage collaboration among students and providing the facility of breakup rooms where students can present their work in small groups and get feedback. Tasks should have strict timelines as it is very easy to get distracted in online learning, and expectations should be communicated to students to encourage them to do better. These are in line with the findings of the present study. Online learning can become an enriching and rewarding experience for all stakeholders involved if these measures are adopted by HEIs in a time-bound fashion.

REFERENCES

Abney K. Alexandra, Laurel A. Cook, Alexa K. Fox, and Jennifer Stevens. 2018. Intercollegiate social media education ecosystem, *Journal of Marketing Education*. 41(3): 254–269. DOI: 10.1177%2F0273475318786026

Allen I. Elaine, and Seaman Jeff. 2014. *Grade change: Tracking online education in the United States*. Newburyport, MA: Sloan Consortium.

Anderson Derek, Sandra Imdieke, and Standerford N. Suzanne. 2011. Feedback please: Studying self in the online classroom, *International Journal of Instruction*. 4: 3–15.

Arbaugh. 2005. Is there an optimal design for on-line MBA courses? *Academy of Research in Open and Distance Learning*. 15: 52–70.

Babbie R. Earl. 1995. *The practice of social research*. Belmont, CA: Wadsworth Publishing Company. 121–122.

Brooks Lori. 2003. How the attitudes of instructors, students, course administrators, and course designers affect the quality of an online learning environment, *Online Journal of Distance Learning Administration*. 6. https://www.westga.edu/˜distance/ojdla/winter64/brooks64.htm

Brown John Seely, Collins Allan, and Dunguid Paul. 1989. Situated learning and education, *Educational Researcher*. 25: 34–41.

Buheji Mohamed, Katiane da Costa Cunha, Godfred Beka, Bartola Mavric, Yuri Leandro do Carmo de Souza, Simone Souza da Costa Silva, Mohammed Hanafi, and Tulika Chetia Yein. 2020. The extent of covid-19 pandemic socio-economic impact on global poverty: A global integrative multidisciplinary review, *American Journal of Economics*. 10(4): 213–224.

Chawinga Winner Domnic. 2017. Taking social media to a university classroom: Teaching and learning using Twitter and blogs, *International Journal of Educational Technology in Higher Education*. 14(3): 1–19.

Chen Ching, and Yang Sin. 2006. The efficacy of online cooperative learning systems, the perspective of task-technology fit, *Campus-Wide Information Systems*. 23(3):35–58. www.emeraldinsight.com/1065-0741.htm

Davidson Robyn. 2015. Wiki use that increases communication and collaboration motivation, *Journal of Learning Design*. 8(3): 94–105.

Faizi Rdouan, Afia El Abdellatif, and Chiheb Raddouane. 2013. Exploring the potential benefits of using social media in education, *International Journal of Engineering*. 3(4): 50–53.

Fardin Mohammad Ali. 2020. COVID-19 and anxiety: A review of psychological impacts of infectious disease outbreaks, *Archives of Clinical Infectious Diseases*. 15. DOI: 10.5812/archcid.102779

Frazer Christine, Sullivan Debra, Henline Weatherspoon, and Hussey Leslie. 2017. Faculty perceptions of online teaching effectiveness and indicators of quality Hindawi nursing research and practice, 1–6. DOI: 10.1155/2017/9374189

Ghavifekr Simin, Razak Ahmad Zabidi Abd, Ghani Muhammad Faizal A., Ran Ng Yan, Meixi Yao, and Tengyue Zhang. 2014. ICT integration in education: Incorporation for teaching & learning improvement, *Malaysian Online Journal of Educational Technology*. 2(2): 24–46.

Gillet-Swan Jenna. 2017. The challenges of online learning: Supporting and engaging the isolated learner, *Journal of Learning Design*. 10(1): 20–32.

Graham Cagiltay, Craner Lim, and Duff. 2001. Teaching in a web-based distance learning environment: An evaluation summary based on four courses. Center for Research on Learning and Technology Technical Report No. 13-00. Indiana University, Bloomington.

Graham R. Charles, and Misanchuk Melanie. 2004. Computer-mediated learning groups: Benefits and challenges to using group work in online learning environments, In T. S. Roberts (Ed.), *Online collaborative learning: Theory and practice*. 181–202. Hershey, PA: Idea Group.

Hemschik. 2008. Course design, instructional strategies, and support in K-8 online education: A case study. Statewide Agricultural Land Use Baseline 2015. http://www.westga.edu/distance/ojdla/winter44/yeung44.html.

Huang Qiang. 2018. Examining teachers' roles in online learning, *The EUROCALL Review.* 26(2): 3. DOI: 10.4995/eurocall.2018.9139

Jacobs. 2014. Engaging students in online courses, Research in Higher Education Journal. 26. Retrieved from http://search.proquest.com/docview/1637636685?accountid=458

Jaques David, and Salmon Gilly. 2007. *Learning in groups: A handbook for face-to-face and online environments.* Abingdon, UK: Routledge.

Kebritchi Mansureh, Lipschuetz Aagie, and Santiague Lilia. 2017. Issues and challenges for teaching successful online courses in higher education: A literature review, *Journal of Educational Technology Systems.* 46(1): 4–29. DOI: 10.1177/0047239516661713

Kenyon. 2007. *Academic success: Are virtual high schools working in Georgia?* (Doctoral dissertation, Capella University).

Khati Kabita, and Khem Raj Bhatta. 2020. Challenges of online education during COVID-19 pandemic in Nepal, *International Journal of Entrepreneurship and Economic Issues.* 4(1): 45–49.

Konetes G. Daniel. 2011. Distance education's impact during economic hardship: How distance learning impacts educational institutions and businesses in times of economic hardship, *International Journal of Instructional Media.* 38: 7–15.

Koole. 2014. Identity and the itinerant online learner, *The International Review of Management Learning & Education.* 4:135–149.

Li, Chi-Sing, and Irby Beverly. 2008. An overview of online education: Attractiveness, benefits, challenges, concerns, and recommendations, *College Student Journal, Part A,* 42: 449–458.

Luyt Ilka. 2013. Bridging spaces: Cross-cultural perspectives on promoting positive online experiences, *Journal of Educational Technology Systems.* 42: 3–20.

Maggio A. Lauren, Daley J. Barbara, Pratt D. Daniel, and Torre M. Dario. 2018. Honoring thyself in the transition to online teaching, *Academic Medicine.* 93(8):1129–34. DOI: 10.1097/ACM.0000000000002285

Mayes R. Luebeck, H. Ku, Akarasriworn Yu, and Korkmaz. 2011. Themes and strategies for transformative online instruction, *Quarterly Review of Distance Education.* 12(3): 151–166.

Morris, V. Libby, Haixia Xu, and Finnegan L. Catherine. 2005. Roles of faculty in teaching asynchronous undergraduate courses, *Journal of Asynchronous Learning Networks.* 9: 65–82. Retrieved from http://search.proquest.com/docview/1637636685?accountid=458

Nambiar Deepika. 2020. The impact of online learning during COVID-19: students' and teachers' perspective, *The International Journal of Indian Psychology.* 8(2):783–793. DOI: 10.25215/0802.094

Santrock W. John, and Halonen S. Jane. 2010. *Your guide to college success strategies for achieving your goals.* 6th ed. Boston, MA: Wadsworth Cengage Learning.

Singh, Swetha, Rylander H. David, and Mims C. Tina. 2012. Efficiency of online vs. offline learning: A comparison of inputs and outcomes, *International Journal of Business, Humanities, and Technology.* 2(1): 93–98.

Sun Anna, and Chen Xiufang. 2016. Online education and its effective practice: A research review, *Journal of Information Technology Education.* 15:157–190.

Tyler-Smith Keith. 2006. Early attrition among first-time e-learners: A review of factors that contribute to drop-out, withdrawal, and non-completion rates of adult learners undertaking e-learning programmers, *Journal of Online Learning and Teaching.* 2:73–85.

Williams Brett, Onsman Andreys, and Brown Ted. 2010. Exploratory factor analysis: A five-step guide for novices, *Australasian Journal of Paramedicine.* 8(3): 1–13.

Yueng Davey. 2001. Toward an effective quality assurance model of web-based learning: The perspective of academic staff. *Online Journal of Distance Education.* IV.

6 Ramifications of Digitalization in Higher Education Institutions Concerning Indian Educators

A Thematic Analysis

Karishma Jain and Swasti Singh

CONTENTS

DOI: 10.1201/9781003132097-6

6.1 INTRODUCTION

India has a heritage of education. Previously, there were institutions like Taxila, Kanchipura (Conjeevaram), Nalanda, Odantapuri (or Uddandapqa), Sri Dharryakataka, Kashmira, and Vikramashila, which were known to provide quality education to students across the globe. Then came the era of British rule, where a well-known 'minute' of Macaulay was written in 1835, in favor of studying English as a language (Seshadri et al. 2017). Since then, a lot of universities, education policies, tie-ups with foreign universities, information technology, etc., have come into the picture. Still, with time, it was essential to make changes in higher education (Laxman and Hagargi 2015). From having a gurukul system to digital classrooms with proper infrastructure, India is growing fast. India needs better infrastructure, collaboration from other universities, and a multidisciplinary approach (Sheikh 2017).

Digitalization has taken place in many sectors apart from higher education. Technology has made life more comfortable and faster; the education system must adapt this to remain relevant in the competitive world. India's IT firms tie up with lots of academic institutions for developing young minds in this field. First came the printing press, and then Educomp Solutions Ltd., which has innovated ways and brought changes in the digital space. Although India's higher education system is the third-largest in the world in terms of student numbers (The Wire 2019), it still needs lots of improvement. Digitalization covers the use of technology to renew, simplify, and enhance processes, tasks, and products (Tømte et al. 2019). Digitalization in education has several facets of quality, ranging from technological infrastructure and organizational difficulties to pedagogical approaches (Bates 2015). It has multiple benefits in higher education; its usage can be varied from the simplest form, like the inclusion of PowerPoint presentations in lectures, to the complex structures of Information and Communication Technologies (ICT)-based learnings.

Digitalization in institutions has taken place in two forms: one is the transformation in service and another one in its operations (Fedirko 2019). Service transformation includes digital communication between students and teachers through online classes, digital notes, and quizzes. Operational change comprises digitalization of the process from students' admission, registration of courses, and teachers' attendance through biometrics to controlling the examination. Higher education institutions (HEIs) are making investments in the Internet of Things (IoT), which is reducing operational costs and easing the process. Cloud computing also has a significant impact on teaching. It can be said that digitalization can be a supplement for online teaching (Jha and Shenoy 2016). The Ministry of Education with NEP 2020 has proposed to upgrade the digital infrastructure of the institutes and promote online education and better learning outcomes (Ministry of Education 2020). The fourth Sustainable Development Goal (SDG) also talks about equitable quality education and promoting lifelong learning opportunities for all (United Nations General Assembly 2015). Digitalization supports equitable quality education by providing access to online video lectures and resources across boundaries.

It can enhance students' experience, and they can access lectures and notes of some foreign universities and can work in collaboration through SaaS products. If the scope of online education increases, we may see digital campuses around us with increased digital literacy, which might be suitable for students, teachers, and organizations.

Information and Communication Technologies (ICT) have made life more comfortable and have broadened the boundaries of learning. To sustain globally and create a position, HEIs need to compete with international standards for which digitalization is a must (Khalid et al. 2018). During the testing time of COVID-19, where every student was stranded in their home, digitalization in education became a necessity. Educators have moved to online teaching through various platforms such as Zoom, Google Meet, and Google Classroom for sharing notes and assignments. This transformation is expected to continue even after the reopening of colleges.

6.2 TECHNOLOGY IN EDUCATION

The role of digital technology in education is not a new phenomenon; it has been there for decades. Students of the current generation have a different style of learning from the previous ones. They are quite facile with virtual and digital technologies (Proserpio and Gioia 2007). There have been profound changes both in learning and teaching styles in the last three decades. Technological and social changes in the environment can significantly impact learning and teaching pedagogies (Proserpio and Gioia 2007; Gioia and Brass 1984). A progressive shift in learning has been there, from verbal to visual and then to a virtual learning approach (Proserpio and Gioia 2007). The 'sage-on-the-stage' teaching approach was there a few decades before, and it still has its importance. But with changing time and generation, the teaching style has also evolved with students of TV-Generation: visual materials have been introduced in teaching, and preference is given to observational learning rather than completely theory based (Gioia and Brass 1984). Later, with the advent of personal computers and then the Internet in the late 90s, the virtual generation came into the picture. Internet, computer simulations, and communication software, have brought a significant change in the learning styles of students. Due to the shift in the learning approach, there is a need for change in the teaching approach to optimize learning. Online teaching has gradually blended with face-to-face teaching without changing the basic classroom model. This type of education method where online teaching is used as a supplement to traditional teaching is known as 'blended learning' (Bates 2015; Moser 2016). Educators have been moderately introducing online study elements in presentations, PDFs, and online discussion forums. Recently, 'flipped classroom' model has been introduced, in which recorded lectures can be viewed by students in their own time, and classroom time can be used for interactive sessions (Bates 2015). Optimal teaching and learning occur when the teaching approach is aligned with learning styles (Proserpio and Gioia 2007). Figure 6.1 shows an integration of technologies into classroom teaching for better alignment between teachers' and students' styles.

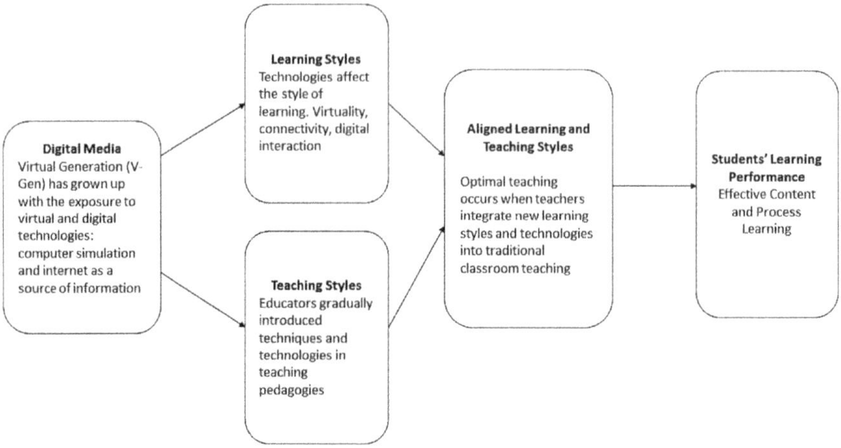

FIGURE 6.1 Alignment between teaching and learning styles. (*Source:* Proserpio and Gioia, 2007.)

6.3 DIGITAL LEARNING

Digital learning is a combination of digital content, technology, and interaction. There is a significant shift in the learning and teaching approaches due to the advent of the Internet and digital technologies. Lai (2011) suggested the use of digital technologies to enhance the quality of the learning experience of students and adopt constant changes in teaching pedagogy. Learning with technology eases students to have control over time, space, and pace. Students can learn any time of the day, from anywhere beyond the classroom's boundaries and according to their speed. Digital learning has better positive learning motivation and learning outcomes than traditional learning (Lin et al. 2017). There are various forms of digital learning; some of them are listed below.

6.3.1 Online Learning

Also known as e-learning or distance learning, the term online learning implies learning from a distance, where a learner is at a different place from an educator. The learner can access learning material and interact with the educator and other learners through any form of technology (Ally 2006).

6.3.2 Gamification

Gamification is an approach to enhance students' motivation and engagement towards learning. It includes regular activities, and rewards are given to students in the form of digital badges. The collection of badges motivates students to attempt the difficult things in their studies.

6.3.3 Blended Learning

The collaboration of digital learning with traditional classroom learning is known as blended learning or hybrid learning. In blended learning, online learning supplements traditional learning (Bates 2015). It comprehends the merits of both approaches of learning methodology.

6.3.4 M-Learning

M-Learning or mobile learning is any form of learning that takes place through mobile devices (Herrington et al. 2009). M-learning is flexible and can be done from any location or space. The educational mobile applications track students' performance, which encourages them to do better.

6.3.5 Flipped Classroom

The flipped classroom approach is also a type of blended learning which is the reverse of the traditional classroom learning approach. In a flipped classroom approach, students are provided with recorded lectures or reading material for studying before class, and discussion on practical and live problem solving is done in a classroom.

6.4 GOVERNMENT INITIATIVES FOR PROMOTING DIGITALIZATION IN EDUCATION

As per the study by KPMG (2017), the online education market is going to overgrow from USD 247 million in 2016 to USD 1.96 billion by 2021. The online channel provides quality education at a low cost due to economies of scale as there are many students and infrastructure cost is less. The Government of India is also taking initiatives to promote digital education. The Government launched the 'Digital India' campaign in July 2015 to empower the country digitally in the technology field by improving IT infrastructure. The campaign was established to provide better Internet facilities in rural areas, providing government services digitally, and promote universal literacy. Under the Digital India initiative, the Government has initiated a massive open online courses (MOOCs) platform SWAYAM (Study Webs of Active Learning for Young Aspiring Minds), to make quality education accessible to many people anytime and anywhere. Other platforms like the National Digital Library of India, National Academic Depository (NAD), Shodhganga, and Global Initiative of Academic Networks (GIAN) are there to promote digitalization in higher education.

The Ministry of Education, erstwhile Ministry of Human Resource Development, has recently come up with New Education Policy (NEP) 2020. NEP 2020 has replaced the education policy of 1986. The purpose of NEP 2020 is to address the growing developmental imperatives of our country. It has released new policies for school education and HEIs. They aim to provide holistic and multidisciplinary learning in higher education. As per NEP 2020, institutions will have the option to run open distance learning (ODL) and online programs, provided that they are accredited to do so, to enhance their offerings. NEP's objective is to increase the gross enrolment ratio (GER) in higher education, including vocational education, from 26.3% (2018) to 50% by 2035. The policy has recommended enhancing and upgrading the digital infrastructure and promoting online education. To promote digital learning, the National Education Technology Forum (NETF), an autonomous body, will be established to exchange ideas on technology to boost learning, planning, assessment, and administration. The NETF will be responsible for organizing conferences, workshops, and seminars, and various software are to be developed for both teachers and students at both school and higher education levels. E-content will be developed in all regional languages by all states. The permanent task of NETF will be to notice new disruptive technologies that will transform the educational system (Ministry of Education 2020).

We discussed NEP with participants, and most of the participants stated that NEP might bring immense substantial transformation in higher education sector. As far as digitalization is concerned, participants were unsure about its impact, and they said that it is too early to comment on how it is going to impact digital learning. One of the participants also mentioned that there were already digitalization policies, but few institutions are not following the regulations in practice.

6.5 LITERATURE REVIEW

Digitalization is a process of improving business processes through digitization. The term 'digitalization' in contrast with computerization was first mentioned by Robert Wachal in 1971. Wachal (1971) discussed the implications of digitization on society in the context of considering objections to and the possibility of computer-aided humanities research. The two conceptual terms, 'digitization' and 'digitalization' are interrelated, but they are different. While the term digitization is the process of converting analog information into digital bits, digitalization is a way of transforming social life from digital communication and media infrastructure (Brennen and Kreiss 2016). Digitalization has changed the way businesses run. It has an impact on every sector, including HEIs. Digital learning, i.e. using technology in learning, has enriched the quality of students' learning experience (Lai 2011; Lin et al. 2017). A survey on students' opinion on digital learning found that digital learning has more positive learning motivation and outcome than traditional learning. Students agreed with the collaboration of digital learning with subject learning. The evolution in education is not only because of technology, but they are more due to societal changes (Baumöl and Bockshecker 2017). The learning style keeps on changing over the generations (Proserpio and Gioia 2007); the learning expectations of generations Y and Z are different from the former generations (Baumöl and Bockshecker 2017).

Education has paramount importance in the development and advancement of any nation. Digitalization in HEIs should be seen from a broader perspective as a part of the digital economy (Komljenovic 2020). Tømte et al. (2019), in their study, suggested that digitalization within HEIs developed along with two approaches: one is external, which is influenced by government agencies, and another is internal, which comprises of both administrative-led initiatives and initiatives taken by individual stakeholders from academic staff. The higher education sector in India is also adapting the developments to meet the challenges globally (Laxman and Hagargi 2015). There are two schools of thought on the impact of digitalization in HEIs: those who foresee it as a bright future, equal accessibility of education and of better content at a reduced cost, and those who consider digitalization as dehumanizing the learning process, removing cultural diversity, and increasing social inequality (Gaebel et al. 2021). The current pandemic of COVID-19 has boosted the overall digitalization in the higher education sector, and HEIs have to switch from traditional classroom to digital learning mode. Digital learning is a difficult task, and to make it effective, educators must design the course, prepare materials assessments, etc., in advance (Young 2010). The lack of digital skills is the biggest barrier in the way of digital learning. Educators face trouble in accepting educational innovation and show resistance (Andres 2019). A survey on European HEIs found that for effective digital learning, active participation is required from both students and teachers along with digitally equipped infrastructure (Gaebel et al. 2021). Thoring, Rudolph, and Vogl (2017), in their qualitative interview study, found three categories that students expect for IT support namely, study organization, online literature, and software provision. Students consider university systems as old fashioned, complex, and less intuitive. They wished to have a centralized platform to access all relevant information and services and easy access to literature and software so that they will be able to focus on content. Kirkwood and Price (2014), in their thematic analysis study on technology-enhanced learning (TEL), have observed that innovative use of technology is required to teach students. Extant literature shows that digitalization has transformed the education sector from traditional classroom learning to digital learning. The purpose of this study is to identify challenges encountered by HEIs educators through thematic analysis.

6.6 RESEARCH METHODOLOGY

Digitalization is the need of the hour, and it is essential to adapt to this change. This study examines emergent themes from experiences and challenges faced by HEIs educators due to digitalization and to identify measures to alleviate those concerns.

6.6.1 Data Collection

As mentioned by (Braun and Clarke 2013), there are no rules for sample size in qualitative research. Fugard and Potts (2015) recommended 6–10 interviews for small projects. For this study, data collection was done through semi-structured interviews; 12 educators were approached, and 2 reminders were sent within a span of 15 days. Eight educators (female: 7, male: 1) agreed to a telephonic interview, leading to an obtained response rate of 66.67%. Interviews were taken from educators teaching in private institutions of North India. Out of the eight educators, two educators are associated with Delhi institutions, two educators are associated with institutions in Madhya Pradesh, and four educators are working in institutions in Uttar Pradesh. Participants were contacted with a summary of the study, and prior consent was taken for recording before commencing the interview. Telephonic interviews were preferred for this study. Interviews were audio recorded and ranged from

30 minutes to 40 minutes in length. Sturges and Hanrahan (2004) suggested that telephonic interviews can be successfully used for a qualitative study, and there is no difference in data quality between face-to-face and telephonic interviews. In a telephonic interview, a respondent can be more relaxed because of anonymity and be in a more personal space as compared to a face-to-face interview (Sturges and Hanrahan 2004). NVivo 12 software (© QSR International) was used for the organization and coding of interview transcripts.

6.6.2 Thematic Analysis

Thematic analysis is a method to do qualitative research. It is a method for identifying, analyzing, and outlining themes within data (Braun and Clarke 2006). It involves searching the entire data set from interviews, web pages, or a range of texts, etc., to find repeated patterns of meaning (Patton 2002). Thematic analysis is a method that works both to reflect reality and to unravel the surface of 'reality' (Braun and Clarke 2006). Themes identified through thematic analysis represent patterned response or meaning within the data set (Braun and Clarke 2006). For this study, we followed an inductive or 'bottom-up' way (Pigden and Jegede 2020). An inductive way of identifying themes and patterns is a data-driven approach, i.e. themes identified are strongly linked with the data themselves (Braun and Clarke 2006). The themes were recognized at a semantic level rather than a latent one (Pigden and Jegede 2020). The semantic approach identifies themes within explicit or surface meanings of the data (Braun and Clarke 2006); codes were made based on what respondents have said without identifying the underlying idea or assumption behind the responses (Patton 2002; Braun and Clarke 2006). For applying thematic analysis, we have followed the six-step framework of Braun and Clarke (2006), shown in Figure 6.2. This is one of the most used approaches in social sciences as it has a logical and clear framework for applying thematic analysis.

Before data analysis, all interviews were transcribed verbatim by the authors. The transcripts were rechecked against the original recording for 'accuracy'. We maintained confidentiality by assigning numbers to each participant. During the first phase of analysis, each transcript was read multiple times to get familiar with the data and to identify initial codes. Codes are attributes of data related to research questions and are meaningful to a researcher. During the initial reading, we highlighted the portions of the interview that described the experiences of the participants due to digitalization. At the end of this phase, we got familiarized with the data and had an

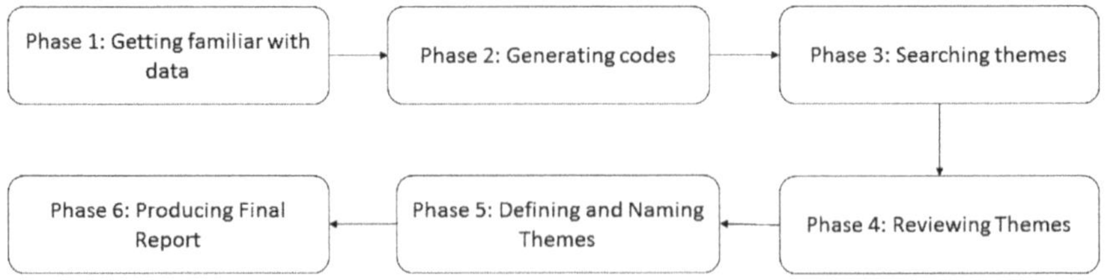

FIGURE 6.2 Six-phase framework for doing thematic analysis. (*Source:* Braun and Clarke, 2006.)

initial list of ideas concerning the data. In the next phase, all the initial ideas and data extracts were coded into a meaningful group. For example, a few participants mentioned that when students put their cameras on during online classes, it reduces the bandwidth and connectivity problems arise. Others have mentioned the problem of the Internet for students in rural areas. Both were coded under the 'Network issues' code. These codes were noted in individual transcript files and a master sheet having a complete list of codes arising from the transcript.

The third phase involves sorting the codes and assigning them to potential themes. Themes were identified based on how often they were occurring, both within a single participant's transcript and across participants' transcripts, and how unique they were. For example, as mentioned earlier, participants faced connectivity issues during online classes, and several participants felt they were lacking resources and required trained for the implementation of digitalization. The authors believed that these codes were related and fall in the main overarching theme of 'challenges in digitalization'. This phase ended with a collection of themes and subthemes. Themes were reviewed and refined to ensure that enough data was present to support every theme in the fourth phase. Themes that had similar codes were collapsed into a single theme. Data within the theme cohered together meaningfully, while there was a clear distinction between themes. In the fifth phase, themes were defined clearly. The essence of each theme cohering with overall themes was identified and named based on data upon each theme capture. The final phase involved the final analysis of themes and reporting them.

6.7 RESULTS

6.7.1 Textual Analysis

The analysis of the text was performed by Word Frequency Analysis in NVivo. Figure 6.3 shows the word cloud outlining words based on their usage. The size of a word in the word cloud represents the frequency of words that occurred in transcripts. The

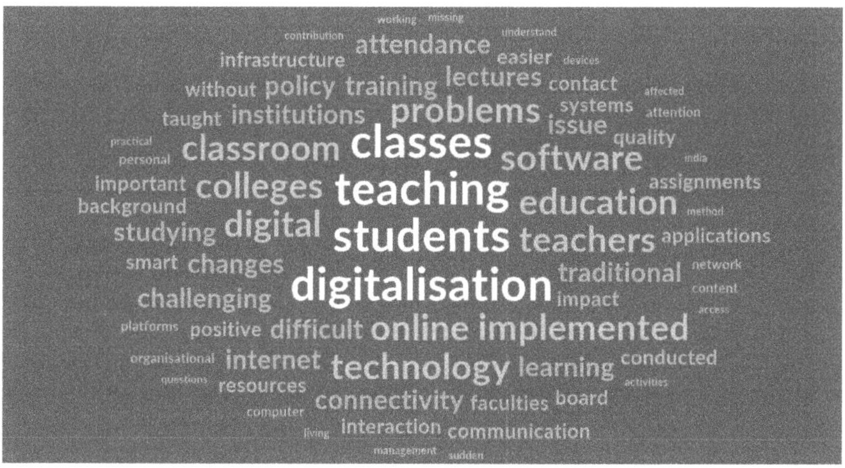

FIGURE 6.3 Word cloud. (*Source:* NVivo.)

higher the frequency of a word, the bigger the size of a word in the word cloud. For Word Frequency Analysis, 70 frequently used stemming words were taken with a minimum of 5 letters.

6.7.2 Emergent Themes

The qualitative analysis revealed five overarching themes: advantages of digital learning, benefits of traditional learning over digital learning, challenges of digitalization, measures to alleviate challenges, and prospects in digitalization. Each theme comprises multiple subcategories, as provided in Table 6.1. This study aimed to explore the overarching themes from interview data, verbatim quotations that represent each theme, and subcategories taken from interviews. The qualitative analysis

TABLE 6.1
Emergent Themes and Subcategories

S. No.	Overarching Themes	Subcategories
1.	Advantages of digital learning	• Better presentation • Access to updated resources • Greater reach • Convenient medium

2.	Benefits of traditional learning over digital learning	• Human interaction
		  • Learning ability
		  • Educators' understanding of students
		• Students' monitoring
		• Concentration retained

3.	Challenges of digitalization	• Connectivity issues • Lack of resources • Security threats • Complexity in teaching • Indiscipline • Health issues • Resistance

4.	Measures to alleviate challenges	• Responsiveness
		  • Organization
		  • Educators
		• Digital infrastructure
		• Government initiatives
		• Training for educators and students
		• Upgradation in regulations

5.	Prospects in digitalization	• Approach toward digitalization • Blended learning • Students' development

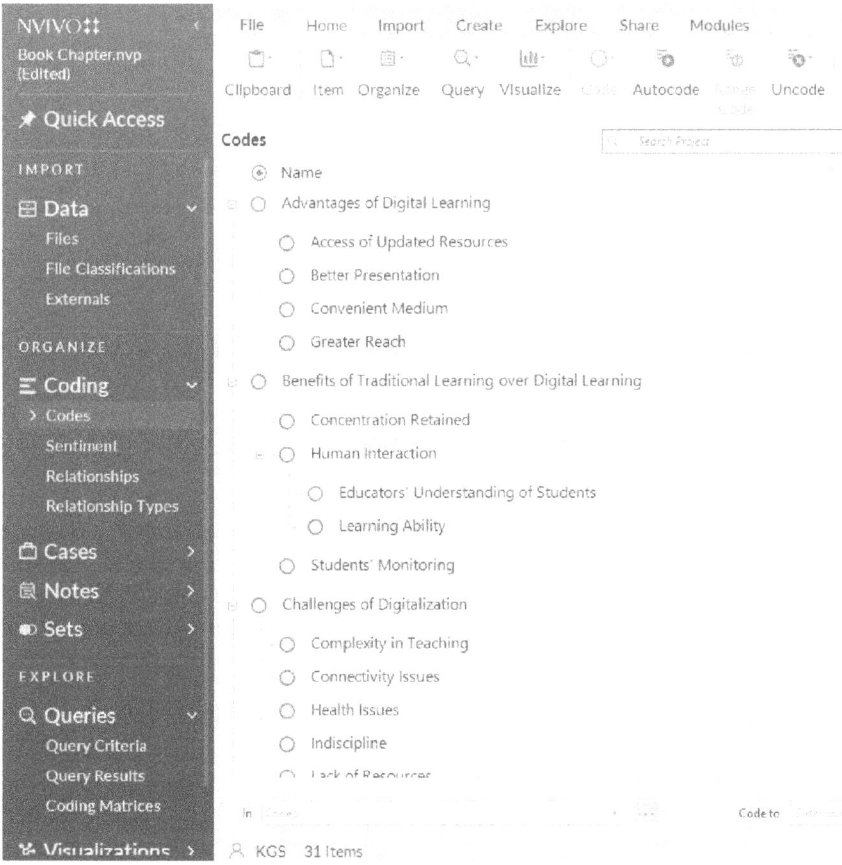

FIGURE 6.4 Image of the coding framework in NVivo. (*Source:* Self-developed.)

software NVivo 12 was used for the coding of interview transcripts. The main themes were represented as a parent node in NVivo and sub-themes as a child node. Figure 6.4 shows the coding framework in NVivo.

6.7.2.1 Advantages of Digital Learning

The overarching theme of advantages of digital learning has four subcategories, namely access to updated resources, convenient medium, better presentation of data, and greater reach. Several participants mentioned that due to digitalization, educators and students have access to updated and latest information and resources. They can access e-books and videos related to the latest topics in their subjects.

> *In digitalization, there are so many resources to learn new things and remain updated … We can teach from the latest sources, books, we can access the latest and updated software.* (P2)

Participants stated that with digital tools, they could present their data to students more creatively and innovatively, which may help students understand easily.

A few of them mentioned that practical aspects that cannot be taught through traditional classrooms could be taught through videos. Using technologies, things can be explained through 3D view also.

> *In traditional classroom teaching, we can only explain things verbally but through technologies, things can be shown better through a 3D view.* (P2)

Educators discussed the convenience of managing their work through digital tools. From teaching to evaluation to managing other academic activities, digitalization makes things easier and saves the time and effort of educators.

> *There are various platforms such as Google Classroom, the submission, evaluation, and record maintenance of assignments is quite easy.* (P1)
>
> *Good thing is, students who are participating in online classes, can interact with them easily. In a traditional classroom, we lose 5 to 10 min in taking classes while in online classes we can simply download the attendance showing when the student entered or exited the class.* (P4)

Educators also indicated that digitalization may increase the reach of education. Students who are in rural areas can access the lectures through digitalization, and in case of connectivity problems, they can watch recorded lectures. Students can learn from renowned professors through their video lectures because of digitalization.

> *Rural background students will have inclusion to study through this way.* (P3)
>
> *We can send the video lectures to the students who are living in remote areas as they are not able to attend the classes due to connectivity issues.* (P4)

6.7.2.2 Benefits of Traditional Learning over Digital Learning

The second overarching theme of the benefits of traditional learning over digital learning comprises three subcategories: human interaction, concentration retained, and students' monitoring. Numerous participants have mentioned that the biggest benefit of traditional learning is human interaction. Human interaction helps educators to understand the students better and teach according to their understanding level. Participants stated that in traditional classroom teaching, students interact with other students of different ideologies, which may improve their intellectual level and communication skills. A few participants believed that classroom teaching improves the learning ability of the students.

> *We can understand the student better, how strong, or weak a student is in a subject and further teach those students accordingly.* (P2)
>
> *In the traditional classroom, students got to meet students of different ideologies which help them to improve their intellectual level.* (P5)

Participants stated that students are more attentive in a traditional teaching setup compared to online classes. Their attention deviates when they are using phones, laptops, and other digital tools to attend classes.

Traditional teaching involves eye contact, and we can notify students in the classroom. We can call out such students and ask them to be mentally present in the class, this way the whole class once again becomes attentive. Teachers' presence makes students attentive. (P8)

We are the traditional teachers; in a classroom, we have eye contact with the students, and we know they are concentrating on whatever we are teaching. (P4)

Participants mentioned that they could not monitor students while teaching them through digital modes. Educators cannot monitor students through a camera while teaching. And sometimes, during online classes, teachers ask students to put their cameras off for better bandwidth and connectivity. Educators can expect that students are studying, but they cannot monitor all the students.

You cannot monitor students through online teaching when you are teaching a good number of students. (P1)

When students' camera is off whatever students are doing, we are not aware of. We expect they are listening and taking all the materials that we are providing but we are not sure that they are studying. (P4)

6.7.2.3 Challenges of Digitalization

Subcategories within the overarching theme of challenges of digitalization included complexity in teaching, connectivity issues, health issues, indiscipline, lack of resources, resistance, and security threats.

Several participants felt that digital teaching takes more effort than traditional classroom teaching. A few participants faced trolling during their online class experience, and a few of them were not comfortable teaching in front of a camera. Older educators are facing difficulties in learning new software. There are few subjects in engineering that require lab sessions and cannot be taught through digital mode. Similarly, management studies involve management games and other activity-based learning, which cannot be conducted through online modes.

It was difficult to understand the know-how of software .... but some subjects cannot be taught digitally like chemistry. You cannot teach practical aspects digitally. (P2)

In management, we do some exercises and classroom activities in class regularly. Those activity-based learning is a real-time activity and not possible through online modes ... Online teaching requires more effort in teaching. It is more tiring. (P4)

Few experienced teachers haven't used technology before, so it is very challenging for them. (P6)

Almost every participant mentioned that the biggest challenge that they faced in the implementation of digitalization is connectivity issues. Especially people in the rural areas bear the problem of network and connectivity during online classes. A few statements that exemplify this challenge:

The biggest con of digitalization is the internet connectivity problem in India. (P1)

... major issue was the network. Students in rural areas who are good at studies are facing problems because of internet connectivity or might not be having internet at all. (P8)

Another subcategory of the challenges of digitalization is health issues. Numerous participants stated that they had faced multiple health issues due to digitalization. Students are losing their social life due to digital classes, as personal interaction is reduced among students. Eyesight problems, back pain, and headaches are common because of using gadgets for longer hours.

> *There will be no social life. Personal interaction will not be there.* (P2)
>
> *We as a teacher also are not acquainted with phone and computers, so sitting in front of phones and laptops for hours at age of 35 we are getting back problems and spondylitis problem. We are not used to taking class sitting at a place for 7 to 8 hours like IT professionals, we are having a practice of moving in the classroom while teaching. UV rays are harmful, so the eyesight problem is also there.* (P4)

Indiscipline is another subcategory; various participants have faced indiscipline activities of students during online classes. Students write offensive comments by changing their names, and notorious students disturb the class. Participants also mentioned network issues as an escape route for not attending classes. A few examples in this subcategory include:

> *... it also creates the problem of indiscipline. It is difficult to manage notorious students as maintaining discipline is problematic during online classes.* (P1)
>
> *Physical interaction is missing so the issue of discipline is raised. Students find an escape route by telling it is a network issue or connectivity issue.* (P3)

Several participants mentioned that resources were lacking mostly with the students. Students who are not from financially strong backgrounds could not afford laptops and smartphones. Apart from this, a few participants stated there was no uniform software for taking online classes.

> *... we don't have a uniform platform for teaching.* (P3)
>
> *... if you do not have a good mobile phone and laptop so that was also an issue.* (P5)
>
> *All students didn't have a laptop and all the facilities.* (P6)

The sixth subcategory of this overarching theme is resistance from both educators and students. Educators, especially the older ones, are not equipped with these new technologies, and students are comfortable with traditional classes; they resist to adapt the online teaching environment.

> *... students are not adaptive towards this change, not ready to accept the change, and interested in face-to-face teaching.* (P3)
>
> *We are used to that whiteboard teaching and diagram. We are not comfortable with teaching on Microsoft Teams.* (P4)
>
> *The resistance of both students and teachers was there earlier. Understanding technology was difficult for everyone.* (P6)

Various participants faced security issues such as hacking of email ids. As email ids and phone numbers are required in every digital platform, educators were troubled by fake calls and fishy emails.

We highly depend on foreign software which is a threat to our security. Lots of email ids are hacked, their phone numbers are hacked, people receive fraud calls, fishy emails especially female faculties. (P3)

6.7.2.4 Measures to Alleviate Challenges

The fourth overarching theme consists of five subcategories under the theme of measures to alleviate challenges. These subcategories are digital infrastructure, government initiatives, the responsiveness of both organizations and educators, training of educators and students, and upgradation in regulations. Various participants highlighted the importance of digital infrastructure that must be set up by the organizations for smooth functioning.

Institutions should have better classrooms with all the facilities like mikes, projectors, systems. A better Wi-Fi system will help. (P8)

As during these online classes, there is less expenditure on infrastructure, colleges should provide financial help to the students to acquire those resources which are necessary for digitalization. (P5)

Digitalization can effectively seep into the system with the help of government initiatives. Participants stated that the government needs to come forward in providing better services to the nation for the growth of digitalization.

Even the Government should support providing better internet facilities. Every corner of India should have better internet connectivity as digitalization is not possible without it. (P7)

The Government also needs to work for better internet coverage. (P1)

Even educators need to take up the responsibility to adopt this change. The responsiveness of educators for doing fair work and upgrading themselves will help the students to learn.

Students and teachers both need to be honest and interested in this online teaching otherwise this may not work. (P2)

Participants mentioned that training for both students and teachers is a must to make them use these technologies easily. Students and teachers need to learn these tools to make use of these resources effectively.

Institutions should provide proper training to both students and teachers for using these digital tools, techniques, and software. (P5)

Proper training should be given to both students and teachers about how to use the technology. (P6)

Upgradation in regulations is required so that the implementation of digitalization is easy. Once rules and regulations are formed, organizations will have to follow them.

Universities need to form groups to frame a policy on how these things should go down. (P3)

The faculties at all levels need to be trained to use these technologies with proper examinations and certifications. As certifications in India is proof of our knowledge. (P8)

6.7.2.5 Prospects in Digitalization

The fifth overarching theme, prospects in digitalization, comprises three subcategories: an approach toward digitalization, blended learning, and students' development. A few participants mentioned the approach toward digitalization that is required for digitalization to grow in the educational field. A few excerpts are given here.

> *We have a large number of smartphone users and so the pace of digitalization will increase. Digital books and the digital library will increase.* (P3)
>
> *We can take classes through digital means which should be a part of our education system as students can be taught by good teachers.* (P6)
>
> *Students are already studying in traditional ways but when they will get content in a different form, they will be more interested.* (P8)
>
> *Quality of education will be impacted in a positive way who are little tech savvy but some of them will be impacted badly as they will misuse the technology like not paying attention to the classes, not listening to teachers, maybe using headphones to listen to music.* (P7)

Participants also mentioned that a successful pathway toward digitalization could be possible through blended learning, i.e. teaching through traditional and digital modes.

> *There is a need for blended learning, online teaching should be continued with traditional one.* (P1)
>
> *Colleges are introducing smart classes to create an environment of blended learning.* (P2)

Most of the participants said that students could be taught using technology which is essential for their development in the future.

> *It is a way to develop students professionally as this process of digitalization makes students learn by themselves which they can apply professionally.* (P7)
>
> *Students from school level only will have a good knowledge of computers and so up-gradation at higher studies will be seen. The digital presence of all people will increase.* (P3)

6.8 CONCLUSION

The objective of this study has been to illuminate the challenges faced by Indian educators due to digitalization and to suggest some measures to alleviate those challenges. The qualitative interview method was applied in this study to obtain diverse perspectives on the research questions. For data analysis, thematic analysis was used to study the detailed phenomenology of educators' experiences. The qualitative data analysis tool NVivo 12 (QSR International) was used to organize the data. We have extracted five overarching themes and various subcategories after sorting out seven overlapping themes. The emergent themes, as shown in Table 6.1, are advantages of digital learning, benefits of traditional learning over digital learning, challenges of digitalization, measures to alleviate the challenges, and prospects of digitalization.

The findings of this study indicate the multiple benefits of digital learning over traditional ones, such as better presentation of data in the form of 3D view and videos. Digitalization provides access to the updated resource and has greater reach (Gaebel et al. 2021); students can access lectures of subject experts from top universities from anywhere in the world to get better and in-depth knowledge. Students in rural areas may also be able to access lectures of educators from premier institutions and can gain knowledge. This will help the nation in achieving the fourth SDG for inclusive and quality education for all. Apart from all these benefits, there are few drawbacks of digital learning as well, such as the lack of human interaction that is present in traditional classroom learning (Gaebel et al. 2021). Face-to-Face human interaction helps the educator to understand the learning ability of students better. It is also difficult to monitor students in online classes, and students need to be self-motivated to study in a digital learning scenario.

Participants discussed the challenges they faced due to digitalization; the major challenge was connectivity issues during classes. India is the second-largest online market, but still, the Internet penetration rate was only 50% in 2020 (Statista 2020); there is a need for a better Internet penetration rate to implement digitalization. Educators who teach science and engineering subjects experienced problems in teaching numerical and lab-based subjects through a digital medium. Institutions also need to upgrade their digital infrastructure to enhance digitalization in organizational culture. Gaebel et al. (2021) in their study on European HEIs indicated that institutions' investment in equipment and infrastructure acts as an enabler for digitalization. Findings also show that students lack resources. Within the same institution, there is a division: a few students have access to the Internet and other resources while other students do not, which makes it difficult to impart equal opportunities to students (Marinoni, Land, and Jensen 2020). Participants agreed that digital medium and tools save their time and effort, but they also underwent some health issues due to the harmful effects of gadgets. A threat to security and data breach is always there when it comes to using a network. People face problems of getting their email ids hacked and data stolen from digital platforms. There is also resistance from both students and educators toward adopting the digital medium of learning. Faculties face difficulties in accepting educational innovation (Andres 2019), and for utilizing the digitalization optimally, it is important to have proactive participation from both students and educators (Gaebel et al. 2021). However, a few issues exist in both the traditional and digital media of teaching, such as indiscipline activities of students and complexity in teaching, but the extent of these challenges are more in a digital medium.

To overcome these challenges, it is mandatory to provide training to both students and educators to make them aware of the digital tools and software packages. Institutions also need to upgrade their IT infrastructure to support the digital learning medium. Universities should also make regulations for maintaining the IT infrastructure. The National Assessment and Accreditation Council (NAAC), which is responsible for maintaining quality and excellence in HEIs, already has policies for having digital resources, e-library, and maintaining IT infrastructure. Students need to be self-motivated for studying through digital mode as it is difficult for educators to monitor every student during class. This study suggests the blended learning approach as an optimal approach to learning. Blended learning is when traditional learning is supplemented with an online learning approach (Bates 2015; Moser 2016; Gaebel et al. 2021). With the blended learning approach, people will gradually get used to using digital media of learning.

6.9 IMPLICATIONS OF THE STUDY

This study offers insights into the issues and challenges faced by Indian educators working in private institutions. The findings of this chapter have significant implications that could be valuable to regulatory bodies, policymakers, HEIs, and faculty members who are interested in the implementation of digitalization. The present study suggests that digitalization is the future of higher education and Indian private institutions need to upgrade their IT infrastructure to sustain in the competitive world. The changes are needed on behalf of both organization and educators. The organization should provide proper digital infrastructures such as digital classrooms and Wi-Fi and LAN connection for better connectivity, and educators need to become more aware of digital tools and technologies. Though various initiatives have been taken by the Government to promote digitalization, such as SWAYAM, which was launched to promote online open courses, commonly known as MOOCs. But better Internet penetration is required; India still has a low Internet penetration rate even when it is the second-largest online market. Initiatives should be taken by the Government in this direction, and special consideration should be given to rural areas. Apart from this, training is mandatory for both students and teachers at all levels to enhance the scope of digitalization in the organization. Educators who are relatively senior and quite experienced are not well versed with these technologies and software. Thus, proper training will help them learn new tools and technologies. Amendments could be made in university regulations to upgrade digital infrastructure that colleges will have to follow. This research study suggests that in the current scenario, a blended learning approach is an optimal approach, and it may help to implement digitalization progressively in the higher education sector.

6.10 LIMITATIONS AND DIRECTIONS FOR FUTURE RESEARCH

The major limitation of this study was having telephonic interviews, as non-verbal cues could not be noticed. Data were collected from eight HEIs educators working in private institutions in North India. These educators' opinions might vary as they lack infrastructural facilities in comparison to those working at premier institutions. Each respondent has shared unique information with us, but they might not want to discuss their flaws or share the organization's lack of support with us. Interviews were transcribed verbatim by authors, but there might be a problem of cohesion as there is a slight chance of misinterpretation. In qualitative research, there is always scope for finding and adding new information to the current themes with a new dataset. Further studies can be done with a large sample covering more geographic regions. Moreover, we suggest studies to be done with premier institutions of India as they might have better infrastructural facilities.

REFERENCES

Ally, Mohamed. 2006. “Foundations of Educational Theory for Online Learning.” In *Theory and Practice of Online Learning*. edited by Terry Anderson and Fathi Elloumi, 111–136. Canada: Athabasca University. https://books.google.com/books?hl=en&lr=&id=RifNwzU3HR4C&oi=fnd&pg=PA343&dq=related:OrgqTVMknoAJ:scholar.google.com/&ots=Sf7rIfH_rw&sig=vM8mWLFZedZUrcuJhTjKSO_KnrQ.

Andres, Hayward P. 2019. “Active Teaching to Manage Course Difficulty and Learning Motivation.” *Journal of Further and Higher Education* 43 (2). Routledge: 220–235. doi: 10.1080/0309877X.2017.1357073.

Bates, A. W. (Tony). 2015. “Teaching in a Digital Age: Guidelines for Designing Teaching and Learning.” Quarterly Review of Distance Education, 16.

Baumöl, Ulrike, and Alina Bockshecker. 2017. “Evolutionary Change of Higher Education Driven by Digitalization.” *16th International Conference on Information Technology Based Higher Education and Training (ITHET)*, 1–5.

Braun, Virginia, and Victoria Clarke. 2006. “Using Thematic Analysis in Psychology.” *Qualitative Research in Psychology* 3 (2): 77–101. doi:10.1191/1478088706qp063oa.

Braun, Virginia, and Victoria Clarke. 2013. *Successful Qualitative Research: A Practical Guide for Beginners*. SAGE Publications. https://books.google.com/books?id=EV_Q06CUsXsC&pgis=1.

Brennen, J. Scott, and Daniel Kreiss. 2016. “Digitalization.” *The International Encyclopedia of Communication Theory and Philosophy*: 1–11. doi:10.1002/9781118766804.wbiect111.

Fedirko, Danylo. 2019. “8 Top Trends of Digital Transformation in Higher Education.” ELearning Industry. https://elearningindustry.com/digital-transformation-in-higher-education-8-top-trends.

Fugard, Andrew J. B., and Henry W. W. Potts. 2015. “Supporting Thinking on Sample Sizes for Thematic Analyses: A Quantitative Tool.” *International Journal of Social Research Methodology* 18 (6). Routledge: 669–684. doi:10.1080/13645579.2015.1005453.

Gaebel, Michael, Thérèse Zhang, Henriette Stoeber, and Alison Morrisroe. 2021. *Digitally Enhanced Learning and Teaching in European Higher Education Institutions*. https://www.eua.eu/downloads/publications/digihe survey report.pdf.

Gioia, D. A., and D. J. Brass. 1984. “Teaching the TV Generation: The Case for Observational Learning.” *Organizational Behavior Teaching Review* 10 (2): 11–18.

Herrington, Jan, Anthony Herrington, Jessica Mantei, Ian Olney, and Brian Ferry. 2009. “Using Mobile Technologies to Develop New Ways of Teaching and Learning.” In *New Technologies, New Pedagogies Mobile Learning in Higher Education*. Vol. 9. Australia: University of Wollongong. https://ro.uow.edu.au/cgi/viewcontent.cgi?article=1092&context=edupapers.

Jha, Nivedita, and Veena Shenoy. 2016. “Digitization of Indian Education Process: A Hope or Hype.” *IOSR Journal of Business and Management* 18 (10): 131–139. doi:10.9790/487x-181003131139.

Khalid, Jamshed, Braham Rahul Ram, Mohamed Soliman, Anees Janee Ali, Muhammad Khaleel, and Md Shamimul Islam. 2018. “Promising Digital University: A Pivotal Need for Higher Education Transformation.” *International Journal of Management in Education* 12 (3): 264–275. doi:10.1504/IJMIE.2018.092868.

Kirkwood, Adrian, and Linda Price. 2014. “Technology-Enhanced Learning and Teaching in Higher Education: What Is ‘enhanced’ and How Do We Know? A Critical Literature Review.” *Learning, Media and Technology* 39 (1): 6–36. doi:10.1080/17439884.2013.770404.

Komljenovic, Janja. 2020. “The Future of Value in Digitalised Higher Education: Why Data Privacy Should Not Be Our Biggest Concern.” *Higher Education*, 0123456789. Netherlands: Springer. doi:10.1007/s10734-020-00639-7.

KPMG. 2017. *Online Education in India*: 2021. https://assets.kpmg/content/dam/kpmg/in/pdf/2017/05/Online-Education-in-India-2021.pdf.

Lai, Kwok Wing. 2011. "Digital Technology and the Culture of Teaching and Learning in Higher Education." *Australasian Journal of Educational Technology* 27 (8): 1291–1303. doi:10.14742/ajet.892.

Laxman, Rajnalkar, and Anil Kumar Hagargi. 2015. "Higher Education of India: Innovations and Challenges." *The Eurasia Proceedings of Educational & Social Sciences (EPESS)* 2: 144–152.

Lin, Ming Hung, Huang Cheng Chen, and Kuang Sheng Liu. 2017. "A Study of the Effects of Digital Learning on Learning Motivation and Learning Outcome." *Eurasia Journal of Mathematics, Science and Technology Education* 13 (7): 3553–3564. doi:10.12973/eurasia.2017.00744a.

Marinoni, Giorgio, Hilligje Van Land, and Trine Jensen. 2020. *The Impact of COVID-19 on Higher Education Around the World: IAU Global Survey Report*. https://www.iau-aiu.net/IMG/pdf/iau_covid19_and_he_survey_report_final_may_2020.pdf.

Ministry of Education. 2020. "National Education Policy 2020 Government of India." 33–49. https://www.education.gov.in/sites/upload_files/mhrd/files/NEP_Final_English_0.pdf

Moser, Karin S. 2016. "The Challenges of Digitalization in Higher Education Teaching." In Zimmermann, T, Jütte, W and Horvath, F (ed.), *University Continuing Education: Facts and Future*. Vol. 147, pp. 93–100. (Arenen der Weiterbildung) Bern, Switzerland: HEP Verlag.

Patton, Michael Quinn. 2002. *Qualitative Research & Evaluation Methods. Qualitative Inquiry*. 3rd ed. SAGE Publications. http://books.google.com/books/about/Qualitative_research_and_evaluation_meth.html?id=FjBw2oi8El4C.

Pigden, Louise, and Franc Jegede. 2020. "Thematic Analysis of the Learning Experience of Joint Honours Students: Their Perception of Teaching Quality, Value for Money and Employability." *Studies in Higher Education* 45 (8): 1650–1663. doi:10.1080/03075079.2019.1661985.

Proserpio, Luigi, and Dennis A. Gioia. 2007. "Teaching the Virtual Generation." *Academy of Management Learning & Education* 6 (1): 69–80. doi:10.1177/1080569907305305.

Seshadri, C., S. B. Menon, S. C. Shukla, Poonam Bhushan, and N. K. Dash. 2017. "Indian Higher Education: The Legacy." In Higher Education: Reteospects and Prospects. http://egyankosh.ac.in/bitstream/123456789/8433/1/Unit-3.pdf.

Sheikh, Younis Ahmad. 2017. "Higher Education in India: Challenges and Government's Initiatives." *Learning Community-An International Journal of Educational and Social Development* 6 (1): 19. doi:10.5958/2231-458x.2014.00011.6.

Statista. 2020. "Internet Usage in India – Statistics & Facts." https://www.statista.com/topics/2157/internet-usage-in-india/#:~:text=It was estimated that by, access to internet that year.

Sturges, Judith E., and Kathleen J. Hanrahan. 2004. "Comparing Telephone and Face-to-Face Qualitative Interviewing: A Research Note." *Qualitative Research* 4 (1): 107–118. doi:10.1177/1468794104041110.

The Wire. 2019. "India's Higher Education Needs a Paradigm Shift." https://thewire.in/education/indias-higher-education-needs-a-paradigm-shift.

Thoring, Anne, Dominik Rudolph, and Raimund Vogl. 2017. "Digitalization of Higher Education from a Student's Point of View." EUNIS 2017–Shaping the Digital Future of Universities (2017): 279–288.

Tømte, Cathrine Edelhard, Trine Fossland, Per Olaf Aamodt, and Lise Degn. 2019. "Digitalisation in Higher Education: Mapping Institutional Approaches for Teaching and Learning." *Quality in Higher Education* 25 (1). Routledge: 98–114. doi:10.1080/13538322.2019.1603611.

United Nations General Assembly. 2015. "Sustainable Development Goals." https://in.one.un.org/page/sustainable-development-goals/sdg-4/.

Wachal, Robert. 1971. "Humanities and Computers: A Personal View." *The North American Review* 256 (1): 30–33. http://www.jstor.com/stable/25117163.

Young, Suzanne. 2010. "Student Views of Effective Online Teaching in Higher Education." *American Journal of Distance Education* 20 (2): 65–77. doi: 10.1207/s15389286ajde2002_2.

7 Digital Education and Society 5.0

Manju Amla

CONTENTS

7.1 INTRODUCTION

The world has been struck by a digital tsunami of change. This digital transformation wave is not going to end soon and is dramatically changing the different aspects of society that include public administration, industries, jobs, and our personal lives too. Digital technology is revolutionizing not just the way we communicate with each other but the patterns of entertainment, industrial work, and education also. It has given us the freedom to live our lives to the fullest by providing ways to overcome a variety of societal and technological challenges. To better harness the ability of digital technologies, Japan envisaged a future society called Society 5.0 within the context of Sustainable Development Goals (SDGs) 2016 (General Assembly 2017) about education, it portrays an essential role in the advancement, integration, and development of society. Education encourages healthy, self-regulatory,

DOI: 10.1201/9781003132097-7

compassionate, versatile, and decision-making citizenship as suggested by European Commission (2018).

Society 5.0 relies on individuals whose ambition is to use technology to make a sustainable future, with better career prospects and change in mentality, economics, and geopolitics as per (Keidanren, 2016) paper. According to European Commission, 2018) a "super-intelligent" culture entails cultural development and ethical considerations and recognition by all those who are concerned. In an attempt to do this, the existing structure attempts to knock down the barriers of (a) governments, with national policies focused on the Internet of Things (IoT) and the "think tank" structure; (b) the judicial system, in which bureaucratic digitization is promoted; and (c) technology, a consistent quest for the creation of an "information basis," cybersecurity, innovative technologies, and more. The present chapter explains digital education and Society 5.0 with its theoretical framework and literature review along with findings. This chapter also covers recommendations for educational institutions to keep up with the process of change, future challenges, and managerial implications along with emerging issues and the conclusion of the study.

7.2 OBJECTIVES

1. To explore the concept of digital education with its theoretical framework.
2. To explore the concept of Society 5.0 with its theoretical framework.
3. To explore the previous studies on related concepts.
4. To suggest possible recommendations according to the findings of the study.
5. To identify the future challenges along with the managerial implications of the study.

7.3 DIGITAL EDUCATION

In recent times, education has become one of the sectors that are going under disruptive changes worldwide. Digital education is proving to be the magic bullet for this educational anathema around the globe. The different ways of communication cover a wide area, which makes it easier and affordable even for the farthest places of the earth. Talking about digital education, it is a creative utilization of digital tools and technologies in pedagogy, also known as technology-enhanced learning (TEL) or e-learning and, more commonly, digital learning. It is defined as a collection of technology-enhanced strategies used to facilitate teaching and education, including evaluation, learning, and instruction. There is a range of digital platforms to sustain e-learning. The Internet, for instance, is used as a networking platform that at one time frees the student from physically attending classes and at the same time connects it to the others while building up the learning communities.

Digital education provides great opportunities both in terms of accessibility and efficiency. It is changing the educational landscape as a lot of students are seeking online learning courses nowadays. Institutions of higher learning are also adopting the efficiency of Internet classes and rapidly implementing online programs to meet the needs of students around the world. Research by Lundberg et al. (2008) found that the growth in the number of online courses offered by universities has been quite significant during the last few years.

The global market size of online learning was around US$187.87 billion in the year 2019 and is projected to reach US$350 billion by 2025 worldwide. Advanced artificial intelligence (AI)-driven platforms are also playing a critical role in the growth of this market across nations. The top two nations in the international online education industry are the United States and China, driven by the rising Internet penetration, increasing disposable income per capita, and the availability of online courses.

There are multiple online learning services in the market, including Udemy, Coursera, Lynda, Skillshare, and Udacity, which support millions of people. Skillshare is mainly for creative people and provides courses on animation, lifestyle, and photography. Coursera provides academic courses at various universities. Top-tier colleges are now democratizing learning by making classes available online. Stanford and Harvard are providing access to online courses in the fields of engineering, computer science, business, art, mathematics, and personal development.

All in all, it took just one generation for digital learning to be born, develop, and become the standard for higher education. Today, students expect courses to have at least one online component, if not a fully online course. There are, of course, benefits and drawbacks associated with digital learning. It is up to the teachers to help students optimize the former and minimize the latter. With a strong understanding of digital learning principles and vocabulary, teachers have to prepare themselves to help students excel.

7.3.1 Theoretical Framework of Digital Education

Contrary to consensus, online education is not a modern phenomenon. While there are ample instances of the use of computers and technology in education across history, e-learning in today's world is a very new approach. However, since the 1950s, the use of TV-based courses and slide projectors has been there. Even so, one of the early instances of online learning in the world can be dated back to 1960 at the University of Illinois, United States. Although the Internet had not yet been invented back then, students used to study from computer terminals that were interconnected to create a network. The University of Toronto offered the first fully online course in 1984 and developed the Electronic University Network in 1986. In 1989, Phoenix University became the world's first educational institution to offer both Bachelor's and Master's degrees, which inaugurated a completely digital college setup. This was the rise of a movement whose possibilities were largely unexplored by the public back then.

In the early 1990s, the Worlds' first Open University was set up by Britain to begin online distance learning. Presently, India's Indira Gandhi National Open University has around four million students enrolled, and most of them are currently attending classes online; this makes it the largest university in the world. Innovation has undoubtedly helped to improve the pace and affordability of correspondence courses as overseas students can now take classes from anywhere.

7.4 SOCIETY 5.0

A dream of a society in which technology would be there in every course of life. This society of the future will sustainably build new values and programs to help and stabilize society as a whole. Japan has already started using the resources of the 4th Industrial Revolution, including all emerging technologies like AI, robotics, sharing economics, big data, and IoT, to create a modern society: Society 5.0. It seeks to solve a range of problems not just by moving toward digitalization at all levels of society but the (digital) transformation of society itself. It aims to upgrade medical services by incorporating robots, to eliminate the regional and linguistic differences by automated translation, and to free people from the inconvenience caused by age or any kind of disability with the use of AI and robotics.

As stated by the Cabinet Office (2017b) "the vision of future society toward which the Fifth Basic Plan proposes that we should aspire, will be a human-centered society that, through the high degree of merging between cyberspace and physical space, will be able to balance economic advancement with the resolution of social problems by providing goods and services that granularly address manifold latent needs regardless of locale, age, sex or language to ensure that all citizens can lead high quality lives full of comfort and vitality."

Another definition of Society 5.0 is that it is a community in which the various needs of the population are precisely distinguished and satisfied by providing sufficient goods and services to the people in adequate amount and in which all people can receive a superior quality of services and live a comfortable, prosperous life and arrangements for inequality among age, sex, region, or culture are made (Harayama, 2017).

Japan aims to "generate new concepts by integrating and collaborating with a variety of different frameworks and plans to standardize file format, templates, system design," etc. and improve the requisite personnel (Hayashi et al., 2017). In addition to it, amendments to intellectual property growth, international standardization, IoT device construction techniques, big data and artificial intelligence technologies and so on are expected to foster Japan's productivity in the "super-smart society" with Society 5.0.

Compared to the knowledge society, where information is gathered through the network and processed by humans, in Society 5.0, the production of information is to be achieved by computers. The best possible outcomes of adding new values to businesses and society can be obtained in this way, which was previously not achievable (Granrath, 2017; Önday, 2019).

7.4.1 Theoretical Framework of Society 5.0

Human society's past has been marked by emancipation from constraints and the development of liberty via increased functionality derived from new methods and techniques. Heading from the Hunting Society to the Agricultural Society, mankind acquired the capacity to produce food that liberated them from starvation. In the Industrial Society, they had increased manufacturing capacity by motivating the employees. The rich list of Society 5.0 can be seen from the history of the development of society (Harayama, 2017), notably Society 1.0 (Hunting Society), recognized as a community of individuals who compete and live together in tune with nature; Society 2.0 (Agrarian Society) defined as a creation specifically based on intensive farming, organizational change, and national development; and Society 3.0 (Industrial Society), a social system that encourages industrial development via the development of humanity. Society 5.0 is founded on Society 4.0 (Information Society), which aims at a people-centric approach. These stages or the list of historical transformations starting from hunting to industrial revolutions have introduced not only greater comfort and advances to society but also systemic improvements to it.

Upon this description, a new community has been developed which recognizes the types of skills that need to be developed and improved, the constraints that need to be simplified, the opportunities to be obtained, and the principles that need to be followed in this new society. Society 5.0 involves a rich imagination to recognize issues and problems that have been there from a long time in the society and also ingenuity to come up with solutions using emerging technology and data sets. Pairing digital transformation with the ingenuity and innovation of various communities would make it easier not only to solve challenges but also to create value that will contribute to a better world.

Digitalization has strengthened the ability of information and communication technology transmission, which has greatly expanded independence of dissemination of knowledge and connectivity of all kinds and further exploring the different options for Web services. Society 5.0 portrays a sustainable socio-economic structure in which human capital is no longer used for data processing and collection, and instead, big data, AI, IoT, robotics, and other advanced technologies would be used. As human productivity is in limited form, there is a major need to explore such tools and techniques which could break through the current sense of inactiveness and create a new generation of society.

7.5 LITERATURE REVIEW

The goal of the review of the literature is to establish a theoretical and conceptual background for the research. It includes the location, documenting, and assessment of relevant studies, which gives the investigator an appreciation of the prior work that has been already performed in this field of expertise. The literature is divided into three sections: studies on digital education, studies on Society 5.0, and studies on digital education and Society 5.0.

7.5.1 Studies on Digital Education

Abel (2005) summarized around 21 research articles on Internet-based learning and narrowed down four major areas that need major attention: encouragement and leadership; curriculum focus; faculty help and student services; and fixing targets and evaluation. Underwood (2009) reported that digital learning has increased the significance of learners in studying, their conviction in practicing skills, and their time on informal learning. Wikramanayake (2005) explained the process of creation of, development of, and acquiring knowledge through technology. The study explained the use of information and communications technology (ICT) to manage and organize explicit knowledge and describe ways to access and apply that knowledge. The study investigates how these technologies have been used in academic achievement and their impact in particular. Coldwell-Neilson et al.(2005) presented the findings of a survey on Deakin University students in 2005. The research examined their experiences of learning in the online world and showed that students were excited about learning in such an atmosphere. The key benefits were the versatility it offers and the opportunity to research according to their preferences. The drawbacks include technological problems such as accessibility and the need to interact regularly. Condie and Bob (2007) found that digital learning rendered science more appealing, reliable, and necessary to students and displayed more time for post-experiment consultation.

Broadley (2007) examined e-learning methods among three schools of the Curtin University of Technology in Western Australia. Samples were collected through interviews and the observation method from the teachers and ICT coordinators. The findings of the study explored the problems related to the introduction of e-learning by educators, such as skill development, shifts in their positions, and the teaching methods they use. The results of the study suggested four main factors: ICT infrastructure and leadership along with support and training programs and the ability of teachers to successfully implement the e-learning environment. Power (2008) presented a study explicitly based on on-field findings and documented case studies and recommended a mix of blended and online learning atmosphere and examined its influence on higher education, incorporating potential positive effects on the handling of educational design and technology.

Qiao and Nan (2009) studied the perspective of educators of using ICT at Capital Normal University. Totally, four areas were examined, and the results of the study showed that most participants were required to have computer skills in graphic design software, the Learning Management System (LMS), and digital educational materials, and a few required to acquire basics such as e-mail writing and the use of the Internet. Besides, participants decided to learn how to incorporate technology effectively and efficiently in classroom instruction. Lee et al. (2009) figured out the activities of 15- to 16-year-old elementary students in the United States and their standardized literacy test scores related to the usage of the computer. Learners were inquired about the number of hours they spent on the system for school and other tasks. Students who utilized systems for a minimum of one hour a day for both school and other tasks were found to have improved reading test scores and better supportive teacher ratings of their classroom efforts than any other group. Yurumezoglu (2009) reviewed the effect of information technology on teaching the basic principles of astronomy for 11- to 13-year-old schoolchildren in Turkey. The respondents were requested to perform considerations about an astronomical event, such as the objectives of the seasons or the phases of the moon. The software was adopted to model and present the results of the reviews. Thereafter, they were requested to describe and generate judgments about their perceptions. The investigation supported that instruction promoted by anecdotes and computer design was greatly helpful in reinforcing the theoretical learning and awareness of the topic. Schneider (2010) performed a mixed-method analysis on the comprehension of student experience, expectations, and attitudes of online learning at one of the colleges of Ontario. A sample of 279 respondents was collected using both quantitative and empirical approaches. The findings proved that most of the students were happy with the online learning experience, and accessibility and the ability to monitor were the key benefits of online learning. Even though the students were happy with their online learning experience, they still chose hybrid/face-to-face programs over a completely online program, the explanation for the same being the socialization that took place in the classroom. Goodwyn (2011) supported the combination of teacher and technology and claimed a "Digi-teacher" (teachers who can incorporate ICT into daily learning) is someone who has a higher desire to communicate with their learners' activities and have standardized digital technology in the classrooms; however, they are mostly self-trained. The writer also talked about the need for appropriate time and training for teachers to develop their practice. Online education also provides flexibility in attending the classes and accessing the course content of any college from anywhere on the earth (Salcedo, 2010). Also, online learning provides a forum for students who do not usually take part in a class or hesitate to talk; they can now express their thoughts and problems more freely without really being heard or evaluated, which improves the overall performance of the classroom (Driscoll et al., 2012).

Mnyanyi et al. (2010) studied the experience of students and educators after incorporating online learning along with its obstacles and possible opportunities. The research results emphasized managing the e-learning content along with the technological aspects of the production of courses and the strategies for the development of skills among its educational stakeholders. The study discussed connectivity issues, ICT facilities, software, training, logistics, low expenditures, economies of scale among learners, and attitudinal factors among social members that need to be addressed during implementation. The study suggested making informed decisions on the design, production, and implementation of ICT components to help learners. Güven et al. (2012) figured out the role of computer-enhanced learning in science and technology coursework on the framework and properties of materials like periodic tables, chemical bonding, and chemical reactions for 13- to 14-year-old children in Turkey. The study formed a notable difference in performance tests between the group of learners who were taught using the computer-enhanced teaching technique and the group who were taught using conventional teaching programs. Hung, Huang and Hwang (2012) inspected the impact of handling multimedia systems in science teaching in Taiwan's middle school. They requested the respondents to accomplish a digital storytelling proposal by selecting pictures with digital cameras, building up a photo-based story, generating a photo-based film, incorporating captions and background, and showing the story. The results identified that respondents in the empirical community preferred a project-based lesson design and regarded it as interesting for the obvious digital storytelling material. Urban-Woldron, (2013) favored long-term integrated learning compared to a face-to-face training session to boost the ability to incorporate technology into education. Lysenko et al. (2014) analyzed the use of two interactive reading comprehension processes for elementary school kids (6–8 years of age) in Quebec, Canada. The data revealed that in classrooms where both programs (multimedia and web-based electronic portfolio) were applied simultaneously throughout the entire school year, learners performed well in both vocabulary and reading awareness. Cheok and Luan (2015) proved organizational support to be the most important factor, which lets teachers to adopt innovation in teaching methods. Dua et al. (2016) addressed the development opportunities and various problems of digital education in India and proposed the empowerment of a dynamic learning classroom model. The emerging developments in interactive education involve digital classrooms, video-based learning, game-based learning, and so on. The study figured out the distinct concept of digital education in India and recommended strategies to rectify the same. The research also suggested some perpetual changes for schools and teachers for the advancement of online education in India. Palvia et al. (2018) evaluated emerging technologies and e-learning patterns for teachers' education and career growth from the perspective of emerging trends and the latest happenings in the Asia-Pacific region. Findings have shown that by focusing on even more recent techniques, e.g., (1) transformation of knowledge distribution to online materials and coursework, (2) development of web-based lecture halls, (3) participation in learning platforms, and (4) developing awareness of development in knowledge-building communities. The research proposed that technical advances should follow social and conceptual improvements and in that educators should play a critical role as they act as stakeholders.

Sousa and Rocha (2018) explored how to assess the effects of digital learning in higher education. The purpose was to bridge the gap between academic and practice-oriented studies on online education. The study was qualitative, and content analysis was used to achieve the goals of the study. The results proved online education platforms could increase the productivity of students, and the training structure, operations, and learning environments powered by smartphones, tablets, and phone apps were found to be immensely popular with students. Kumari (2019) examined digital education and its advantages, challenges, and future direction in the Indian context. The study talked about the digitization of education in the 21st century and proved beneficial as it combined the elements of both school learning and online learning methods. Rolling hand in hand, both serve as a helping network for each other, giving progressive students a foothold. Online examination portals prohibited the unnecessary use of papers, explicitly restricting the chopping down of forests. Digital education gives the ability to connect with teachers, address peers, and access learning resources and gives freedom to complete projects from anywhere on the web (Richardson and Swan, 2003).

7.5.2 Studies on Society 5.0

Prasetyo and Arry (2017) focused on obtaining the right resource to apply in the Community Management System for the Smart Society Platform. This research suggested that the computational architecture of community management systems has important benefits to help Society 5.0 in the context of a smart social network. Saxena et al. (2020) examined the Vision of Industry 4.0 for Society 5.0. The vision will help prepare a smart society with a modern version of 5.0 of education and instructors. The New Education Policy (NEP) 2020 in India focuses on the principles of Education 5.0. The study also discussed some new knowledge bases for the education system. Rodriguez-Abitia et al. (2020) looked at the problems of innovation and implementation faced by seven institutions of higher education (three from Spain and four from Mexico). For data collection, direct observations and in-depth interviews were conducted. The results summarized three components: technological, methodological, and administrative, which need to be focused on for growth and availability of technology. The findings also showed that cultural differences play a decisive role in the ability of the organization to prosper from technology that supports the educational process and ensures its performance. Supendi and Nurjanah (2020) studied Society 5.0 and its readiness among students for the realization of society. The study summarized that students should have three specialized qualifications: problem-solving skills, analytical thinking, and innovation. Airawaty and Widarjo (2021) examined the effectiveness and implements of e-learning in education in Industrial Revolution 4.0 and Society 5.0. They applied the analysis method to examine the consent of 125 accounting graduates and to assess the learning outcome using the structural equation modeling (SEM) and the Mann-Whitney test. The investigation of classroom instructions using two separate approaches did not declare any considerable disparities in discoveries. This research reinforced the claim that a varied teaching method using advanced e-learning technologies can enhance Indonesian education.

7.5.3 Studies on Digital Education and Society 5.0

Mathews and Gondkar (2017) analyzed IoT in education and narrowed down that some features like digital content were easy to share and were accessible to every student. Abbasy and Quesada (2017) examined IoT and its application in higher education. It was found that IoT was very useful and more efficient compared to the older techniques of education, with the reason being it can work on a bigger database and tons of resources. The same proved to be very helpful in better engagement of students as students felt more attentive, curious, motivated, and passionate about learning through means of videos rather than just reading from a textbook. Abdel-Basset, et al., (2019) summarized the advantages of IoT in the education sector. It enables advanced attendance tracking systems, which reduces the proxy attendances in the institute. It can help provide advanced security measures to restrict the illegal bunking of classes on the campus. However, the availability of a large amount of faculty and student data comes with multiple security and privacy challenges that need to be dealt with. The introduction of all these devices requires a huge investment as capital, which puts up another challenge. Proper administration of these devices requires a lot of accuracy and efficiency.

Sudibjo et al. (2019) evaluated the learning characteristics of the Jakarta Private University in the times of Industry 4.0 and Society 5.0. The qualitative approach was used to get the results. The findings showed that the learning characteristics in the era of Industrial Revolution 4.0 and Society 5.0 are evolving concepts. They also found that IoT and AI were rapidly changing the scenario. The most appropriate teaching approach used in the digital age is student centered and teacher directed. Project-based learning and collaborative learning are the learning models that suit the digital age. The pedagogical strategies that suit the modern age are combined with hybrid learning and e-learning.

7.6 MATERIALS AND METHODS

The qualitative analysis approach has been used to provide a comprehensive description of the context to explain the actions and inter-relationships of digital education and Society 5.0. The research is typically based on secondary sources, and for that, a vast volume of published information has been used, including books, magazine papers, scholarly journals, and websites articles.

7.7 FINDINGS AND RECOMMENDATION

The findings of the study revealed the importance of ICT infrastructure to manage and organize explicit knowledge among students (Wikramanayake 2005; Broadley 2007). Previous studies (Schneider 2010; Lin, Chen and Liu 2017; Sudibjo et al., 2019) have also focused on a mixed method of education as students do not want to compromise on the socialization part that takes place while attending the classes through offline mode. The importance of software training to teachers has to be realized so that they will be able to transform knowledge distribution to the online medium, assist in developing web-based lecture halls, and participate in various learning platforms. Online education has been found to be very helpful in studying difficult subjects and increasing students' practicing skills (Condie and Bob 2007; Bigelow, 2009). The usage of computers has been found to increase reading scores along with theoretical learning and awareness.

Yurumezoglu (2009) supported that instruction promoted by anecdotes and computer design was greatly helpful in reinforcing the theoretical learning and awareness topic.

The findings also talked about the importance of teachers' knowledge of basic computer skills and computer software such as LMS s (Qiao and Nan 2009). The more computers become familiar with the learning behavior, the more effective the specific assignment, assessment, and creation of new content is expected to be in the future days. Cheok and Luan (2015) proved organizational support to be the most important factor, which lets teachers adopt innovation in teaching methods. Kumari (2019) talked about the benefits of online learning methods and also drew attention to conducting online exams as it prohibits the unnecessary use of papers, explicitly restricting the chopping down the forests.

Findings also suggested the development of digital classrooms and video-based and game-based learning to raise the interest of learners in subjects. Project-based and collaborative learning models were found to be the best learning models that suit the digital age. Online education proves beneficial for teachers as it helps in monitoring the performances of students.

The usage of modern technology like IoT, AI, and big data is on the rise for future education. Society 5.0 demands students to prepare for critical thinking and be equipped with problem-solving skills and, above all, good command over soft skills (Supendi and Nurjanah 2020). To face the difficulties that emerge in Society 4.0 and Society 5.0, societies should have the opportunity to observe things such as critical thinking to resolve issues; the group must also have the confidence to try new things to compete with other societies, as well as the ability to cope well. These three skills can be accomplished by thinking to a greater extent. Current developments allow the student to play active participation in the learning process, with daily input and evaluations.

At present, almost every country is dealing with the lack of self-guided, independent, and fundamental academic abilities at all levels of learning and needs to focus on human capabilities more. Based on the findings, there are certain recommendations to be followed by educational institutions and countries while making the national curriculum for digital education and to keep in pace with the changes. These recommendations are suggestive and not exhaustive.

1. *Equal and Optimized Learning*: Individual learning that focuses on realistic research and development should be undertaken with usual activities like sports, volunteer work, and other activities. These activities enhance cultural competence among students and provide a deep awareness. Every student must learn basic academic skills e.g., simple reading comprehension, knowledge competence, and mathematical reasoning, etc.
2. *Focus on Fundamentals*: Focus on gaining basic academic skills, including a structural understanding of sentences, arithmetic, vocabulary knowledge, mathematical thought, and reading. For this purpose, learning support can be enhanced by intensifying the preparation of teaching materials, ICT environments, and EdTech (educational technology) that will enhance instructional methods. Data science and statistical education should be improved through primary, lower, and upper secondary schools. Instructional protocols in schools and the credential system for teaching should be strengthened to ensure that basic academic skills are continually mastered.
3. *Surpass the Gap Between Humanities and Science*: The aim should be to provide futuristic education e.g., STEAM (science, technology, engineering, art, and mathematics), design thinking, and online labs, etc. and for that, educational institutions should make smooth arrangements to transit from one stream to another e.g., liberal arts to data science.
4. *Prioritize Liberal Arts:* Society 5.0 will require us to be more flexible with the grade, leaving behind the conventional age-grade-progress approach and also seeking a more cross-disciplinary approach to subjects and a more elaborate understanding of liberal arts. Japan seemed to view liberal arts as synonymous with philosophy and ethics up until a few years ago. However, now it is believed that the liberal arts curriculum aims to improve students' ability to think critically and shape normative decisions through the study of a wide range of topics, such as humanities, social sciences, and natural sciences so that students can learn basic skills to identify and solve problems and to formulate and design them.
5. *Foster Entrepreneurial Education:* Entrepreneurship as a model offers a solution-oriented education with sustainable economic outcomes. Instead of training students to work for others, which will not exist in the next ten, or even five, years, we should teach them how to solve problems and build

their employment effectively. There are many entrepreneurs today who are changing the whole scenario of online education, and with EdTech on the rise, it is becoming easier and easier to self-educate.

6. *Use Mixed Methodology:* Studies have shown that personalized, blended programs aim to improve students' knowledge persistence and learning skills. Digital learning courses can be paired with other learning platforms, such as videos, podcasts, and even multimedia courses, to enhance their learning.

As the future belongs to AI and other technologies which will take most of the jobs, we require human skills to make a meaningful connection.

Although teachers have learned the technological and operational needs of digital classrooms, virtual classrooms have also opened up a world of opportunities to reimagine what teaching could look like in the years to come. Today, we are living in what we expected was a distant future. Although the method of teaching has changed over time, the role of the teacher has stayed essentially the same: a coordinator of learning in the minds of the learners and a stimulus for the chemistry of understanding in the hearths of the mind. Existing and upcoming technology like AI, machine learning, and big data is proving to be the most sought-after technologies that assist teachers in understanding the learning habits of students in a better way. These innovations are prepared to change the potential of learning.

7.8 CHALLENGES

1. Technology-based classrooms are effective, but in countries like India, where, due to the slow Internet speed most of the time, parents have to suffer downloading online videos for their children, it is a challenging task. Therefore, getting proper bandwidth is a major concern for implementing the concepts of Society 5.0.
2. There is a generation gap between the children and their parents. Due to the lack of knowledge on technology, they are unable to assist kids with the proper supervision that is needed, and that may hinder their studies.
3. Many of us are scared of the digital transition. There is increasing concern that AI-powered development would lead to mass layoffs. Even McKinsey has predicted that up to 800 million jobs will be lost by 2030. To a certain degree, the traditional education system strengthens these stereotypes, refusing to educate students about the incredible possibilities presented by technological progress.

7.9 MANAGERIAL IMPLICATIONS

Even though technology is the driving force behind technological revolutions, the subject of focus in Society 5.0 is human. Hence, education has the responsibility for this phase of transition. It is the first and foremost concern for educational institutions to educate individuals who are capable of responding and taking advantage of new technologies and satisfying the needs of modern society. Secondly, highly skilled individuals must be educated to meet the requirements of a technology-intensive labor market.

> *"In the Google generation, people don't need to remember every single reality. Most functions are better done by machines nowadays." "Thus the concentration must be on interpersonal skills like communication, teamwork and resilience, and also enthusiasm, understanding and reading ability." stated the Education Minister of Japan, Yoshimasa Hayashi.*

Computers can manage computerized knowledge much better than humans can, but we excel at self-expression, and this makes us special. In this scenario, learners are required to learn not to obtain unnecessary, memorized data but to analyze data and use it to solve real-life problems. This enables the use of education policies, approaches that make it easier for individuals to access knowledge during an interactive teaching phase. There is a lot of highly priced technology, though there are many ways in which higher education institutions can save money on technology. Money invested in e-readers, for example, will lead to long-term savings through reducing or removing the cost of purchasing and upgrading textbooks.

Besides, learners can use technology to browse free assignments and teaching materials. Technology-enabled persons would easily find a job as the future belongs to digital skills. There are platforms that offer easy and free access to a variety of coursework, namely MOOCs (massive online open courses). It allows students from the most remote parts to access the finest academic material at a very affordable price. Other than study materials, students can take part in live interactions that allow them to explain their questions and communicate with professors from top universities and colleges.

Society 5.0 will need more schools and universities to teach students how to build opportunities for themselves and how to solve challenges, inspire others, and bring society together. There are already a variety of locations that are welcoming the change in education, but the future belongs to the online environment.

7.10 EMERGING ISSUES

As per the industry requirements, universities such as Harvard, Georgia Tech, and Duke University are offering open certification programs. These online programs are developed in collaboration with EdTech companies such as 2U, Udacity, and Simplilearn to deliver predominantly *outcome-based learning programs*, which provide hands-on practical learning resources that allow students to be future ready.

Gamified learning is a concept that makes the learning fun and creative and encourages a positive, competitive nature among teammates within an organization that inspires them to appreciate their training accomplishments accompanied by exciting rewards.

Another recent trend in corporate training is the use of bite-sized teaching methods. *Microlearning* requires brief, unique bursts of material with a simple learning framework. Cutting lengthy, in-depth lessons into small, digestible chunks is the ideal way to keep workers interested in longer, traditional e-learning courses.

7.11 DISCUSSION

Even a year ago, nobody would have imagined that more than 1 billion children worldwide would be forced out of schools in 2020, and as an immediate response to the pandemic, all educational institutes have closed, which has let them redesign their classrooms, enabling children to study from their homes. COVID-19 has sparked an unparalleled use of technology in the teaching and learning process, which has proven to be the most sought-after option for this attempt to continue education. Technology has played a critical role in different social contexts, making it easy for skills and competencies to become redundant.

The present study suggests that a mixed method of education is a promising model as students do not want to compromise on the socialization part that takes place while attending the classes in offline mode. Knowledge of software training has proven to be essential for teachers as it helps in transforming education material to online platforms and can assist in developing web-based lecture halls. Certain platforms are offering easy and free access to a variety of coursework, namely MOOCs, allowing students to access course material from any part of the world. Students would likely to pursue quality education without sacrificing working and family and spending time on traveling, and online education provides this kind of freedom as it liberates students from being at a particular place at a particular time and allows connecting to any college across the world, which once was restricted to a single college.

Online education has been found very helpful in studying difficult subjects and increase students' practicing skills. It further emerged in the study that knowledge of basic computer and software such as LMS proved to be very beneficial for teachers. The more computers become familiar with the learning behavior, the more effective the specific assignment, assessment, and creation of new content is expected in the future days. The study also suggests developing a digital classroom and video-based and game-based learning to raise the interest of learners in subjects. Project-based and collaborative learning models were found to be the best learning models that suit the digital age. The study also confirmed that critical thinking, problem-solving skills, and, more importantly, good command over soft skills are the requirements for Society 5.0. Challenges like getting proper bandwidth, lack of technical knowledge, and resistance to change have also been discussed.

7.12 CONCLUSION

e-education or online education is modifying our view toward classroom learning. Improvements in overall education systems are significant and revolutionary as they have reduced classrooms, offices, cafes, dormitories, and library space. It has become a feasible alternative and a desirable need for business executives, households, and other related populations. As institutions and universities keep up with the new trends, a very holistic learning environment has received maximum maximized attention among scholars, educators, administrative staff, policymakers, publishing companies, and corporations. To plan for and confront the age of Society 5.0, students must be able to think objectively, share ideas, and have strong problem-solving skills. These talents are found in higher-level thinking skills and are considered the most critical competency in Society 5.0.

Thus, it can be inferred that higher-order thought is a capacity that must be possessed by all people, particularly students. In recent times, we are witnessing the extensive use of modern devices in almost every area of our lives. Additionally, educational institutions are responsible for increasing the skilled workforce. The chapter concludes with the impact of Society 5.0 on classroom teachings and learning practices and how that can benefit the next generation. Besides, big data analysis and AI in education processes are expected to be simplified in the future by formulating courses, scheduling and organizing educational resources, and using sophisticated algorithms that comply with student expectations. In line with the principles declared by the Japanese Government, various activities have emerged in Japanese academic circles and industry. And while Society 5.0 originates from Japan, it is not intended solely for the advancement of one nation. There is no question that the systems and innovations built here will connect to the solution of sustainability issues worldwide. It is therefore up to us to transform technical advances into possibilities.

7.13 LIMITATIONS OF THE RESEARCH

The present study was qualitative, which requires more time, and interpretation is limited as it restricts the verification of results.

The concept of Society 5.0 is rather new, so not much research work has been carried out on the same. This gives the scope for further research on the topic.

Also, only a few studies have been performed on the relationship between Society 5.0 and digital education, which made it a little problematic in formulating the scope of the research.

7.14 SCOPE FOR FUTURE RESEARCH

The present study explores the relationship between digital education and Society 5.0; nonetheless, there can be other variables also that affect both the concepts in today's world that could be taken into account for further research.

Similar research could be done by using primary sources such as interviews, questionnaires, and personal observations to gain a wider perspective of the concepts.

A comparative study can be done on the situation before and after COVID-19 in the education sector, how digital education has changed the perspective of learners toward education, what the requirements are for Society 5.0, and how it is going to affect the world.

REFERENCES

Abbasy, Majid. Bayani & Enrique V. Qesada. 2017. "Predictable influence of IoT (Internet of Things) in the higher education." *International Journal of Information and Education Technology*, 7(12), 914–920.

Abdel-Basset, Mohamed Manogaran, Gunasekaran Mohamed Mai & Rushdy Ehab. 2019. "Internet of things in smart education environment: Supportive framework in the decision-making process." *Concurrency and Computation: Practice and Experience*, 31(10), e4515.

Abel, Rob. 2005. "Achieving Success in Internet-Supported Learning in Higher Education: Case Studies. Illuminate Success Factors, Challenges, and Future Directions." *Alliance for Higher Education Competitiveness*. Retrieved from www.ahec.org/research/study_reports/IsL0205/TOC.html.

Bigelow, Cale. 2009. "A comparing student performance in an online versus a face-to-face introductory turfgrass science course-a case study." *NACTA J*, 53, 1–7.

Broadley, T. 2007. Implementation of e-learning: A case study of three schools, in Jeffery, R., Shilton, C. & Davies, M. (ed), Proceedings of the International Educational Research Conference, Nov 25–29 2007. Fremantle, WA: AARE.

Chaudhary, Asiya. 2020. "The Divide in Digital Education." The Hindu. May 30, 2020. https://www.thehindu.com/opinion/open-page/the-divide-in-digital,education/article31710304.ece, 2020.

Cheok, Mei Lick & Wong Su Luan. 2015. "Predictors of e-learning satisfaction in teaching and learning for school teachers: A literature review." *International Journal of Instruction*, 8(1), 75–90.

Coldwell-Neilson, Jo, Annemieke Craig & Goold, A. 2006. Student Perspectives of Online learning. 97–107.

Cosmas BF, Jabiri Bakari & Tolly S. A. Mbwette. 2010. "Implementing E-learning in Higher Open and Distance Learning Institutions in Developing Countries: The Experience of The Open University of Tanzania." MIT LINC 2010 Conference May 23–26, https://linc.mit.edu/linc2010/proceeding

Driscoll, Adam, Karl Jicha, Andrea N. Hunt, Lisa Tichavsky, & Gretchen Thompson. 2012. "Can Online Courses Deliver in-Class Results? A Comparison of Student Performance and Satisfaction in an Online versus a Face-to-Face Introductory Sociology Course." *Teaching Sociology*, 40(4), 312–331. https://doi.org/10.1177/0092055X12446624

Dua, Shikha. Wadhawan, Seema & Gupta, Sweety. 2016. "Issue, Trends & Challenges of Digital Education: An Empowering Innovative Classroom Model For Learning" *International Journal of Science Technology and Management*. Vol. No. 5, Issue No. 05. ISSN 2394-1537.

European Commission/EACEA/Eurydice. 2018. "Teaching Careers in Europe: Access, Progression and Support." Eurydice Report; Publications Office of the European Union: Luxembourg.

General Assembly. 2017. Resolution Adopted by the General Assembly on 6 July 2017. 71/313. Work of the Statistical Commission Pertaining to the 2030 Agenda for Sustainable Development; United Nation: New York, NY, USA, 2017; pp. 1–25.

Goodwyn, Andrew. 2011. English Teachers in the Digital Age – a Case Study of Policy and Expert Practice from England. English in Australia, May 2011.

Goold Annegret, Coldwell Neilson, Jamie Mustard & J. 2008. "Perceptions of roles and responsibilities in online learning: A case study." *Interdisciplinary Journal of E-Learning and Learning Objects*, 4. 205–223. 10.28945/375.

Granrath, Lorenz. 2017. "Japan's Society 5.0: Going Beyond Industry 4.0." Retrieved from https://www.japanindustrynews.com/2017/08/japans-society-5-0-going-beyond industry-4-0.

Güven, Gökhan, Yusuf Sülün & Türk Fen. 2012. "The Effects of Computer-Enhanced Teaching on Academic Achievement in 8th Grade Science and Technology Course and Students' Attitudes towards the Course." *Journal of Turkish Science Education (TUSED)*, March.

Harayama, Yuko. 2017. "Society 5.0: Aiming for a new human-centered society." *Collaborative Creation through Global R&D Open Innovation for Creating the Future*, 66(6) August

2017. Hitachi Review. Pp., 2017. http://www.hitachi.com/rev/archive/2017/r2017_06/pdf/p08-13_TRENDS.pdf

Hayashi, Hisanori, Hisashi Sasajima, Yoichi Takayanagi & Hirco Kanamaru. 2017. "International standardization for smarter society in the field of measurement, control and automation." *56th Annual Conference of the Society of Instrument and Control Engineers of Japan (SICE).* https://doi.org/10.23919/sice.2017.8105723

Hiroaki, Nakanishi. 2020. "Modern Society Has Reached Its Limits. Society 5.0 Will Liberate Us." World Economic Forum. https://www.weforum.org/agenda/2019/01/modern society-has-reached-its-limits-society-5-0-will-liberate-us/.

Hung, Chun-Ming, Gwo-Jen Huang & I. Hwang. 2012. "A project-based digital storytelling approach for improving students' learning motivation, problem-solving competence and learning achievement." *Journal of Educational Technology & Society*, 15(4), 368–379.

Keidanren (Japan Business Federation). Toward realization of the new economy and society. Reform of the economy and society by the deepening of "Society 5.0." 2016, http://www.keidanren.or.jp/en/policy/2016/029_outline.pdf

Kumari, Umesh. 2019. "Digital education: Scope and challenges." *International Journal of Applied Research*, SP4: 01–03.

Laferriere, Therese, Mary Lamon & Carol Chan. 2016."Emerging e-trends and models in teacher education and professional development." *Teaching Education*, 75–90. 10.1080/10476210500528087.

Lee, S.M. et al., 2009. "Computer Use and Academic Development in Secondary Schools." *Computers in the Schools (Sic)*, January.

Lin, Ming-Hung, Huang-g Chen, & kuang-Sheng Liu. 2017." A Study of the Effects of Digital Learning on Learning Motivation and Learning Outcome.". *Eurasia Journal of Mathematics, Science and Technology Education*, 13(7), 3553–3564. https://doi.org/10.12973/eurasia.2017.00744a

Lundberg, Johan & Merino, David & Dahmani, Mounir. (2008). Do Online Students Perform Better than Face-to-face Students? Reflections and a Short Review of some Empirical Findings. *Revista de Universidad y Sociedad del Conocimiento*, 5, 35–44. 10.7238/rusc.v5i1.326.

Lysenko, Larysa, V. Abrami, & C. Philip. 2014. "Promoting reading comprehension with the use of technology." *Computers and Education*, 75, 162–172.10.1016/j.compedu.2014.01.010.

Mathews, Suja P. & R. Raju Gondkar. 2017. "Solution integration approach using IoT in education system." *International Journal of Computer Trends and Technology*, 45(1), 45–49.

Önday, Özgur. 2019. "Japan's society 5.0: Going beyond Industry 4.0." *Business Economics Journal*, 10(389), 2. Retrieved Jun 22, 2019, from https://www.researchgate.net/publication/330500307

Palvia, Shailendra Aeron, Prageet Gupta, Parul, Mahapatra, Diptiranjan, Parida, Ratri. Rebecca Rosner & Sumita Sindhi. 2018. "Online education: Worldwide status, challenges, trends, and implications." *Journal of Global Information Technology Management*, 21(2018), 233–241.

Power, Michael. 2008. "The Emergence of a Blended Online Learning Environment." Retrieved from http://jolt.merlot.org/vol4no4/power_1208.htm.

Prasetyo, Yuli & Arman Arry. 2017. "Group management system design for supporting society 5.0 in smart society platform." 398–404. 10.1109/*ICITSI*.2017.8267977.

Qiao, Ailing & Wang Nan. 2009. "An investigation of teachers' needs on using ICT in teaching and learning." *International Conference on Computer Engineering and Applications IPCSIT*, 2(2011), Singapore: IACSIT Press, 285–289.

Rodríguez-Abitia, Guillermo & Pérez, Sandra & Ramírez-Montoya, María-Soledad & Lopez-Caudana, Edgar. 2020. “Digital Gap in Universities and Challenges for Quality Education: A Diagnostic Study in Mexico and Spain”. *Sustainability*. 12. 9069. 10.3390/su12219069.

Richardson, Jennifer. C. & Karen Swan. 2003. “Examining social presence in online courses in relation to student’s perceived learning and satisfaction.” *Journal of Asynchronous Learning Networks*, 7, 68–88.

Salcedo, Claudia S. 2010. “Comparative Analysis of Learning Outcomes In Face-To-Face Foreign Language Classes vs. Language Lab And Online”. *Journal of College Teaching & Learning (TLC)*, 7(2). https://doi.org/10.19030/tlc.v7i2.88

Saxena, Abhay, Durgesh Pant, Amit Saxena & Chandrashekhar Patel. 2020. “Emergence of educators for industry 5.0 – An Indological perspective.” *International Journal of Innovative Technology and Exploring Engineering (IJITEE)*, 9(12), 2278–3075, October.

Schneider, Klaus. 2010. “Ontario Colleges in the Digital Age: Understanding the Student Experience, Perceptions and Attitudes of Online Learning at One Ontario College.” A thesis submitted in the Department of Theory and Policy Studies in Education Ontario Institute for Studies in Education University of Toronto, 21.

Sousa, Maria Jose & Rocha, Alvaro. 2018. “Digital Learning Analytics In Higher Education.” Conference: 10th International Conference on Education and New Learning Technologies. https://doi.org/10.21125/edulearn.2018.0282

Sudibjo, Niko, Lusiana Idawati & H.G. Retno Harsanti. 2019. “Characteristics of learning in the era of Industry 4.0 and Society 5.0.” *Advances in Social Science, Education and Humanities Research*, 372, International Conference on Education Technology (ICoET 2019), 276–278.

Supendi, Ahmad & Nurjanah Nurjanah. 2020. “Society 5.0: Is it high-order thinking?” *International Conference on Elementary Education*, 2(1), 1054–1059. http://proceedings.upi.edu/index.php/icee/article/view/716.

Underwood, Jean. 2009. “The impact of digital technology (A Review of the Evidence of the Impact of Digital Technologies on Formal Education).” Becta, 2009.

Urban-Woldron, H. 2013. “Integration of Digital Tools into the Mathematics Classroom: A Challenge for Preparing and Supporting the Teacher.” *International Journal for Technology in Mathematics Education*, 20(3), 116–123.

Wikramanayake, Gihan. 2005. “Impact of Digital Technology on Education. Conference: 24th National Information Technology Conference.

Yurumezoglu, Kemal. 2009. “The effect of 3d computer modeling and observation-based instruction on the conceptual change regarding basic concepts of astronomy in elementary school students.” *Astronomy Education Review*, 8. 10.3847/AER2009006. http://adamasuniversity.ac.in/a-brief-history-of-online-education/

Section II

Understanding Technology in Education

8 MOODLE
Learning Management System

Praveen Srivastava and Shelly Srivastava

CONTENTS

8.1 INTRODUCTION

Learning is a never-ending process. Learning happens at all stages of human life. However, over time, how it is conducted has changed (Ahern and Repman 1994). From traditional chalk and talk to present-day technology-enabled virtual learning, the process has taken a giant leap. Research has shown us how online learning could be seen as a way to act in this social system (Peach and Bieber 2015). The configuration of faculty responsibilities shifts as faculty members are gradually expected to allow complete or limited use of online technologies to provide education (Coppola, Hiltz, and Rotter 2002; Young 2002).

The various online learning platform was changing the wind toward the online learning when from nowhere COVID-19 emerged and ensured that online learning remains the only mode of disseminating knowledge. Since March 2020, academicians around the world are looking forward to this mode of learning only. However, as an academician, they have increased responsibility for learning new educational technology and disseminating the content.

Online teaching or teaching using Information and Communication Technology (ICT) is not the same as classroom teaching and it is believed to be more taxing. A comparative analysis between traditional and online teaching highlights the issues of concentration, control, and dissemination. Though the challenges in online teaching

DOI: 10.1201/9781003132097-8

have been highlighted in several studies (Laczko-Kerr and Berliner 2006; Uren and Uren 2009; Yam and Rossini 2012; Gaskell and Mills 2014; Shava and Ndebele 2014; Islam 2015; Esfijani 2018; Gorman and Staley 2018), the fact that currently, it is the only option cannot be denied. Hence a debate whether we should use this mode or not holds no value in the current scenario.

Currently, the focus of discussion in the academic world is features required to ensure the learning-teaching process continues without hassle. A point to note here is that the teaching process has also shifted its focus from teaching-learning to learning-teaching, ensuring that teaching is designed to keep the learners' requirement (*learning*) in mind. Initially, a Content Management System (CMS) and in-class learning were accepted as one of the best Hybrid Models. However, CMS is a passive technology used primarily for displaying documents. On the other hand, Learning Management System (LMS) is an application that motivates learners to be involved with the program (Mijatovic et al. 2013). Learners can try a quiz, for instance. Creators can create a quiz and monitor students' progress. Given the advantage of the LMS, there is a need to shift from CMS to LMS.

8.2 THEORETICAL FRAME

8.2.1 Moodle

LMS is a platform for e-learning courses of all kinds. It helps to ensure that active learning takes place between the user and the virtual learning platform (Lyndon and Hale 2014; Parsons 2017). LMS helps to coordinate learning, reduces the time needed to set up courses, and assigns courses successfully to students. LMS makes learning interactive as a creator can build a quiz, monitor user progress, and see its results. There are numerous LMS which are providing various features to the users. This chapter aims to explore the basic feature of Moodle, one of the most popular LMSs.

Modular Object-Oriented Dynamic Learning Environment, commonly known as *Moodle*, is one of the most popular LMSs used by academicians and trainers. Moodle is a trademark owned by Moodle HQ (i.e. Moodle® LMS platform, Moodle® Hosting services, MoodleCloud®). Moodle can be installed in the Server of the organization or used as a web-based application, MoodleCloud. Authors are not associated with Moodle HQ, and all the IP rights belong to Moodle HQ. Necessary permission from Moodle IP team has been taken to use the screenshots of MoodleCloud.

Previous literature has explored the Moodle designing with different learning tools (Katsamani, Retalis, and Boloudakis 2012; Limongelli et al. 2016), given insight regarding the security of Moodle (Mudiyanselage and Pan 2020), investigated few features of Moodle (Lu and Law 2012), explored the adoption of Moodle in higher education (Costello 2013), however, there is scarce literature available on MoodleCloud course designing.

8.2.2 Learning Moodle

To give an in-depth understanding, MoodleCloud has been selected to explain the features of Moodle. Since its web-based, it is easily accessible by the audience, and they can explore its features. The web address to reach the home page of MoodleCloud is moodlecloud.com.

8.3 RESEARCH QUESTION

There have been plethora of studies focusing on the student perspective about online education in general and use of Moodle as LMS in particular (Khatib 2013; Damnjanovic, Jednak, and Mijatovic 2015; Horvat et al. 2015; Chang and Lan 2019; Pérez-Pérez, Serrano-Bedia, and García-Piqueres 2020). However, the authors observed sparse research focusing on Moodle as a Cloud-Based LMS and its application. Hence, the literature providing the details of cloud-based Moodle LMS and its detailed operational step is rare. With this backdrop, the present study focused on Cloud-based Moodle, i.e. MoodleCloud and provided detailed steps to ensure that academicians from every generation can avail the benefit of MoodleCloud.

8.4 METHODS

To provide the detailed steps, a visual representation is a must. Hence, the authors used the screenshots of every critical step required to make the reader understand MoodleCloud use. To ensure no copyright issue, the authors contacted the MoodleCloud Team and requested them to provide the necessary permission. Moodle IP Team provided permission to use MoodleCloud screenshots for academic purposes. Every screenshot has been given a particular figure number and its explanation in the steps.

The home page of MoodleCloud gives two options to the audience. A user can log in, if they have already created their account in MoodleCloud, if not, they can sign up (Figure 8.1). The first-time user should sign up and fill in the essential details. Moodle cloud provide free access to the admin where he can register 50 users, however, recently, Moodle has changed the policy for its accounts due to heavy traffic on the

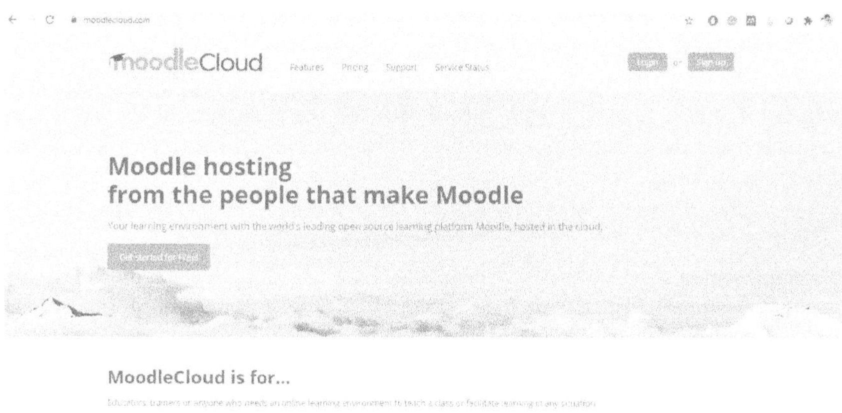

FIGURE 8.1 Home page of MoodleCloud. (*Source:* MoodleCloud.)

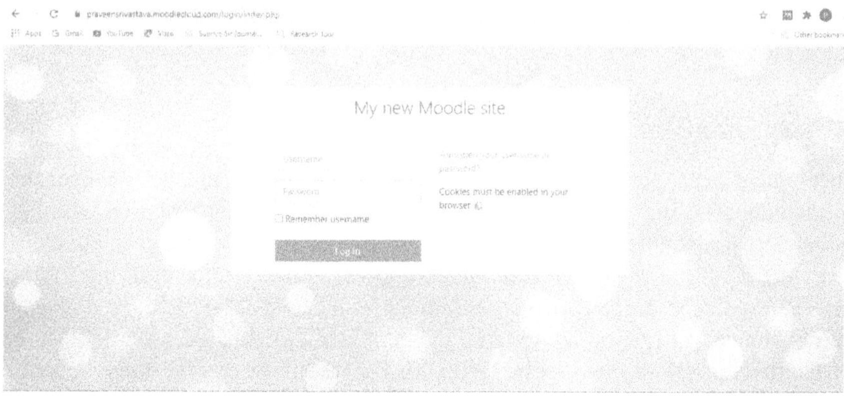

FIGURE 8.2 Home page of praveensrivastava.moodlecloud site. (*Source:* MoodleCloud.)

site. One can have a maximum of 200 users registered with 400 Mb space in the free site, but it will be available for a trial period of 45 days only. Currently, Moodle offers plans under three headings, i.e. free trial, starter, and Moodle for school. More information on the same can be obtained at https://moodlecloud.com/app/en/. To sign up, a plan should be selected, and basic information should be entered, and lastly, the terms and condition is to be accepted. This will give an option to the user to select their site name, which looks like www.xxxxxxxxxxxx.moodlecloud.com. The authors created a MoodleCloud site www.praveensrivastava.moodlecloud.com for reference.

Once a site has been created, the user of the site gets the admin role by default. They can log in and create a course of their choice. The login interface is depicted as shown in Figure 8.2.

The user can log in from their home page. By default, the login Id will be "admin" and password will be the one selected during the signup process (Figure 8.2). Once login, the admin will get to see their Home, where they find various options like Dashboard, Calendar, Private files, Content Bank, My Course, Introduction to Moodle, and above all, Site Administration in the left panel (Figure 8.3). The Home also gives an indication of the total number of registered users on the site and total storage being used. Apart from these options, the site admin's name with a drop-down arrow and a gear icon is also visible on the right side of the home page. This icon has a drop-down too, which gives options for edit setting, filter, backup restore, and more.

8.4.1 Adding a New Course

The current section will include the process of creating a course in Moodle and the next section shall deal with adding the students.

Step 1: To add the Course, the admin needs to navigate to the *Site Administration* option. This will show various sub-menus, and the menu name course should be selected under which option of adding a new course will help build our first Moodle Course (Figure 8.4).

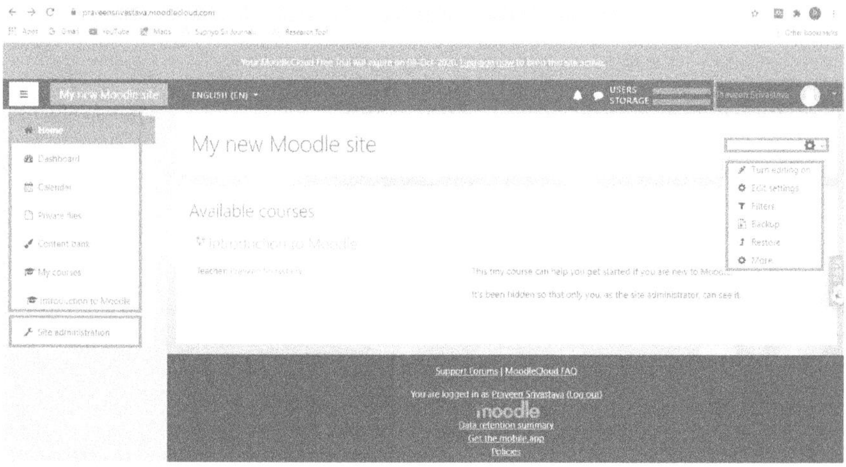

FIGURE 8.3 Home of praveensrivastava.moodle cloud site. (*Source:* MoodleCloud.)

Step 2: Add a new course option will seek various information regarding the Course like General Information (course full name, category, start date, end date, etc.), Description (course summary and image), Course Format (topic, weekly, etc.), Appearance, Files and uploads, Completion tracking, Groups, Role renaming, and Tags (Figure 8.5).

Step 3: In order to explain the course creation, a course name ***check*** was created, and it was mentioned in General option. Once administrator creates the Course, it will be visible under the Home option. Clicking on the Course will show options to either enter the topic (if format type has been set as topic), or else weekly format appears. To change the topic and enter the topic of our choice, we need to first click

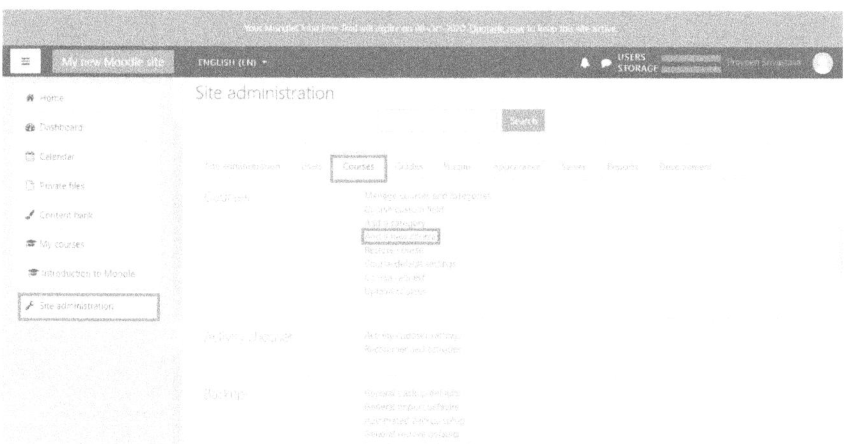

FIGURE 8.4 Adding a course (1). (*Source:* MoodleCloud.)

FIGURE 8.5 Adding a course (2). (*Source:* MoodleCloud.)

on *Turn editing on* which will then change to *Turn editing off.* This will make the MoodleCloud site editable and we can make the changes (Figure 8.6).

Step 4: Now, as an admin, we can enter the topic name as per the Modules. For the purpose of this paper, Module 1, Module 2, and Module 3 have been entered as topic name. We need to write the topic and press enter to ensure that changes are implemented. Now, we need to add activities to these Modules. For this, we need to click on add an activity or resource option visible in front of that module. This will open a new window with three sections, namely All, Activities, and Resources (Figure 8.7). All option displays all the content available under Activities and Resources. However, Activities and Resources will show the content separately.

Step 5: We can now select activity of our choice to add to a particular module. There are numerous activities like Assignment, Lesson, Chat, Choice, Database, Glossary, Quiz, etc. However, for the purpose of current study, we shall take *adding*

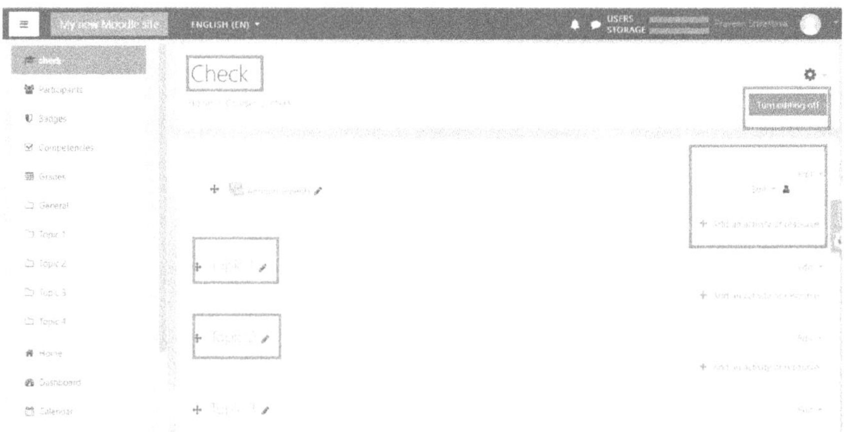

FIGURE 8.6 Turn editing on (3). (*Source:* MoodleCloud.)

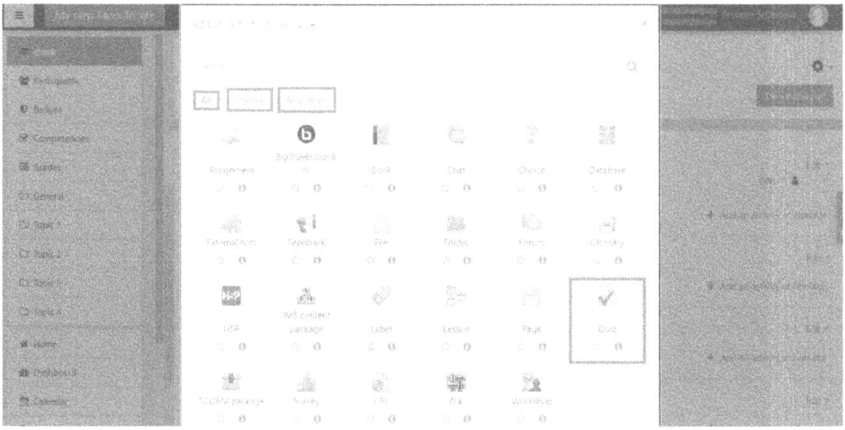

FIGURE 8.7 Adding a resource (4). (*Source:* MoodleCloud.)

a Quiz to the Moodle site. Once we select Quiz, we get several options like *Timing* (it gives us an option to start and end the Quiz), *Grade* (an option to give grade method, attempts allowed and Grade to pass), *Layout* (Questions starts in a new page or several questions in one page), *Question Behavior* (shuffle within question), *Overall Feedback*, *Safe Exam Browser*, *Tags* along with *General Setting* of name and description (Figure 8.8). After filling the necessary entries, admin can click on *Save and Display*. This will create the Quiz and will give an option to *Edit the Quiz* i.e. to add questions to the Quiz created.

Step 6: We can now add a question to the Quiz (Figure 8.9). Moodle gives us options to add different type of questions to the Quiz. The option to *Edit Quiz* will be available as soon as we click on *Save and Display*. Clicking on that option, we get to a screen where we can add a new question, add a question from question bank, or add any random question. Since it is a new course, we have no question added to it.

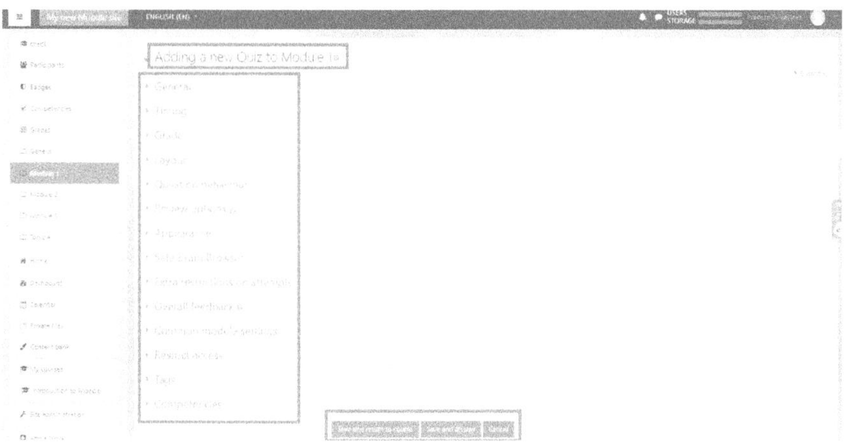

FIGURE 8.8 Adding a new quiz (5). (*Source:* MoodleCloud.)

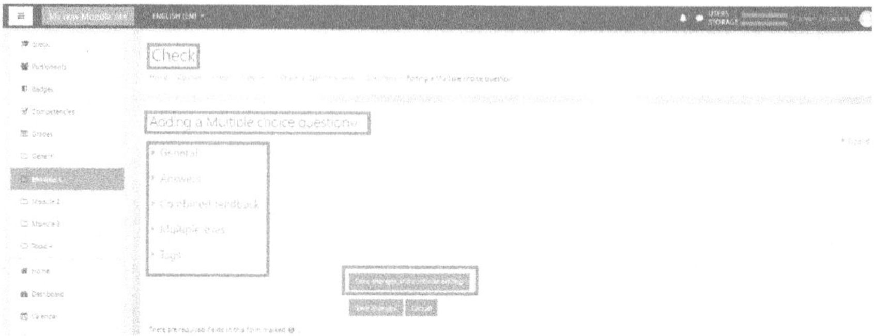

FIGURE 8.9 Adding a new quiz (6). (*Source:* MoodleCloud.)

Hence, we shall select the option to *add a new question*. This will further give us a plethora of choices for Question Type. The choice includes Calculated, Calculated Multiple Choice, Calculated Simple, Drag & Drop into text, Drag & Drop markers, Drag & Drop onto image, Embedded Answers, Essay, Matching, Multiple choice, Numerical, Random short answer matching, Select missing words, Short answer, and True/False. For the purpose of the present paper, we shall select Multiple Type Question only. For that, we need to click on the radio button next to *Multiple Type question* and click on *Add*.

This will open a new page where we need to give details under various heads like General (Question name, Question text, Default marks, single or multiple answers, etc.), Answers (Choice of options, Grades per answer, etc.), Combined Feedback (Feedback for correct, partially correct, and incorrect responses), Multiple Tries (Penalty for each incorrect tries), Tags (The keywords which is associated with the question). Once all the details have been entered admin can *save the changes and continue* to edit or *save the change* and come out from this page.

Step 7: We have created a question, "*What is the Capital of India,*" and gave three wrong and one right answer. For the correct answer, a 100% grade has been given (Figure 8.10). For others, no grade is awarded, and we select *none* for it. This

FIGURE 8.10 Giving grade in quiz (7). (*Source:* MoodleCloud.)

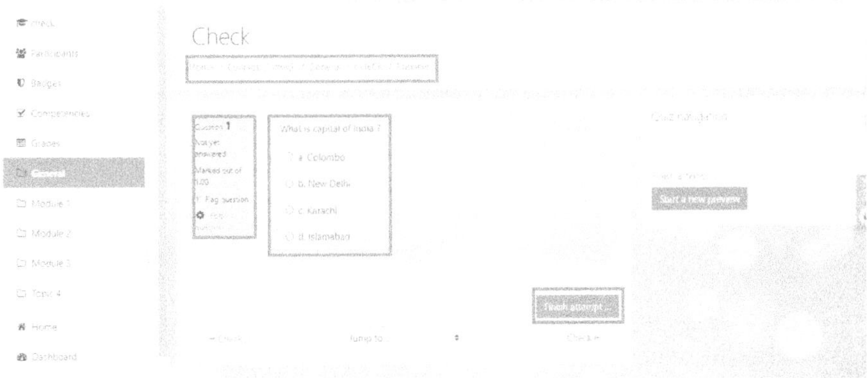

FIGURE 8.11 Giving grade in quiz (8). (*Source:* MoodleCloud.)

implies that if students select the wrong answer, they get no marks, and if they select the correct answer, they get full marks. In case we have more than one correct answer for any question, we can divide the Grade accordingly. Moodle gives us the option to provide different grades to the choices. For two correct, we can provide 50% to both the right choice, and so on. If students select only one correct answer, they will get 50% of the total marks. However, in order to give a question with multiple correct answers, we need to select *multiple answers allowed* under *one or multiple answers.*

Step 8: Once we have created the question and given the Grade for every choice, we need to click on *save changes.* This will ensure that our question has been created with the proper grades for every option. The student will get to see the question in a new window with four options. Students can select the option of their choice by selecting the respective radio button and then click on Finish attempt to submit the response (Figure 8.11). Once the student clicks the finish attempt, they get the option to "Return to attempt" and "Submit all and finish." To change the answer, students can go back to return to attempt and to submit the response, they can click on submit all and finish.

8.5 DISCUSSION

The MoodleCloud is rich in features that can be of great use to the academician. Self-tracking by students; awarding badges to the performers; displaying grades to students; adding an assignment, page, website, video, audio, etc.; creating a question bank for Quiz; taking feedback; Interactive Quiz via H5P; creating glossary; creating numerical Quiz, multiple-choice, drag and drop, essay, missing word, short answer; BigBlueButton are some of the features which are worth exploring. The plug-ins of Moodle are also an added advantage that keeps Moodle updated with the new features as and when it is required. With all these features, Moodle has the upper hand as an LMS.

8.6 CONCLUSION

Authors have made an attempt to explain the basic features of MoodleCloud, which include creating a MoodleCloud site, creating a Course in the site, adding the section topic-wise, adding a quiz, and including a multiple-type question in the Quiz. This will be of help to the academician who is willing to use MoodleCloud as a platform for online learning.

8.7 IMPLICATION

This study explores the features of MoodleCloud. The specific focus toward a single LMS makes this study unique. Screenshot and explanation both have not been presented together in the previous studies focusing on LMS.

The current study will be of help to academicians in more ways than one. Academicians will get a ready reference to prepare their Course in MoodleCloud. Academicians will be able to create their own MoodleCloud site and add different courses. They can add several features to make their Course more interactive. Those academicians who are already using Moodle platform can understand the MoodleCloud feature and can compare to select the best platform for their Course. Lastly, even those academicians who are using other platforms for online learning-teaching can understand the Moodle cloud feature and hence understand the difference between Moodle and other LMS features.

8.8 LIMITATION

The present study has some limitation which is as under:

- This study has explored some of the basic features of MOOC, which will be of use to an academician. However, the student interface of Moodle has not been explained.
- The study does not include a comparison of Moodle feature with other such open platforms.

8.9 FUTURE RESEARCH

The study gives an option to explore different features of Moodle. Hence, researchers in the area may find it interesting to explore the same. Some of the suggestions for future research are as under:

- The researchers may identify other features of Moodle which can be of use to the academician.
- Researchers may also investigate other open-source learning platforms like Gnomio and explore its features.
- Further, a comparison of features available in different learning platforms by researchers will be of interest to the academicians.

REFERENCES

Ahern, Terence C., and Judi Repman. 1994. "The Effects of Technology on Online Education." *Journal of Research on Computing in Education* 26 (4): 537–46. https://doi.org/10.1080/08886504.1994.10782109.

Coppola, N. W., S. R. Hiltz, and N. Rotter. 2002. "Becoming a Virtual Professor: Pedagogical Roles and ALN." *Journal of Management Information Systems* 18 (4): 169–90.

Costello, Eamon. 2013. "Opening up to Open Source: Looking at How Moodle Was Adopted in Higher Education." *Open Learning* 28 (3): 187–200. https://doi.org/10.1080/02680513.2013.856289.

Damnjanovic, Vesna, Sandra Jednak, and Ivana Mijatovic. 2015. "Factors Affecting the Effectiveness and Use of Moodle: Students' Perception." *Interactive Learning Environments* 23 (4): 496–514. https://doi.org/10.1080/10494820.2013.789062.

Esfijani, Azam. 2018. "Measuring Quality in Online Education: A Meta-Synthesis." *American Journal of Distance Education* 32 (1): 57–73. https://doi.org/10.1080/08923647.2018.1417658.

Gaskell, Anne, and Roger Mills. 2014. "The Quality and Reputation of Open, Distance and e-Learning: What Are the Challenges?" *Open Learning* 29 (3): 190–205. https://doi.org/10.1080/02680513.2014.993603.

Gorman, Emily F., and Catherine Staley. 2018. "Mortal or Moodle? A Comparison of In-Person vs. Online Information Literacy Instruction." *Journal of Library and Information Services in Distance Learning* 12 (3–4): 219–36. https://doi.org/10.1080/1533290X.2018.1498635.

Horvat, Ana, Marina Dobrota, Maja Krsmanovic, and Mladen Cudanov. 2015. "Student Perception of Moodle Learning Management System: A Satisfaction and Significance Analysis." *Interactive Learning Environments* 23 (4): 515–27. https://doi.org/10.1080/10494820.2013.788033.

Islam, A. K. M. Najmul. 2015. "The Moderation Effect of User-Type (Educators vs. Students) in Learning Management System Continuance." *Behaviour and Information Technology* 34 (12): 1160–70. https://doi.org/10.1080/0144929X.2015.1004651.

Katsamani, Maria, Symeon Retalis, and Michail Boloudakis. 2012. "Designing a Moodle Course with the CADMOS Learning Design Tool." *Educational Media International* 49 (4): 317–31. https://doi.org/10.1080/09523987.2012.745771.

Khatib, Nahla M. 2013. "Students Attitudes towards the Web Based Instruction." *Gifted and Talented International* 28 (1–2): 263–67. https://doi.org/10.1080/15332276.2013.11678421.

Laczko-Kerr, Ildiko, and David Berliner. 2006. "The Effectiveness of." *Architectural Engineering and Design Management* 2: 73–86.

Limongelli, Carla, Matteo Lombardi, Alessandro Marani, Filippo Sciarrone, and Marco Temperini. 2016. "A Recommendation Module to Help Teachers Build Courses through the Moodle Learning Management System." *New Review of Hypermedia and Multimedia* 22 (1–2): 58–82. https://doi.org/10.1080/13614568.2015.1077277.

Lu, Jingyan, and Nancy Wai Ying Law. 2012. "Understanding Collaborative Learning Behavior from Moodle Log Data." *Interactive Learning Environments* 20 (5): 451–66. https://doi.org/10.1080/10494820.2010.529817.

Lyndon, Sandra, and Beverley Hale. 2014. "Evaluation of How the Blended Use of a Virtual Learning Environment (VLE) Can Impact on Learning and Teaching in a Specific Module." *Enhancing Learning in the Social Sciences* 6 (1): 56–65. https://doi.org/10.11120/elss.2014.00019.

Mei-Mei Chang & Shu-Wen Lan (2021) Exploring undergraduate EFL students' perceptions and experiences of a Moodle-based reciprocal teaching application. *Open Learning: The Journal of Open, Distance and e-Learning,* 36:1, 29–44, DOI: 10.1080/02680513.2019.1708298

Mijatovic, Ivana, Mladen Cudanov, Sandra Jednak, and Djordje M. Kadijevich. 2013. "How the Usage of Learning Management Systems Influences Student Achievement." *Teaching in Higher Education* 18 (5): 506–17. https://doi.org/10.1080/13562517.2012.753049.

Mudiyanselage, Akalanka Karunarathne, and Lei Pan. 2020. "Security Test MOODLE: A Penetration Testing Case Study." *International Journal of Computers and Applications* 42 (4): 372–82. https://doi.org/10.1080/1206212X.2017.1396413.

Parsons, Anastassia. 2017. "Accessibility and Use of VLEs by Students in Further Education." *Research in Post-Compulsory Education* 22 (2): 271–88. https://doi.org/10.1080/13596748.2017.1314684.

Peach, Harold G., and Jeffery P. Bieber. 2015. "Faculty and Online Education as a Mechanism of Power." *Distance Education* 36 (1): 26–40. https://doi.org/10.1080/01587919.2015.1019971.

Pérez-Pérez, Marta, Ana M. Serrano-Bedia, and Gema García-Piqueres. 2020. "An Analysis of Factors Affecting Students' Perceptions of Learning Outcomes with Moodle." *Journal of Further and Higher Education* 44 (8): 1114–29. https://doi.org/10.1080/0309877X.2019.1664730.

Shava, G., and C. Ndebele. 2014. "Towards Achieving Quality Distance Education, Challenges and Opportunities: The Case of the Zimbabwe Open University." *Journal of Social Sciences* 39 (3): 317–30.

Uren, Martin, and James Uren. 2009. "ETeaching and ELearning to Enhance Learning for a Diverse Cohort in Engineering Education." *Engineering Education* 4 (2): 84–90. https://doi.org/10.11120/ened.2009.04020084.

Yam, Sharon, and Peter Rossini. 2012. "Online Learning and Blended Learning: Experience from a First-Year Undergraduate Property Valuation Course." *Pacific Rim Property Research Journal* 18 (2): 129–48. https://doi.org/10.1080/14445921.2012.11104355.

Young, J. R. 2002. "The 24-Hour Professor: Online Teaching Redefines Faculty Members' Schedules, Duties, and Relationships with Students." *Chronicle of Higher Education* 38 (48): A.31–A.33.

9 Digital Transformation of Higher Education

Opportunities and Constraints for Teaching, Learning and Research

Viju Mathew, A. I. Abduroof, and J. Gopu

CONTENTS

9.1 INTRODUCTION

The social, cultural and economic scenario that emerged as an outcome of the ongoing pandemic of Covid-19 has forced organizations in resorting to choosing digital platforms for most of their social and business interactions to take place. There is hardly any sphere of life that is untouched by the pandemic and higher education (HE) is no exception where the current situation is acting as a catalyst in the process of digital transformation, which, in fact, had started in the pre-Covid-19 period. Machekhina (2017) observes that many important analysts and forecast experts look at the transition of the process of education into the digital stage as the turnaround idea in the history of teaching and learning. To respond to the multi-faceted

DOI: 10.1201/9781003132097-9

changes taking place in their environment in recent times, almost all HE organizations around the globe consider the digitalization of courseware and course delivery as a top priority. Digitalization can offer the possibility of expanding the scope of learning and acquiring new audiences, and it can add safe spaces of freedom to some communities (DAAD/DIE 2018). The phenomenon of digitalization which offers rich opportunities in HE has a multitude of elements. Kuzu (2020), for example, identifies the Internet, mobile networks and devices, big data, cloud services, social media networks and artificial intelligence as some important components of the digitalization process in HE. All these factors and their interplay with one another in addition to its impact on various stakeholders in HE make the digital transformation an extremely complex process.

It is assumed that the technological shift has been an ongoing process in all sectors of life. Presently, technology is utilized in daily lives that brought changes in the development and delivery of knowledge (Lim and Wang 2016). Several studies discuss the exposure to technology skills and diversity among individuals (Helsper and Eynon 2010). Additionally, studies demonstrate the realities of diverse demographics and technological exposure, a socio-economic challenge due to improper access to the Internet (Hargittai 2010). The use of technology and its application within the learning system, especially in HE has changed significantly in the last few years as a platform for reaching the desired students and delivery of its services. Alavi and Leidner (2001) reveal that traditional whiteboards and handouts have largely been changed by digital ones, e.g. online classrooms, smart boards and online course delivery boards. Predominantly, physical infrastructure like the classroom, library, lecture materials, etc., has been transformed into a digital medium creating new possibilities that tend to continue into even more advanced technology in the HE. The technology-based system has altogether transformed the HE system from a teacher-centered classical form to a learner-centered model (Jani et al. 2018) incorporating ICT tools for an enhanced experience.

Much research has presented that technology had a significant impact on varied aspects of HE (Brahimi and Sarirete 2015). The availability of and accessibility to information and communication technology (ICT) has transformed the higher education institutions (HEIs) as similar to class learning which is based on individual choice and preference (Singh and Kaurt 2016). However, due to the lack of resources and skills among different stakeholders in HE, the enactment of digital tools and technology had a lower impact on the teaching and learning processes in many developing countries in motivating and improving student activity and involvement than that of traditional classroom teaching. The improper management of digital technology has failed to attract the students in effectively adopting the digital/online education platforms. Therefore, HEIs need to provide essential digital resources and training to enable changes (Kirkwood 2014) to fall in mark with emerging and developed countries. Previous research often emphasizes the various challenges that concern the utilization of technology in the classroom. The absence of measuring the effectiveness of e-learning (Aguti, Wills and Walters 2014) is still a question for many researchers. The continuous and rapid digitalization of HE institutions around the world has caused research gaps in fully understanding the pros and cons of digital transformation. This study will contribute to knowledge by providing rich insight into the impact of digital technology in HE and by drawing its implications (Walsham 1995) and challenges and opportunities to HE, teaching-learning process and society as a whole. This study will also give insight into the broader context of HE to gain a better understanding of how HE can adopt new digital technology in teaching, learning, research, collaboration and developing solutions to problems of the society and business.

9.1.1 Research Questions

The research questions for this study are given below:

1. How digital technologies and learning tools are employed in the HE sector?
2. What are the opportunities that await digital technologies in the field of HE?
3. What are the challenges in adapting digital technologies for teaching and learning in HE?
4. What is the role played by adaptation of digital technologies in motivating teachers, students and researchers?
5. How digital technologies in HE are impacting upon the policies and functioning of educational organizations, educational industry, research agencies, government and the community in general?

9.1.2 Digitalization Technology in Teaching and Learning

From the past decade, HE institutions in many countries have increased their investment in digital technology (Kirkwood 2014) and started reflecting on how technology can transform educational practices in developing countries. Technological development is also changing the social context of HE shaping the young mind to take full benefits to achieve quality education. It is also observed that many HEIs are adopting multiple approaches and methods for enhancing the experience of students and teachers helping to achieve academic and research goals. The use of digital tools and technologies in HE has caused a shift from the traditional lecture of one-way communication with the teacher being the dominant figure to students being an active participant in the learning process (Zandvliet & Fraser 2004). Nevertheless, digital or online teaching cannot overcome the advantage of face-to-face teaching with academic-students relationship to understand and express each creating a conducive environment for the teaching-learning process. Meanwhile, the students and teachers need to be well equipped with digital technology and its innumerable aspects to overcome the regular problems to teach and learn efficiently with technology (Kennedy et al. 2008). The inability of the student to adopt the technological updates and skills will directly affect their grade rather than assessing the ability and knowledge of the student for a particular course. Similarly, the teaching processes are also not far from being affected by the use of digital technology (Kirkwood 2014). According to Kirkwood (2014), technological advancement can be used to improve teaching, but the main challenge is to ensure that new teaching methods improve the learning experience. It is also observed in studies that absence of applied knowledge and incompetence to acclimatizing with new technology can discourage teachers (Clegg, Konrad and Tan 2000) showing resistance to adopt and integrate new functionalities for teaching. Therefore, it is very essential to support academics to adopt and use new technological features in the teaching process (Jamieson 2003).

All these factors point out that the digitalization of HE is an elaborate and complex process. In practical terms, it is a restructuring of all spheres of HE-academic, administrative and industrial in terms of digital technologies and new media communication. HE organizations throughout the world are placing a high priority on developing and applying digital tools and technologies in the education and research process realizing its importance in branding and marketing the HEIs in a world of increasing competition. Leading educational organizations were keen to utilize the possibilities of the Internet ever since the World Wide Web gained popularity and globalization gained momentum. Several universities and academic institutes started offering online courses for students spread across the globe. Online learning gradually emerged as a much sought-after business model and several organizations started to popularize it. The model which began by delivering recorded lectures and textual material, soon developed into using animation and graphics, online video lectures, flipped classrooms, virtual reality labs, digital simulation, online assignments and a plethora of other tools making use of the immense possibilities of digital and new media technologies. The components of interactivity, flexibility, personalization, availability of different platforms and the capability of transcending geographical and time zones make digital learning a popular and reliable feature of present-day HE, which no organization can afford to ignore. The abundance of free material available on various websites and social media platforms is also another factor that prompted HEIs to give importance to the digitalization process. This resulted not only in offering online classes of existing courses but also in developing new courses. The online delivery which was once confined to theoretical courses in social science and humanities gradually spread to highly practical areas like engineering. For example, almost a decade ago, Stanford University in California had initiated the process of launching an online course in Masters in Engineering (Kulkarni 2013). In the present times, the process is widespread in many Western, Asian and African HE organizations.

Researchers have revealed that digital technologies have a higher impact on learning (Imhof, Scheiter and Gerjets 2011) than the normal ‘lecture-listening’ process has thanks to the elements of both visualization and sound simultaneously. Additionally, technological adaption also affects the organizations in terms of structural change (Alavi and Leidner 2001; Kolb and Kolb, 2005). It is noted that the technology used has to be adopted in finding new solutions and implementing strategies, e.g. group activity and learning, collaboration, etc., to yield consistent positive conclusions (Andrews et al. 2002; Hartley 2007). Numerous researches have concluded that ICT has a positive and measurable effect on learning. Meanwhile, many researchers have confirmed that technological interventions have a relative impact that is not usually considered (Marzano 1998; Hattie 2008). In this regard, investigational and interpretational evidence does not offer a conclusive case showing the impact of digital technologies on the learning outcomes. It is, therefore, the urgent need to recognize more accurate and articulate clear evidence (Hrastinski and Keller

2007) of the correlation effect of digital technologies and education both in teaching and learning benefits (Schacter and Fagnano 1999). It has been observed that only a few adopt digital technology at the early stages (Rogers 2003; Chan et al. 2006), particularly toward pedagogy and related areas. While the majority are late adopters who need to scale-up and expand aiming use of technology for pedagogical innovation and as a solution for teaching and learning efficacy in HE.

9.2 LITERATURE ON VIEWS OF ADOPTING TECHNOLOGY IN HE

The ecological view of the adoption of technology in HE is necessary to overcome several challenges present today, where the justification of technology adoption is a relative (Zhao and Frank 2003) which needs further enhancement. The organizations and academics have queries while choosing to adopt technologies (Somekh 2007) which later can become a problematic practice. Hence, it is very much necessary to view the existing ecosystem, justifying technology adoption (Zhao and Frank 2003) while replacing the existing technologies to effectively integrate the available resources to support (Luckin 2008) the teaching and learning process fetching effective practices that enable an efficient learning context (Underwood and Dillon 2004).

Unlike the traditional educational system of yesteryears where the premises of teaching and learning was confined to a world which was local and relatively immobile, HE has been transformed into a highly dynamic and mobile process involving international stakeholders with a focus on accelerated learning, ever since the globalization era was ushered in. Limani et al. (2019) observe that accelerated learning increases cost-effectiveness by combining the favorite student-learning styles with digitalization, thus enhancing the learning process. A literature survey indicates that, in general, HEIs the world over, have formulated their strategic plan with ample importance on the digitalization process which indicates that the global HE scenario is on the verge of a digital transformation. Davies, Mullen and Feldman (2017) reported that the technology-enhanced approaches that are in practice in the UK include Massive Open Online Courses (MOOCs), technology-assisted experiments and field trips and redesigning of assessments in addition to flipped learning. They also note that technology-enhanced curriculum design and online learning resulted in considerable positive outcomes in the National Student Survey (NSS) scores at Manchester Metropolitan University. With regard to digitalization in HE, the HEIs in Denmark are closely steered and supported by the government (Tømte et al. 2019). Similarly, a report on the digitalization strategy for HE in Norway argues that the Norwegian Higher Education sector is at the forefront of digital solutions with effective infrastructure solutions (digitalization strategy for the HE sector 2017–2021). It can thus be seen that the HE scenario is getting revamped with digital technologies penetrating all spheres of HE.

There are several factors that necessitated the incorporation of digital technologies into HE. According to Prensky (2001), the traditional design of the educational system is no longer appropriate to teach the digital natives. He further states that the assumption of teachers – with a vast majority of digital immigrants – that the same method which worked for them when they were students will now work for their students is no longer valid. Adapting a different and unique strategy for present-day students who are digital natives is one of the realities that compel the process of digitalization in HE. Makosa (2013) argues that one of the reasons for promoting educational organizations into the digitalization process could be the fashion and fascination of the young generation with electronics. Professionals in education need to master skills that enable them to cater to the needs of the younger generation because teachers are now increasingly realizing the fact that their existing skills are obsolete and inadequate to impart knowledge to their students. With the advent of new media technologies and the penetration of social media into the lives of students, learning is no longer a classroom-confined process. The digital technologies with their participatory nature take the learning process beyond the traditional classrooms. According to Farkas (2012), traditional ideas on authority and expertise are challenged by the popularity of digital technologies. Zandvliet and Fraser (2004) argues that digital technology has caused a shifting of role in the case of students. Rather than mere recipients of information, which they were used to be in the pre-digital era, students are active participants in the learning process. Digital technologies allow them to continue as active participants even when remaining physically separated.

At the same time, the concept of digital transformation goes beyond the process of mere technological change. According to Seres, Pavlicevic and Tumbas (2018), the goal should be in adopting new working methods for delivering user-focused services. They highlight digital services, digitally skilled educators as well as students and decisions that consider available evidence as some important characteristics of successful digitalization process in HE (Seres et al. 2018). Benavides et al. (2020) identify several other dimensions of digitalization in HE, apart from the mere technological dimension. These dimensions span over a vast area which includes but are not limited to education, infrastructure, curriculum, management, research, human resources, extension, digital technology governance, marketing and business process.

9.2.1 Global Tendencies on Technological Usage and Increasing Skepticism

Western countries like the USA, UK, Canada, Germany, France and other developed countries are the pioneers to use digital technologies in numerous areas of education and research. Research shows that technological usage contributes to virtual learning environments (Passey and Higgins 2011), improves pedagogy, positively affects interactive sessions (Higgins, Beauchamp and Miller 2007), and integrates learning systems (Parr and Fung 2000). Published research is focusing on various aspects of teaching and learning where technology can be used but does not give the desired outcome in terms of effectiveness. Pedagogy, assessment, curriculum (Mabry and Snow 2006), collaboration, research and effective delivery are inextricable in HE.

Global tendencies adopted technological usage more widely (Voogt and Knezek 2008) offering new teaching and learning opportunities like simulation, learning analytics, cloud computing, etc. (Chan et al. 2006). Additionally, the Internet service (Cole and Hilliard 2006) in developing countries has a relatively slow speed, having a negative impact on educational resource delivery especially, when it comes to online teaching. Secondly, it is very difficult to identify the impact of the one-to-one provision of technology (Silvernail and Gritter 2007) and the challenges faced in mobile technologies (Naismith et al. 2004). Also, developing effective interaction and collaboration (Liu and Kao 2007) and addressing teachers' concerns effectively (Donovan, Hartley and Strudler 2007) seem challenging at this moment. The third trend is the lack of substantial confirmation of the beneficial impact of technological use in e-learning on students' achievement (Kanuka and Kelland 2008) and the development of skills that contribute to success. Additionally, concern about the negative impact of technology on the health and mental development of young learners (Straker et al. 2005) was growing among various stakeholders. This concern is also related to the social and physical and mental well-being of learners who use technology for a longer period of time. It includes several aspects, such as health issues like obesity, social isolation, personality change, physical and mental fitness, addiction to certain programs, stress, etc.

9.2.2 Digital Advancements and Transformation of HEIs

Many HEIs are forced to transform with the digital advancement of technology in teaching, learning and research, shifting away from the traditional stand to digital platforms bearing new interaction avenues, educational resources and collaboration between and among tutors, trainers, mentors and learners. However, it has been observed and emphasized in the previous researches that the technological changes have triggered pressures and are still in the nascent stage struggling in their efforts to adapt and overcome the impact of the digital advancement (Lonka 2015). Meanwhile, the diverse populace with multiple educational requirements, changing profiles of learners and market completion have posed new challenges to the HEI (Şahin and Alkan 2016) in terms of process and strategy. Other challenges are related to administrative and structural aspects (Odabaşı, Fırat and İzmirli 2010) while adopting new technologies which can be evident in many HEI when they try to cope up with the managerial and organizational issues.

According to researchers, various factors like management systems (Glenn 2008), social networking, competition, internal processes, external relations, educational services, accountability requirements, internationalization of HE, structures of the universities and society are changing with the transformation and progression of technology. Meanwhile, digital technological tools are offering varied solutions to the challenges that HEIs face in the 21st century. Digital tools are used for virtual and augmented reality (Şendağ and Gedik 2015), online social networking, research and innovations, learning management systems (Glenn 2008), admission and registration, delivering and accessing quality educational content, lifelong learning, sustainable development, connectivity among different stakeholders like industry and academia offering solutions for the existing and potential challenges which may exist in HEIs. For taking maximum paybacks of these techno-system tools, the HEIs need to reform the policies and integration process and encourage innovations within the education system, constantly updating the system (Glenn 2008) and properly allocating the resources for effective system management.

Another fundamental element is the 'learners' as a significant factor (Glenn 2008) that needs to be addressed with the adoption of technological tools for educational purposes. The majority of HEIs are overwhelmed with a large number of students every year which poses increased challenges with the change associated with the learners. The change in the learner's profile and requirement (e.g. requirement of specialized course development, technical and managerial skill training, etc.) has created immense pressure on the HEIs. The change in the social and professional skills in the industrial workforce has added to the problem of transforming the qualification and skill requirement in graduates matching with the contemporary job market. The fast-changing digital age has given opportunities to young learners not only to depend on HEIs for skill development and qualification but also to use the information and media technology for life learning and career change. It also provides flexibility and adaptability within the environmental situation and is self-directive to achieve the desired objectives under given socio-cultural factors with the capability to accommodate cross-cultural settings within the HEI and its functional units. Technological transformation can be utilized for ensuring productivity, accountability, optimum utilization of resources, management control and obligation toward the service user. Communication, collaboration and problem solving within a particular time duration are other significant benefits of the technology use.

The second major fundamental element is the 'teacher' playing substantial roles and responsibilities to bring change and develop the learners imbibing them with desired knowledge and skills. Instructors also are required to be equipped with new digital technology, skills and qualifications (Odabaşı et al. 2010) to surpass the challenges within the digital world. The instructor plays a dominant role in delivering the foundation of knowledge to young learners. Even though learners have the access to digital repositories, websites, web libraries, social media and online learning groups and networks, they cannot, nonetheless, replace the role of the teacher with intellectual resources. Additionally, instructors play a major role in exposing learners to fulfill specific learning needs using new sophisticated learning management systems, learning analytics and adaptive learning (Lonka 2015). However, the instructor needs to develop individualized learning perspectives, development of knowledge and assessment of the learner through the use of digital innovations. The primary change in the role of the instructor with the adoption of digital technology in many HEI is from the 'knowledge provider' to that of 'facilitator or mentor for learning'. This paradigmatic shift has transformed the overall scenario of teaching and learning demanding the specific mindset and approach for continuous skill development and successfully aligning to meet specific learning needs ensuring the quality and lifelong learning combined with other attributes required to achieve learning objectives.

The third significant factor that plays a major role is the 'learning and teaching environment' that directly impacts the knowledge structure in any HEI. The digital revolution is shaping the learning and teaching environment shifting from the traditional process to the collaborative knowledge structure of education in the digital age. With the pedagogical shifts, informal learning (Lonka 2015) is considered to be more effective in shaping the learning activities than the customary formal method. Therefore, the current emerging techno innovation emphasizes developing collective social and organizational practices keeping learners' needs as a priority. According to Collins and Halverson (2009), the adoption of technological innovations has directly affected the pedagogical and organizational environment to realize the desired changes among the learners. The digital technologies adapted environment help learners to match the industrial needs with the learning specialization outcome necessary for the job market. Standardized learning with desired quality assurance can be achieved with the change in the learning environment allowing adaptive and subtle learning.

Even though incorporating technology into education is an age-old tradition, nothing has impacted education in a far-reaching and profound way as the process of digitalization had. If we take the examples from the last two centuries, we have the Magic Lantern of the 1870s, the radio revolution that started in 1920, and the overhead projector and headphones which were started to be used in education in 1930 and 1940, respectively. A decade later, videos became a popular tool in education which was followed by the introduction of a photocopier after ten years. The instructional television initiated in the US way back in 1939 reached its zenith during the 1960s. But it was the introduction of computers that revolutionized the use of technology in HE thanks to the popularization of personal computers by IBM in the early half of the 1980s. The same decade witnessed another development in HE technology with the arrival of laptops. But the real revolution came when the digital era was ushered in as a result of the popularization of the Internet after the development of the World Wide Web and HTML in the early 90s. Today, 'digital transformation' of commercial processes, procedures and opportunities had an enhanced impact on society and educational institutions throughout the world. Digital transformation has become a high priority in the sphere of HE during the last decade. It would remain so and evolve into new forms in the years to come.

9.3 RESEARCH METHODOLOGY AND TOOLS

This study states various methods that can provide a strong foundation to comprehend digitalization and its significance in HE. Therefore, mixed methods were used to explore official discourse and to examine several aspects at various levels of the transformation process of HE. The combination of in-depth document analysis, investigation, interpretation and consultations at multiple levels provided substantial pieces of evidence to explore the digital transformation of HE and its significance at the international, national and institutional levels.

Research has been conducted within the structural framework to seek and establish a comprehensive insight into digital reforms and the development over the past decade exploring the challenges faced and identifies frequent changes that build opportunities, fixed counteractive mechanisms and measures adopted to overcome serious issues. Additionally, qualitative information using thematic focusing method was obtained through self-evaluation and experience-sharing mechanisms to analyze and provide recommendations for advancing the digitalization of HE.

9.4 MAJOR CHALLENGES AND SOLUTIONS OF DIGITAL TECHNOLOGY USAGE IN THE HEI

Various factors that affect the HEI create challenges while adopting new technologies in the HEIs. These challenges are as follows:

1. Integration of technology in daily life: The complete integration of technology in daily teaching and learning environment and pedagogy seems a puzzling task. According to Ertmer et al. (2012), additional effort and a dedicated process are needed to entirely overcome these challenges. This challenge needs to be tackled from the initial stages itself with a concrete strategy without which the organization may probably end up suffering high cost and ineffectiveness in the delivery process.
2. Insufficient equipment and connectivity (Habibu, Mamun and Clement 2012): The major challenge any HEI faces with the development of the technological change is the insufficient equipment and connectivity within the internal system. Without adequate number and standardized equipment and fast connectivity, the educational technology remains ineffective, which can create constraints and incompetency in utilizing the technology to its full potential?
3. Training of human resources: Adequate training to different stakeholders must be provided focusing on skill development for using educational technology without which implementation is not feasible. Human competency is considered the cardinal resource in any development process. Inadequate training related to technology and professional skills creates challenges for HEIs. Professional development (Pearce et al. 2011) of the activities can be easily conducted and coordinated by the use of digital technologies in their work that is helpful for effective delivery and achievement of desired results.
4. Accessibility (Dintoe 2018): Regular and reliable access to various components is the fundamental requirement for any system. Access and integration of required equipment, resources and connectivity to all stakeholders are essential for the successful conduct of an educational program. It is observed that the majority of students, academics and personnel in HEIs in developing countries have no access to the basic necessity for implementing digital technological adoption which acts as the greatest challenge for improvement. Unreliable access to connectivity and equipment makes it extremely difficult for teaching, learning and research.
5. Lack of technological skills: It has been observed that lack of technology-adaptive skills affects and acts as a constraint for the implementation of educational goals. According to Ertmer et al. (2012), inadequate professional development and training are the most commonly mentioned reason for the proper implementation of teaching technology in most of the HEIs. New technologies are regularly developed and used for teaching and research by academics. Lack of additional training to keep their technological skills (Rienties, Brouwer and Lygo-Baker 2013) updated with technological

innovations can cause inadequate professional development and insufficient familiarity and availability of necessary resources, which in turn, can act as a barrier in deriving the benefits of technology adaptation.

6. Inadequate technical support and administrative/peer support: Lack of technical and administrative support to academics affects the optimal use of technology. It is revealed that technology-based application acts as a barrier and consumes a large amount of time without adequate support for educational technology. With the regular and extended support from trained professionals of educational technologies, the HEI can reduce the burden considerably and can achieve quality delivery of courses and research outputs.
7. Attitude and beliefs toward technological adaptation: Attitude of academics and researchers toward the adaptation and usage of digital tools plays a vital role in determining the effectiveness and implementation of programs. The HEI must promote and develop positive attitudes among its academics, researchers and staff for regular usage and optimizing technological utilization. Now that technology is inevitable (Ertmer et al. 2012), it is essential to update and regularly maintain the skills for determining the effectiveness.
8. Self-reliance and flexibility: Assuming that the technical implementation is essential for achieving the organizational goals (Digital Dawn 2015), it is, therefore, the need of the day to be self-reliant, developing flexibility to be adapted at different levels for multiple purposes and capable of the new addition with the change in the environment.
9. Availability of guidelines and regular feedback (Al-Bashir and Kabir 2016): Without handy guidelines and instruction on various issues of technology, it is challenging for different stakeholders to operate and adopt changes. Guidelines will act as a key for regular and proper usage without daily interruptions. Regular feedback must be taken from the users to understand the difficulties and complications faced by stakeholders at different levels and point while adopting, implementing and using new technologies.
10. Technology transformation must fulfill strategic priorities: Digital transformation signifies a deep alteration of HEI's activities and processes. Digital technology shifts (Perkin and Abraham 2017) must fulfill short and long-term strategic priorities and should be fully exploited to achieve competence within the institution accelerating its impact across society. The HEIs should explore the huge opportunities to mix digital technologies, resources, people and innovations to present, promote and position its capabilities to develop quality and capability to be built in the future.
11. Policy reformation: Fast transition of technology and individualized learning has raised the pressure on HEIs for transforming the technological infrastructure. The conversion of the traditional form of teaching to technology-based teaching has seen strong resistance and increased expectation from students and parents. These factors and practices contradict the very nature of the traditional system to consider individual and group differences. To avoid these contradictions between the systems, the HEIs need to reform policies and procedures to update the education process in a staged manner irradiating the confronting factor.

12. Utilization of internal and external resources: The optimum use of internal and external resources like human, financial, infrastructural, etc., must be accomplished to maximize the output and achieve effectiveness in the HEI. The ability of resource utilization by the HEI will ultimately reduce the organizational cost and risk associated with the adoption of technology. The organization needs to look at micro functional level and resources utility of technological application for achieving the goals.
13. Innovations: The attitude of innovation and change must be inculcated among diverse stakeholders throughout the educational processes. Necessary changes in strategies and support from the top management must be incorporated at all levels. However, if innovation falls short to achieve its mark in the educational system, the institute must evaluate the approach to identify the deficiency for further action.

Just like in any other area, digitalization offers an array of opportunities in HE. As students can access the courses and learning material remotely, education can be brought to geographically remote areas in a cost-effective manner. On the one hand, digitalization can ensure cost-effectiveness, while on the other hand, it can increase productivity by saving time and energy because of its easy accessibility. Makosa (2013) observes that the most important advantages of digital education are the increase in the effectiveness of education and the equalization of educational opportunities. In addition to being environment friendly, digitalization also ensures prolonged preservation of knowledge in multitudes of forms in cyberspace which ensures possibilities of upgradation from time to time. The recording facility of online lectures and its round the clock availability will be useful for students for future referencing purposes. It will enhance self-directed learning skills in students which will result in strengthening critical thinking, analytical skills and reasoning. It can also motivate students to learn new skills and share information among student groups, thereby creating a new learning culture. Once course delivery becomes online, it not only transcends geographical boundaries but also bypasses cultural constraints. While the latter opens up opportunities for intercultural communication and imparting education across various cultures, it also poses certain challenges as both educators and learners will now need to develop cultural sensitivity. Digitalization has its own challenges in many other areas as well.

Presently, many of the educators who belong to the analog era generation are not well trained to handle the rapid digital transformation. Imparting time to time training and self-improvement of skills can be a challenge to many professionals. In addition to this, some other challenges await HEIs. They should realize the relevance and power of big data. Successful business houses are capable of rightly analyzing the vast data available to them and formulate business strategies based on the insight derived by data analysis. Unless HEIs acquire capabilities for developing data analytics, they are not going to gain a strategic edge in the increasingly competitive market place. In order to ensure enhanced stakeholder experience, HEIs need to rationalize their operations based on big data analytics to face the challenges posed by the realities of funding and student expectations. Yet, another challenge inherently involved with the digital premise is related to the security and privacy of individuals. Institutions have to address this issue wisely to retain the trust of their public vested in them.

With the development of the digitalization of HE in GCC (Gulf Cooperation Council countries) and the rest of the countries in the Middle East, Oman has seen the fast growth than ever before. The HE institutions are using the IT tools from the admission of students to the exit. Withstanding the effort, Oman faces several challenges that have been recognized and streamlined which result in the growth of digitalization in coming years. These roadblocks include the lack of technological skills with the shortage of trained employees handling the vertical situation of teaching and learning support. Secondly, as rest of the world, Oman also faces inadequate technical support due to fast growth and shortage of technical adaptation. Additionally, the country has seen a big shift in attitude and beliefs toward technological adaptation which enables the introduction of technological adaptation from the school level to the HE. Thirdly, due to the scarcity of trained individuals, full utilization of internal and external resources cannot be adopted at the ground level. Also, lack of innovation in the HE technological adaption restricts the institution to fully utilize the resources available to its optimum.

9.5 OPPORTUNITIES FOR TECHNOLOGICAL ADVANCEMENT IN HEI

1. Improvement of teaching and learning (Selwyn 2016): Despite several challenges, the universities and colleges have accelerated the process of digitalization in HE which helps them to the pervasive use of information technology for supporting and improving teaching and learning (Hanna 2016). The technological tool enables the students to search and access the content of their program given by the teachers. They also have easy access to other resources, such as library books, digital knowledge banks and software with minimum effort and time during the course.
2. Cost-cutting effect: Often, the question of cost has been the major problem for HEI due to its nature of non-profitability service toward the society. Digitalization of system and process of HE universities (Bowen 2013) will transform from the traditional time consuming to speedy digital system reducing the cost drastically. With the environmental and competitive situation existing today, cost-cutting will be an advantage that can be added to improve and provide quality education. Also, a reduction in the cost factor is equally applicable to the students who can simultaneously pay lesser for their education. Like modern businesses in a competitive world, universities must seek all means to drive efficiency and cost-saving. To achieve the cost-saving objective, the HEIs should reduce the budget of manual and time-consuming operations like the admission and registration process and searching books in the library and add budget on technological advancement for making things easier and faster with lesser cost-creating advantage center.

3. Overcome the techno barrier among young learners: Most of the courses other than information technology and communication engineering do not expose the students to technological development in the practical world. The induction of the technology for educational purposes in schools, colleges and universities helps the young learner to interact with and adopt the technology on a day-to-day basis. These newly adopted in-depth skills will be helpful in future for building their career in any sector apart from the IT sector. Additionally, the extent of the effect may be influenced by the capability of teachers to use digital learning tools and resources effectively to achieve improved learning outcomes (Bates and Sangrá 2011).
4. Reducing inequalities and promoting opportunity for all: Previous studies show that many HEIs in developed countries have the evidence of the use of digital technology tools and resources helping to reduce gaps and inequality among areas like gender, subject selection, positions, research and attainment of effective implementation of strategies. It has also been observed that the support needs toward the teachers and learners will improve creating a conducive environment for academics and research in the university.
5. Improving learner's transitions into employment: Digital technology improves skills, competencies and collaboration among the learners linking the industry and job market. This support is very crucial for beginners who are going to enter the job market hunting for jobs. The advantage of digital tools builds network and collaboration among the learners and industry before the completion of the program, enhances knowledge and indulgence in career paths and working environments for a smooth transition to employment.
6. Engagement of multiple participants: It is evident that the digital tools are capable of handling multiple participants who can monitor, review, analyze and provide feedback for further improvement. Students, peers, teachers, management staff, industrial representatives, parents, advisors, functional departments (Gaebel et al. 2014) and technical staff can be involved simultaneously to understand, support and overcome certain challenges to improve the effectiveness and achieve the goals and performance of learners facilitating the information to all engaged.
7. Assessment and evaluation (Siemens et al. 2015): Studies have already presented that digital technology can be a game-changer in HE. Digital tools can be used for assessing different components linked to the outcomes, check attendance, students' past details and progression, preferences, etc., which can be useful to support the students for learning, develop cognitive and life skills and provide guidance from time to time. Secondly, on-the-job training, internship, part-time works assignment, etc., can be monitored and evaluated in a timely manner by various evaluators from academics as well as industry. Even though the students are far from the HEIs for undertaking various assignments or internships, the faculty can monitor and guide them even from geographically remote places for achieving the program objectives. Feedback on the assessment by using technological tools can help the students to improve grades and overall performance.

8. Research and collaboration: Nowadays, technology acts as a tool to conduct high-quality research and analysis. Many HEIs are looking forward to and investing in huge amounts of money on technology – both hardware and software – that enables the researchers to conduct complex studies and present the finding to overcome unsolved problems. Technology in the field of medicine and engineering research has exposed as a lifesaving component in the present day. Additionally, research and collaboration among scientists, academics and researchers from all fields can be shared, collaborated and utilized for overcoming the complex challenges and reducing the cost of repetition and errors.
9. Efficiency and effectiveness of education process: The technological tools can be used beginning from developing the course material, assessment and evaluation to estimating the achievement of outcomes and attributes imbibed by the learners during the program. HEIs need to invest in instructional technology that can be directly linked to an application for academic staff (Nworie, Haughton and Oprandi 2012) that is valuable for determining the effectiveness of the program with regard to other elements and showing the efficiency in creating or adding value.
10. Curriculum redesign: The most potential opportunities of technology to renovate and redesign curriculum in order to incorporate the requirements of present and future job market directly affect the processes of teaching, learning and research. Technology can be potentially used to enable great improvement in outcomes that enable the students to be equipped with knowledge and skill for the job market. With the curriculum redesign, the major component is the pedagogic approaches; assessments, etc., of the course which can be altered to fit the need to support learning. In this regard, it becomes necessary to rethink the use of technology and its possibilities to achieve better learning outcomes at a lower cost to take advantage of the digital technological capabilities the HEI owns.
11. Cooperative teaching and learning for the vulnerable: Technological tools can be used to support and provide quality education to the vulnerable having difficulty to pay tuition fees cannot leave their home due to family situation, people with disability to travel employed citizens aged and others who are not in a position to spend a considerable amount of money for further education. Digital technology can reach them in their homes at a very low cost. Additionally, cooperative learning, peer-to-peer learning and distant interface can be easily conducted by effective use of technology.
12. Quality assurance (Austen et al. 2016): With the use of technology, quality can be assured in the education process and progression. Technology-enhanced learning initiatives will lead to excellent teaching that monitors and ensures the quality of teaching, learning, course material; organizational issues, pedagogical approaches, etc. (Selwyn 2016) and resources provided for delivery and research. Based on the technological predetermined indicator, corrective action can be taken without any deviation from the quality path.

13. Sustained learner engagement: With the use of technology, HEIs have the opportunity of sustaining student engagement even after they pass out from the institution. Alumni can serve as important resources for various purposes, such as industry linkage, sponsorship, job market feedback, curriculum review and development, student on-the-job training, etc. Community and social development activities can also be an added advantage for sustaining learners' engagement. This type of formal and informal engagements among the alumni and the community has a long-term effect on building reputation and relationship. Digital technology acts as the standing platform for sustaining learners' engagement that reflects overall learning progress.
14. Digital tools as learning platform: During the current situation of the Covid-19 pandemic, technology-based digital platforms act as a space for delivery of knowledge, engaged learning and instruction activities. Technology tools carry out high-impact interactions that have the capacity to replace the face-to-face teaching and learning system. The tools offer flexible learning platforms that are student-centered, collaborative and adapted for learning.
15. Database management and analytics: Technology provides a great opportunity for maintaining and analyzing data of the graduates, undergraduate and other learners that can be used for strategic decision making. The data from the HEIs can then be used to measure the students' engagements as learners, individual preferences and learning behavior, predict the grades and the outcome expected by the completion of the program, comparative analysis within and other sections, dropout rates, admission and registration management, overall individual performance in that year, etc. The database analytics will help the HEI to reduce the overall grievance and disengagement to support students in achieving the desired performance.
16. Increasing retention and success rate: Database analytics provide opportunities and enable HEIs to track, engage and reduce dropout rates among the students to attain a higher progression rate and addressing potential challenges that come across during the program. The technological advancement provides early warning and predicts the students at risk of dropping out or repeating the course. This process will simultaneously indicate the corrective measures to be taken and reduce the students' attrition rate for the HEI.
17. Provide support system to teachers and students: Technological tools offer valuable insights for supporting the teachers and students from potential constraints and errors and intervene for achieving designated path such as selection of courses – students may select wrong courses – change in programs, academic advising, grade prediction and calculation, etc. The digital tools offer solutions and support for students at risk for improving the grades and can also identify the academic weak students or students struggling with certain courses to provide individual or group support.
18. Control system: Technology-enhanced financial system may act as a regulatory tool for controlling university spending and accessibility to funds for its daily requirements. The techno tool will provide support for managing

and allocating funds, assessing the risk, and later will help develop systematic datasheets for auditing and strategic decision making.

19. Behavioral insight of teachers and students: Technology-enhanced internal system provides an opportunity for understanding the behavioral pattern of all its stakeholders, especially teachers and students. The techno-tools help to predict the behavioral pattern in the selection of courses, intention for continuing with the opted courses, health data, interest data, etc. This behavioral data analysis is very helpful for HEIs for providing sports-related courses, armed force training program, etc., where physical and mental endurance is highly needed which make students drop out of the course.
20. Knowledge management (Mathew 2009): Technology can be of great use and provide opportunities for capturing, storing, sharing and generation of new knowledge from time to time. Universities and colleges have a large number of academics and students who generate knowledge and this knowledge can be captured and reused by the future faculty to improve and develop newer knowledge that can be useful for creating a competitive advantage (Mathew 2010a).
21. Team-based approach (Austin, 2006), as well as personalized learning: Technology with flexible tools can provide a personalized learning experience or team-based approach which cannot be achieved in the classroom setting. Each student can select or delete the courses during the semester from the pool of courses that he or she has to complete during the program. This experience provides students to manage their activities and priorities both individually and with groups in their choice to overcome challenges and score desirable grades.
22. Internationalization and flexible educational programs (Conole 2014; O'Connor 2014): Technology-enhanced education has the capacity for widespread reach and attracts a large number of students around the world with flexible educational programs.
23. Data security system and collaborative studies: Technological tools can be used for considering data security and research through collaborative action (Khalid et al. 2018) across the campus or with other HEIs. Detection of plagiarism, protecting data theft, hacking, and data fraud, etc., can be prevented and taken care of with the help of specific hi-tech tools. HEIs have opportunities for better collaboration for providing services like library services, sports services, training programs, research and professional development (Gibbs, Knapper and Piccinin 2008), etc.

Furthermore, technological development and digitalization require adequate competencies (Rienties et al. 2013) and strategic investment for achieving desired outcomes. It includes external processes (Zawachi-Richter and Naidu 2016) and internal processes (Zawacki-Richter and Latchem 2018). These factors affect the overall strategies, require systematic implementation and support from the academic, management, IT specialists and individuals for bottom-up initiatives (Selwyn 2016) or 'top-down' processes (Nworie et al. 2012).

The changing environment and globalization have also posed opportunities to the Gulf region, especially to the GCC countries as they are developing at a faster rate in adopting ICT in daily life. Oman's fast growth in the technological arena has received its due credit in the field of business, education, hospitality and other sectors. In HE, Oman has a substantial advantage with modern equipment and appropriate infrastructural facilities available inside the country. The country has the opportunity for improvement of teaching and learning at both graduate and post-graduate levels with a continuous focus on research and collaboration. With a vast geographical spread of students in Oman, the use of the digital platforms in HE will not only create an efficient and effective education process but also cut costs by engaging multiple participants supporting sustainable learner engagement during the regular studies. Additionally, the digital adaptation supports HEIs and the regulatory authority to ensure rigorous assessment and evaluation and curriculum redesign, provides a support system to teachers and students, controls system and quality assurance of institutions and programs domestically and internationally. Furthermore, the opportunity for internationalization, accreditation and flexible educational programs using digital tools as learning platforms will further strengthen the development of educational reform in the country.

9.6 ANTICIPATING EMERGING TRENDS

The future holds a great expectation and turn around trend related to the use of digital technology tools in HE. Specific skills and interests are growing with the change from the traditional system to digitalization. Few main aspects that deserve specific attention while anticipating the emerging trends are as follows:

i. Skill gap among the young learners to handle the digital tool and technological instrument decreases with the time and the pace of adaptation. Moreover, it is likely that the IT-related occupational gap to handle the technology will reduce in comparison with the previous decade and it is likely to be perpetuated at least for the coming periods.
ii. Initiatives at all levels in both government and private sectors focusing the digital and technological development are expected to increase. A major initiative will be target oriented and directed toward achieving performance.
iii. Database management and analytics could constitute a specific process in the future and the market expands with the use of digitalization.
iv. The future trend of technology utilization will affect our daily routine lives and education system and will change with regard to knowledge distribution and management.
v. Digital transformation will increase specialization and differentiation with regards to occupations and constantly emerge new knowledge and skills among young individuals.
vi. Traditional occupations tend to change with the adoption of technology and therefore suffer a shortage of techno-skilled workers. Additionally, occupations will change and dependence on the knowledge provided will be extremely beneficial and a compelling opportunity for HEIs.

vii. Integrated system with technological efficiency will lead the HEIs in the coming future. Career guidance and education (Cedefop 2019) will be closely linked to a successful future. The students need to identify their capacities, competencies and interests while selecting the courses that match their career path and objectives.

viii. Entrepreneurship with respect to developing newer technology will have better scope in the future. The entrepreneur will use technological tools and gadgets for producing and serving the social needs along with materializing the career objectives and sources of income.

ix. Future holds difficulty that may arise when implementing techno-entrepreneurs in most of the developing countries (Mathew 2010b). The lack of techno-based courses in HEIs and skills among the nascent entrepreneurs will create a gap for a shorter period of time.

x. Applications of artificial intelligence and algorithms (Goksel and Bozkurt 2019) will take a higher place in the HE system, especially in ICT-related courses.

xi. Majority of teachers tend to utilize ICT tools predominantly for making the daily activities simpler and meeting specialized teaching applications needs. The policies can differ from country to country that may affect the implementation of the technological tool in the HE system.

xii. Students tend to have their own devices making them more acquainted and comfortable with techno gadgets. The smartphone will be commonly used as a learning platform. Individual techno-gadget will help the students and teachers to feel more and better use of the spare time that will directly generate positive outcomes for teaching and learning.

xiii. Data analytics will be highly used for measuring and anticipating the future and optimizing learning (Ferguson et al. 2016). Analytics will be useful not only for analyzing data but also for predicting the students' performance, dropout rates, etc., along with the calculation with regard to grades in the future.

xiv. Personalization of learning content (Maseleno et al. 2018) and assessment will be very common in HE programs and curriculum. Students will have a large number of choices to select from the pool and access the learning content which will be helpful for the larger community. This change will influence personal level performance and individual learning behavior which can be named as 'adaptive learning' or 'smart learning' (Peng, Ma and Spector 2019; Kinshuk et al. 2016).

xv. Teaching and learning will be majorly based on virtual reality, augmented reality and other innovative virtual tools (Akçayır and Akçayır 2017). Virtual reality will dramatically change current learning and teaching settings along with the simulator for better reach and delivery of educational content to the community.

xvi. Classrooms and laboratories will be highly equipped and connected in the future for HE purposes. The differential variation will be seen on facilities, networking requirements, digital equipment and digital content from one HEI to another. Cutting-edge ultra-fast networks will be the game-changer in the future scenarios within the classroom situation. Highly equipped and connected classrooms can be seen in the future.

xvii. Wide public acceptance and demand for online programs among the students can be seen in the next decade. The flexible online program will be in great demand that will enable the professionals and working people to go for HE.

xviii. Knowledge-based processes (Schwarz 2010) will improve the opportunities for sharing knowledge, experience, resources, intellectual properties and services.

xix. Digitalization will change the way for international cooperation for academics, researchers, HE institutions and communities. This opportunity will transform the way we make networks to gain knowledge and access to global resources. It will be at the micro-level, individual-level and functional department level and at the level of individual HEIs as a whole.

xx. Technology will enhance graduate employability and cultivate 'enterprising' individuals.

With the adoption of ICT in Gulf regions to implement the strategies to develop HE in countries, the technology-blended teaching and learning have become a new normal with a majority of students' time and focus on the practical aspect related to projects and application-based assessments. Countries like Oman have rightly adopted the technology in the HE with the majority of programs are based on digital tools and their application in day-to-day life. Teaching and learning will be majorly based on virtual reality to reach the common with the high standard of delivery and assessments. Most of the classrooms and laboratories are equipped and connected with ultra-fast networks giving wide acceptability and flexibility for both institutions and students. These growing trends in the Oman HE system will likely expand further to personalize education and enhance graduate employability.

9.7 DISCUSSION

Globally, the HE scenario and its proportions are continuously altering with the change in competition. The growing complexity in competition and related activity raised several issues of survival and sustainability questions in the recent past of HE. However, there are now high priorities to digitally transform of key HE elements, such as teaching, learning and research around the world. The development of HE around the world has recognized digital technology and its application in teaching, learning and research purposes. Presently, HE aiming for international dimensions has used the technological know-how as an opportunity to dominate all aspects of education and its management. Meanwhile, many countries are opening up to adopt digital tools to attract incoming students offering internationalized education experiences and several new are in consideration. Digital technology integration also creates challenges which need to be addressed to improve teaching-learning experience. In this regard skills of staff, students and researchers in digital tools need to respond with the technical and specialized professional with clear procedures for operation.

In this chapter, we have identified opportunities and challenges that come across the HE institutions during the digital transformation supported by the finding of the previous studies. Few opportunities include improvement of teaching-learning practice, cost advantage, reducing inequalities, promoting opportunity for all, easy and quick assessment/evaluation, research and collaboration and many others. Challenges while adopting the digital technology include integration of technology in routine activities, insufficient equipment and connectivity, accessibility, lack of technological skills and flexibility, low policy reformation, innovations, etc., which need to be responded for achieving effective usage. The study also anticipated trends that hold great expectation in the near future and have the ability to provide effective and affordable education to a large population.

9.8 IMPLICATION

Digital technology continues to grow and remains a vital aspect of HE. Since there are no signs of turn back and losing drive, researchers are exploring deep to bring new methods and indulgence to produce effective digital tools and technology in all aspects of HE. Our study provides a solid picture and indicates that the HE institution has a key role in applying digital tools keeping in opportunities and challenges of digital technology in mind. This chapter will help the scholar, students and HE organization to have a deep understanding of trends that can apparently transform the HE strategic implementation. This chapter will also help the stakeholders gain a deeper understanding of how digital tools can be implemented and it affects a different aspect of HE. Further research and analysis are, however, needed to provide a deeper and full understanding of how various aspects of digital technology and interrelated aspects are associated with impacting the HE and its implementation for better preparation to reduce the future effect of digitalization on HE.

9.9 CONCLUSION

In General, digital technologies include the use of digital gadgets like computers, smartphone phones, etc. (Rice 2003) for specific purposes. The main benefits of technology-based e-learning are cost-saving to the HEI (Arbaugh 2005) compared to traditional teaching having huge investment in infrastructure and resources. With the advancement of digital technology, there is a potential for wide scope in cost reduction and improvement in the HE system (Unwin et al. 2010; Bolu and Egbo 2014); improved learning (Motiwalla 2007) and quality of teaching and learning and accessibility (Moya, Musumba and Akodo 2011). Technology integrated e-learning benefits both the HEIs and the students (Bhuasiri et al. 2012) equally. On the other hand, digital tools help the HEIs to form an international environment (Lee 2010) and deliver knowledge anytime with the use of technological tools (Taylor 2007). It also enables the institution for collaboration for sharing of knowledge, research and joint programs which may result in receiving full benefits of global knowledge provided by specialized academics selected from around the globe making the HEIs achieve a high level of acceptance. Students will have a choice (Hollenbeck, Zinkhan and French 2006) that is flexible and suitable to their individual needs based on their career goals. Additionally, they can undertake studies along with a job (Wisloski 2011) benefiting from not attending physically in the classroom for lectures (Bhuasiri et al. 2012). Nevertheless, HEIs may face various constraints (Stantchev et al. 2014) by the adoption of digital technologies, such as capital investment, resource arrangement and continuous innovation. Future holds bright prospects for the HEIs in terms of technology-integrated learning, making the HE system more flexible and acceptable.

REFERENCES

Aguti, B., Wills, G. B., & Walters, R. J. 2014, An evaluation of the factors that impact on the effectiveness of blended e-learning within universities. In International conference on information society (i-society), pp. 117–121. IEEE. https://doi.org/10.1109/i-Society.2014.7009023

Akçayır, M., & Akçayır, G. 2017, Advantages and challenges associated with augmented reality for education: A systematic review of the literature. *Educational Research Review*, 20, 1–11. https://doi.org/10.1016/j.edurev.2016.11.002

Alavi, M., & Leidner, D. 2001, Research commentary: Technology-mediated learning – a call for greater depth and breadth of research. *Information Systems Research*, 12(1), 1–10. http://dx.doi.org/10.1287/isre.12.1.1.9720

Al-Bashir, Md. Mamoon, & Kabir, Md. Rezaul. 2016, The value and effectiveness of feedback in improving students' learning and professionalizing teaching in higher education. *Journal of Education and Practice*, 7(16), 38–41.

Andrews, R., Burn, A., Leach, J., Locke, T., Low, G. D., & Torgerson C. 2002, A systematic review of the impact of networked ICT on 5-16 year olds' literacy in English. In: *Research Evidence in Education Library*. Issue 1. London: EPPICentre, Social Science Research Unit, Institute of Education.

Arbaugh, J. B. 2005, Is there an optimal design for online MBA courses? *Academy of Management Learning & Education*, 4, 135–149. https://doi.org/10.5465/AMLE.2005.17268561

Austen, Liz et al. 2016, Digital capability and teaching excellence: An integrative review exploring what infrastructure and strategies are necessary to support effective use of technology enabled learning (TEL), QAA 2016.

Austin, R. 2006, The role of ICT in bridge-building and social inclusion: theory, policy and practice issues. *European Journal of Teacher Education*, 29(2), 145–161.

Bates, T., & Sangrá, A. 2011, *Managing technology in higher education: Strategies for transforming teaching and learning*. New Jersey: Jossey Bass.

Benavides, Lina M.C., Johnny A. Tamayo Arias, Martín D. Arango Serna, John W. Branch Bedoya, and Daniel Burgos. 2020. "Digital Transformation in Higher Education Institutions: A Systematic Literature Review" *Sensors* 20, no. 11, 3291.

Bhuasiri, W., Xaymoungkhoun, O., Zo, H., Rho, J. J., & Ciganek, A. P. 2012, Critical success factors for e-learning in developing countries: A comparative analysis between ICT experts and faculty. *Computers & Education*, 58(2), 843–855. https://doi. org/10.1016/j.compedu.2011.10.010.

Bolu, C. A., & Egbo, K. 2014, The role of higher education institutions in the development of ICT professionals for innovation in Nigeria. *International Journal of Engineering Innovations and Research*, 3(1), 1.

Bowen, W. G. 2013, *Higher education in the digital age*. Princeton, NJ: Princeton University Press.

Brahimi, T., & Sarirete, A. 2015, Learning outside the classroom through MOOCs. *Computers in Human Behavior*, 51, 604–609. http://dx.doi.org/10.1016/j.chb.2015.03.013

Cedefop. 2019. Over-qualification rate (of tertiary graduates): Skills panorama. https://skillspanorama.cedefop.europa.eu/en/indicators/over-qualification-rate-tertiary-graduates

Chan, T., Roschelle, J., Hsi, S., Kinshuk, Sharples, M., Brown, T., Patton, C., Cherniavsky, J., Pea, R., Norris, C., Soloway, E., Balacheff, N., Scardamalia, M., Dillenbourg, P., Looi, C., Milrad, M., & Hoppe, U. 2006, One-to-one technology-enhanced learning: An opportunity for global research collaboration. *Research and Practice in Technology Enhanced Learning*, 1(1): 3–29.

Clegg, S., Konrad, J., & Tan, J. 2000, Preparing academic staff to use ICTs in support of student learning. *International Journal for Academic Development*, 5(2), 138–148. http://dx.doi.org/10.1080/13601440050200743

Cole, J. M., & Hilliard, V. R. 2006, The effects of web-based reading curriculum on children's reading performance and motivation. *Journal of Educational Computing Research*, 34(4): 353–380.

Collins, A., & Halverson, R. 2009, *Rethinking education in the age of technology: The digital revolution and the schools*. New York: Teachers College Press.

Conole, G. 2014, A new classification schema for MOOCs. *INNOQUAL: International Journal for Innovation and Quality in Learning*, 2(3), 65–77.

DAAD/DIE 2018: Digital Transformation: Higher Education and Research for Sustainable Development. Position paper. Bonn.

Davies, S., Mullen, J., & Feldman, P. 2017, Rebooting learning for the digital age: What next for technology-enhanced higher education? Higher Education Policy Institute, HEPR Report 93. Retrieved from https://www.hepi.ac.uk/wp-content/uploads/2017/02/Hepi_Rebooting-learning-for-the-digital-age-Report-93-20_01_17Web.pdf

Digital Dawn. 2015, *Hays recruiting revue*, 8(46), 18–26.

Dintoe, Seitebaleng Susan. 2018, Information and communication technology use in higher education: Perspectives from faculty. *International Journal of Education and Development using Information and Communication Technology (IJEDICT)*, 14(2), 121–166.

Donovan, L., Hartley, K., & Strudler, N. 2007, Teacher concerns during initial implementation of a one-to-one laptop initiative. *Journal of Research on Technology in Education*, 39(3): 263–269.

Ertmer, P. A., Ottenbreit-Leftwich, A., Sadik, O., Sendurur, E., & Sendurur, P. 2012, Teacher beliefs and technology integration practices: A critical relationship. *Computers & Education*, 59, 423–435.

Farkas, M. G. 2012, Participatory technologies, pedagogy 2.0 and information literacy. *Library Hi Tech*, 30(1), 82–94.

Ferguson, R., Brasher, A., Clow, D., Cooper, A., Hillaire, G., Mittelmeier, J., Rienties, B., Ullmann, T., & Vuorikari, R. 2016, Research evidence on the use of learning analytics: Implications for education policy. Joint Research Centre. http://dx.doi.org/10.2791/955210

Gaebel, M., Kupriyanova, V., Morais, R., & Colucci, E. 2014, E-learning in European Higher Education Institutions. Results of a Mapping Survey Conducted in October–December 2013 (Brussels, European University Association). Available at: http://www.eunis.org/wpcontent/uploads/2015/02/e-learning-survey2.pdf (accessed 22 March 2019).

Gibbs, G., Knapper, C., & Piccinin, S. 2008, Disciplinary and contextually appropriate approaches to leadership of teaching in research-intensive academic departments in higher education. *Higher Education Quarterly*, 62(4), 416–436.

Glenn, M. 2008, *The future of higher education: How technology will shape learning*. London: Economist Intelligence Unit.

Goksel, N., & Bozkurt, A. 2019, Artificial intelligence in education: Current insights and future perspectives. In S. Sisman-Ugur & G. Kurubacak (Eds.), *Handbook of research on learning in the age of transhumanism* (Vol. 9, pp. 224–236). Hershey, PA: IGI Global. https://doi.org/10.4018/978-1-5225-8431-5.ch014

Habibu, Taban, Mamun, Md Abdullah Al, & Clement, Che. 2012, Difficulties faced by teachers in using ICT in teaching-learning at technical and higher educational institutions of Uganda. *International Journal of Engineering Research & Technology*. 1(7), 1–9.

Hanna, N. 2016, Mastering digital transformation: Towards a smarter society, economy, city and nation (First edition). *Innovation, technology, and education for growth*. Bingley: Emerald Publishing.

Hargittai, E. 2010, Digital natives? Variation in internet skills and uses among members of the 'next generation'. *Sociological Inquiry*, 80(1), 92–113.

Hartley, J. 2007, Teaching, learning and new technology: A review for teachers British. *Journal of Educational Technology*, 38(1), 42–62.

Hattie, J. 2008, *Visible learning: A synthesis of over 800 meta-analyses relating to achievement*. London: Routledge.

Helsper, E. J., & Eynon, R. 2010, Digital natives: Where is the evidence?. *British Educational Research Journal*, 36(3), 503–520.

Higgins, S., Beauchamp, G., & Miller, D. 2007, Reviewing the literature on interactive whiteboards. *Learning, Media and Technology*, 32(3), 213–225.

Hollenbeck, C. R., Zinkhan, G. M., & French, W. 2006, Distance learning trends and benchmarks: Lessons from an online MBA program. *Marketing Education Review*, 15(2), 39–52. https://doi.org/10.1080/10528008.2005.11488904

Hrastinski, S., & Keller, C. 2007, An examination of research approaches that underlie research on educational technology: A review from 2000 to 2004. *Journal of Educational Computing Research*, 36(2), 175–190.

Imhof, B., Scheiter, K., & Gerjets, P. 2011. Learning about locomotion patterns from visualizations: Effects of presentation format and realism. *Computers & Education*, 57(3).

Jamieson, P. 2003, Designing more effective on-campus teaching and learning spaces: A role for academic developers. *International Journal for Academic Development*, 8(1-2), 119–133. http://dx.doi.org/10.1080/1360144042000277991

Jani, J., Muszali, R., Nathan, S., & Abdullah, M. S. 2018, Blended learning approach using frog VLE platform towards students' achievement in teaching games for understanding. *Journal of Fundamental and Applied Sciences*, 10(5S), 1130–1141.

Kanuka, H., & Kelland, J. 2008, Has e-learning delivered on its promises? Expert opinion on the impact of e-learning in higher education. *Canadian Journal of Higher Education*, 38(1), 45–65.

Kennedy, G., Judd, T., Churchward, A., Gray, K., & Krause, K. 2008, First year students' experiences with technology: Are they really digital natives. *Australasian Journal of Educational Technology*, 24(1), 108–122.

Khalid, J., Ram, B. R., Soliman, M., Ali, A. J., Khaleel, M., & Islam, M. S. 2018, Promising digital university: A pivotal need for higher education transformation. *International Journal of Management in Education*, 12(3), 264–275.

Kinshuk, Chen, N.-S., Cheng, I.-L., & Chew, S. W. 2016. Evolution is not enough: Revolutionizing current learning environments to smart learning environments. *International Journal of Artificial Intelligence in Education*, 26(2), 561–581. https://doi.org/10.1007/s40593-016-0108-x

Kirkwood, A. 2014, Teaching and learning with technology in higher education: Blended and distance education needs 'joined-up thinking' rather than technological determinism. *Open Learning: The Journal of Open, Distance and E-Learning*, 29(3), 206–221. http://dx.doi.org/10.1080/02680513.2015.1009884

Kolb, A., & Kolb, D. 2005, Learning styles and learning spaces: Enhancing experiential learning in higher education. *Academy of Management Learning & Education*, 4(2), 193–212. http://dx.doi.org/10.5465/amle.2005.17268566
Kulkarni, K. G., 2013, Digitalization in Higher Education: Costs and Benefits. www.researchgate.net
Kuzu. O. H. 2020, Digital transformation in higher education: A case study on strategic plans. *Vysshee obrazovanie v Rossii = Higher Education in Russia*, 29(3), 9–23. https://doi.org/10.31992/0869-3617-2019-29-3-9-23
Lee, W. J. 2010, Online support service quality, online learning acceptance, and student satisfaction. *Internet and Higher Education*, 13, 227–283. https://doi.org/10.1016/j.iheduc.2010.08.002
Lim, C. P., & Wang, T. 2016, A framework and self-assessment tool for building the capacity of higher education institutions for blended learning. In C. P. Lim, & L. Wang (Eds.), *Blended learning for quality higher education: Selected case studies on implementation from Asia-Pacific*, (pp. 1–38). Bangkok: UNESCO Bangkok Office.
Limani, Y, Hajrizi, E., Stapleton, L., & Retkoceri, M. 2019, Digital Transformation Readiness in Higher Education Institutions (HEI): The Case of Kosovo, www.sciencedirect.com
Liu, C.-C, & Kao, L.-C 2007, Do handheld devices facilitate face-to-face collaboration? Handheld devices with large shared display groupware to facilitate group interactions. *Journal of Computer Assisted Learning*, 23, 285–299.
Lonka, K. 2015, *Innovative schools: Teaching & learning in the digital era*. Brussels: European Union. Retrieved from http://www.europarl.europa.eu/studies
Luckin, R. 2008, The learner centric ecology of resources: A framework for using technology to scaffold learning. *Computers & Education*, 50, 449–462.
Mabry, L., & Snow, J. Z. 2006, *Laptops for High-Risk Students: Empowerment and Personalization in a Standards-Based Learning Environment. Studies in Educational Evaluation*, 32: 289–231.
Machekhina, O. N. 2017, Digitalization of higher education as a trend of its modernization and reforming. Revista ESPACIOS. ISSN 0798 1015, Vol. 38 (N° 40) Año 2017, Revista ESPACIOS.com
Makosa, Pawel. 2013, Advantages and disadvantages of digital education. Retrieved from https://www.researchgate.net/publication/264419797_Advantages_and_disadvantages_of_digital_education
Marzano, R. J. 1998, *A theory-based meta-analysis of research on instruction*. Aurora, Colorado: Mid-continent Regional Educational Laboratory. Available at: http://www.peecworks.org/PEEC/PEEC_Research/I01795EFA.2/Marzano%20Instruction%20Meta_An.pdf
Maseleno, A., Sabani, N., Huda, M., Ahmad, R., Azmi Jasmi, K., & Basiron, B. 2018, Demystifying learning analytics in personalised learning. *International Journal of Engineering & Technology*, 7(3), 1124–1129. https://doi.org/10.14419/ijet.v7i3.9789
Mathew, V. 2009, Knowledge Management in Higher Education: Implementation agenda in Distance learning. Proceeding of International Conference on Distance Education and Open Learning (DEOL 2009) at NEC (Nanyang Technological University) Singapore. DOI: 10.1109/ICDLE.2010.5606015.
Mathew, V. 2010a, Service delivery through knowledge management in higher education. *Journal of Knowledge Management Practice*, 11(3).
Mathew V. 2010b, Women entrepreneurship in Middle East: Understanding barriers and use of ICT for entrepreneurship development. *International Entrepreneurship and Management Journal*, 6(2), 163.
Motiwalla, L. F. 2007, Mobile learning: A framework and evaluation. *Computers & Education*, 49(3), 581–596.

Moya, M., Musumba, I., & Akodo, R. 2011, Management attitude, support and integration of information communication technologies in higher education in Uganda. Journal of Modern Accounting and Auditing, USA.

Naismith, L., Lonsdale, P., Vavoula, G., & Sharples, M. 2004, Literature Review in Mobile Technologies and Learning. Bristol: Futurelab. Retrieved from http://www2.futurelab.org.uk/resources/documents/lit_reviews/Mobile_Review.pdf

Nworie, J., Haughton, N., & Oprandi, S., 2012, Leadership in distance education: qualities and qualifications sought by higher education institutions. *American Journal of Distance Education*, 26(3), 180–199.

O'Connor, K. 2014, MOOCs, institutional policy and change dynamics in higher education. *Higher Education*, 68(5), 623–635.

Odabaşı, H. F., Fırat, M., & İzmirli, S. 2010, Küreselleşen dünyada akademisyen olmak. *Anadolu Üniversitesi Sosyal Bilimler Dergisi*, 10(3), 127–142.

Parr, J. M., & Fung, I. 2000, A review of the literature on computer-assisted learning, particularly integrated learning systems, and outcomes with respect to literacy and numeracy. Auckland: University of Auckland.

Passey, D., & Higgins, S. 2011, Learning platforms and learning outcomes–insights from research. *Learning, Media and Technology*, 36(4), 329–333.

Pearce, N., Weller, M., Scanlon, E., & Kinsley, S. 2011, Digital scholarship considered: How new technologies could transform academic work. *Durham Research Online*, 16(1), 72–80.

Peng, H., Ma, S., & Spector, J. M. 2019, Foundations and trends in smart learning. Proceedings of 2019 International Conference on Smart Learning Environments. https://www.worldcat.org/oclc/1090447802

Perkin, N., & Abraham, P. 2017, *Building the agile business through digital transformation*. London: Kogan Page Stylus.

Prensky, M. 2001, Digital natives, digital immigrants part 1. *On the Horizon*, 9(5), 1–6. https://marcprensky.com/writing/Prensky%2020 Digital%20 Natives,%20Digital%20 Immigrants%20-%20 Part 1.pdf (accessed 15 December 2020).

Rice, M. F. 2003, Information and communication technologies and the global digital divide: Technology transfer, development, and least developing countries. *Comparative Technology Transfer and Society*, 1(1), 72–88.

Rienties, B., Brouwer, N., & Lygo-Baker, S. 2013, The effects of online professional development on higher education teachers beliefs and intentions towards learning facilitation and technology. *Teaching and Teacher Education*, 29, 122–131.

Rogers, E. M. 2003, *Diffusion of innovations*, 5th ed. New York: Free Press.

Şahin, M., & Alkan, R. M. 2016. Yükseköğretimde değişim dönüşüm süreci ve üniversitelerin değişen rolleri. *Eğitim ve Öğretim Araştırmaları Dergisi*, 5(2), 297–307.

Schacter, J., & Fagnano, C. 1999, Does computer technology improve student learning and achievement? How, when, and under what conditions? *Journal of Educational Computing Research*, 20, 329–243.

Schwarz, M. 2010, Social Innovation. Innovation, (0), 78.

Selwyn, N. 2016, *Education and technology: Key issues and debates*. London: Bloomsbury.

Şendağ, S., & Gedik, N. 2015, Yükseköğretim dönüşümünün eşiğinde Türkiye'de öğretmen yetiştirme sorunları. *Eğitim Teknolojisi Kuram ve Uygulama*, 5(1), 72–91.

Seres, L., Pavlicevic, V., & Tumbas, P. 2018, Digital Transformation of Higher Education: Competing on Analytics, Proceedings of INTED2018 Conference, Valencia, Spain, 5–7 March 2018, Research Gate, DOI: 10.21125/inted.2018.2348

Siemens, G., Gašević, D., & Dawson, S. 2015, Preparing for the Digital University: A Review of the History and Current State of Distance, Blended and Online Learning. Athabasca University Press, Athabasca AB Canada.

Silvernail, D. L., & Gritter, A. K. 2007, Maine's Middle School Laptop Program: Creating Better Writers – Research Brief Gorham, Me: Maine Education Policy Research Institute.

Singh, R., & Kaurt, T. 2016, Blended Learning-Policies in Place at Universiti Sains Malaysia. Blended, 103. UNESCO, United States.

Somekh, B. 2007, *Pedagogy and learning with ICT: Researching the art of innovation*. London: Routledge.

Stantchev, V., Colomo-Palacios, R., Soto-Acosta, P., & Misra, S. 2014, Learning management systems and cloud file hosting services: A study on students' acceptance. *Computers in Human Behavior*, 31, 612–619.

Straker, L. M., Pollock, C. M., Zubrick, S. R., & Kurinczuk, J. J. 2005, The association between information and communication technology exposure and physical activity, musculoskeletal and visual symptoms and socio-economic status in 5-year-olds. *Child Care Health and Development*, 32(3), 343–351.

Taylor, P. S. 2007, Can clickers cure crowded classes? *Maclean's*, 120(26–27), 73.

Tømte, C. E., Fossland, T., Aamodt, P. O., & Degn, L. 2019, Digitalisation in higher education: Mapping institutional approaches for teaching and learning. *Quality in Higher Education*, 25(1), 98–114, DOI: 10.1080/13538322.2019.1603611

Underwood, J., & Dillon, G. 2004, Capturing complexity through maturity modelling. *Technology, Pedagogy and Education*, 13(2):213–225.

Unwin, T., Kleessen, B., Hollow, D., Williams, J. B., Oloo, L. M., Alwala, J., & Muianga, X. 2010, Digital learning management systems in Africa: Myths and realities. *Open Learning*, 25(1), 5–23.

Voogt J., & Knezek, G. (Eds.) 2008, *International handbook of information technology in primary and secondary education*. New York: Springer.

Walsham, G. 1995, The emergence of interpretivism in IS research. *Information Systems Research*, 6(4), 376–394.

Wisloski, J. 2011, Online education study: As enrollment rises, institutions see online education as a 'critical part' of growth, Online Education Information.

Zandvliet, D., & Fraser, B. 2004, Learning environments in information and communications technology classrooms. *Technology, Pedagogy and Education*, 13(1), 97–123.

Zawacki-Richter, O., & Latchem, C. 2018, Exploring four decades of research in Computers & Education. *Computers & Education*. 122, 136–152. 10.1016/j.compedu.2018.04.001.

Zawacki-Richter, O. & Naidu, S., 2016, 'Mapping research trends from 35 years of publications in distance education', Distance Education, 37(3), 245–69.

Zhao, Y., & Frank, K. A. 2003, Factors affecting technology uses in schools: An ecological perspective. *American Educational Research Journal*, 40, 807.

10 'Rubrics' as a Tool for Holistic Assessment

Design Considerations and Emerging Trends

Umesh Prasad, Abhaya Ranjan Srivastava, and Soumitro Chakravarty

CONTENTS

10.1 INTRODUCTION

The term 'rubric' has caught the attention of academicians across the globe in recent times, and to date, the term lacks any formal standard definition (Dickinson and Adams 2017; Fraile et al. 2017; Debattista 2018; Tobón et al. 2020). It conveys various meanings from an educational perspective. It is often considered as some kind of secret scoring sheet that remains under the possession of the teacher, which is disclosed only after a student's work has been evaluated. 'Rubrics' can be very flexible. They may convey a fairly detailed scoring and grading logic, or they may also convey qualitative assessment in terms of 'good', 'average', etc. Some Rubrics may rely more upon graphics rather than words, and some may rely more on words and less on graphics and so on (Elsheikh 2018; Byerley and Thompson 2017; Trace et al. 2016). Since its inception and now reasonably wide use, especially in online education, the term has not been clearly defined still.

Researchers have taken an interest in the concept and use of Rubrics in the last few decades, and they have pointed out the attributes of Rubrics, too. Most researchers

DOI: 10.1201/9781003132097-10

opine that Rubrics dominantly refer to a scoring pattern that is used for evaluating the performance of students. However, though, this has been widely agreed upon in terms of meaning and usage; there has been a wide array of meanings associated with the term 'rubric' without any formal definition of the term. It has been observed that many institutions use Rubrics without an operational definition of the term. Researchers have conducted a wide range of empirical studies, but there have been varied interpretations of the term, and a gap in terms of a formal definition remains unaddressed to date.

The current paper aims to look into the conceptual framework associated with Rubrics and provide a theoretical understanding that would facilitate discussions about this mechanism, which has grown in importance, especially in this period of the COVID-19 pandemic, which has brought about online teaching and learning to the forefront in the absence of conventional classroom lectures.

The present study relies upon an extensive review of the literature, covering the last decade or so with a view to synthesize and present it in coherent terms that could facilitate the development of a theoretical framework around 'Rubrics'. It should be clearly be noted that the present inquiry seeks to provide a framework and thereby encourage further research in this emerging trend that has grown in importance in the recent past owing to the pandemic and subsequently increased reliance upon online education.

10.2 THEORETICAL FRAMEWORK

The main objective of the study is to explore and find out more about the assessment mechanisms using Rubrics in modern-day education across different courses with a special emphasis on B-schools in India. Taking into account the present scenario and the restrictions being posed by the pandemic COVID-19, there is a need for ensuring that assessment of students be carried out in the online mode as traditional classes are not taking place. The concept of employing Rubrics for the purpose of facilitating assessment by the students based upon criteria that are communicated to them in advance can facilitate effective assessment in the present times. This can ensure that students become more interested in their assignments as now their participation is also being actively encouraged through the implementation of 'Rubrics'. As very limited information is available on the given area, the present study tries to put forward insights based upon the review of existing literature and discussions with stakeholders, and it also discusses a rubric designed by students at the post-graduate level in a B-school of the region. Focus is given on conveying greater clarity of the term, looking into the various aspects of Rubrics and also putting forward the scope for further research in the given area.

10.3 RESEARCH QUESTIONS

Taking into account the objectives, the present study seeks answers to the following research questions:

1. How can Rubrics play a role in ensuring effective assessment of the students in the online mode of education in the present era of restrictions imposed by the pandemic COVID-19?

2. What are the potential advantages that this approach offers over conventional assessment methods?
3. What are the critical design issues and concerns in the case of Rubrics, and how can the same be managed?
4. What are the likely future trends in assessment using Rubrics?

10.4 METHODS

Taking into account the objectives of the inquiry and also keeping in mind the fact that no standard or universally accepted definition of the term exists to date, there has been considerable research carried out by fellow researchers in the area over the last few decades. For the purpose of this study, we have relied upon an intensive review of the literature with a special emphasis upon the works done in the past decade (the period from 2010 onward). Apart from an intensive review of literature, the views of stakeholders like faculties, students and administrative staff of institutions in and around the study location (Jharkhand) have been incorporated to gain insights relevant for the inquiry.

10.5 RUBRICS IN EDUCATION: EVOLUTION AND THE PRESENT SCENARIO

In the present era, Rubrics are very widely used in education across primary, secondary and even post-secondary levels in schools, colleges and universities around the globe. Research in the 1980s very significantly contributed toward the development of Rubrics as creatively and very carefully designed assessment criteria for students that could be capable of clearly communicating to the students about the parameters based on which they are going to be evaluated (Dickinson and Adams 2017; Ford et al. 2019; Clabough and Clabough 2016; Jönsson and Ernesto 2017). Rubrics that have adequate and clearly specified descriptions regarding the criteria and performance assessment logic are considered to be an effective and also a transparent method for evaluating the work of students and also providing them feedback based on their submitted work. Owing to the growing importance and popularity of this assessment tool, there has been a significant amount of research done in the last few decades on the quality and effectiveness of Rubrics used for educational purposes among students all around the globe (Cooper and Gargan 2009; Luft 1999). The most important contribution of Rubrics for teachers has undoubtedly been the innovation and value addition it offers in terms of evaluating the performance of students. Earlier, the evaluation of performance was limited to observations of the teacher as far as the performance of students in the classroom was concerned, coupled with the marks that they scored in written examinations. However, with the evolution of 'Rubrics' as a tool for assessment, the assessment mechanism now incorporates greater flexibility and more student participation as well. Further, if effectively used, it also incorporates a high degree of flexibility. It helps the teacher evaluate students across various parameters that the teacher sets up, and the students are also well informed about the same.

10.5.1 Rubrics: The Advantages

There are numerous advantages that assessment through Rubrics offers when compared to the conventional assessment carried out by teachers before their wide use in education. Rubrics, if properly used, facilitate the evaluation of a student's performance based upon a wide range of specified criteria rather than only a quantitative score, which often failed to do justice in terms of assessing the creative potential and analytical skills of the students (Wang 2017; Menéndez-Varela and Gregori-Giralt 2018a, b; Sharma 2019; Elsheikh 2018; Gallardo 2020). Further, the authenticity associated with Rubrics, if properly designed, is high, as the criteria for evaluation are made known to the students in advance, and they are aware to a very large extent of the teacher's expectations from a given assignment. It can be inferred with a reasonable degree of certainty that Rubrics have the potential to enhance the effectiveness of direct instruction as well as evaluation.

Further, Rubrics have a very high degree of flexibility as well, and Rubrics may be designed for almost all contents in education irrespective of the stream, such as arts, science or professional courses (Minnich et al. 2018; Junisbai et al. 2016). The following two Rubrics that have been outlined below were designed by a group of post-graduate students of a reputed B-school in Jharkhand as a part of their co-curricular work that was assigned to them. Though the Rubrics are not yet in a fully designed and functional form, they do give us some idea about the assessment criteria and logic. Such Rubrics can be designed easily by instructors in collaboration with students as well if needed.

Model 1: Rubric for assessment of a research proposal submitted by students at the PG level as part of the requirements for assignment evaluation in the semester for the subject Research Methodology.

The written assignment elements that the students chose to judge were:

- The clarity of the specified need for conducting the given study.
- The number of sources referred to and cited for the given assignment.
- Language, grammar and syntax of the write-up.
- The relevance of the study for tackling real-life issues.
- The degree of originality of the proposed study.

4. Excellent

- The clarity in terms of few sentences itself serves the purpose.
- More than ten sources have been referred to and cited from recent literature.
- Very little to no errors in terms of language usage, syntax, etc.
- High degree of relevance in the present times.
- Fairly original in terms of contribution to the existing body of literature.

3. Good

- The clarity is satisfactory but requires too much text and needs condensing.
- More than ten sources have been referred to and cited from literature, but few of the sources are dated.

TABLE 10.1

Criteria	Excellent	Good	Needs Improvement	Poor
Clarity of need	High	Satisfactory but needs condensation	Not evident and needs to be worked on	No clarity
Sources referred to and cited	More than ten and recent	Adequate but few are dated	Most sources dated	Poor and inadequate
Quality of write-up	High	Slight editing needed	Whole write-up needs editing and spell-check	Full rewriting needed
Degree of relevance	High	Limited	Very little	No relevance
Contribution to existing knowledge in the field	Significant	Somewhat notable	Lacks originality	None

- Some phrases need to be edited and spell checked.
- A limited degree of relevance in present times.
- Somewhat original in terms of contribution to the existing body of literature.

2. Needs Improvement

- The clarity is not evident upon a first-time reading.
- Most of the sources are dated.
- The whole write-up needs to be edited and spell checked.
- Very little degree of relevance in present times.
- Lacks originality in terms of contribution to the existing body of literature.

1. Poor

- No clarity in the objectives.
- Referred to and cited literature do not match.
- The whole write-up needs to be rewritten.
- No relevance in present times.
- Lacks any contribution to the existing body of literature.

The Table 10.1 presents Rubric.

Similar Rubrics based on the need for given cases can be designed and put to use for other subjects and areas as well with satisfactory results.

Model 2: Rubric for assessment of an oral presentation for a miniproject submitted by students (designed by students with inputs from faculty).

10.6 FINDINGS

10.6.1 Rubrics: Design Issues and Concerns

Apart from the advantages mentioned above, there are some design issues associated with Rubric design and use as well, with the most notable issue being the large number of options associated with scaling in a given rubric for a specified purpose.

TABLE 10.2

Criteria/Description	Score
High voice clarity throughout the duration of the presentation with proper use of multimedia tools, positive body language with straight and confident posture, satisfactory resolution of all queries and finishing the presentation within the stipulated time frame of 15 minutes.	5
Reasonable voice clarity for the best part of the presentation with proper use of multimedia tools, generally positive body language and posture, resolution of most queries and finishing up the presentation 3–4 minutes earlier than scheduled.	4
Reasonable voice clarity but occasionally referred to written materials, proper use of multimedia and decent body language, resolved only one query and finished the presentation five minutes earlier than scheduled.	3
Voice clarity was just about satisfactory and speaker frequently referred to notes, use of multimedia was inconsistent, no query resolution and presentation got over very quickly.	2
Very poor voice clarity, difficult to understand the words, no multimedia support and no interaction or query resolution.	1
No formal presentation prepared at all.	0

For one teacher, the scale may use numbers; for others, it may use letters, symbols or words and so on. Ultimately, this lack of any standard approach toward scaling can be a cause of concern and create confusion both among teachers as well as students. Wrong choices of scale, especially in the scenario of no standards, may result in unproductive Rubric design. Further, there is no consensus regarding the number of options in a given scale that should be considered as an ideal. Four or five options are generally considered ideal, but again, it becomes a very subjective choice, and a wrong choice for a given purpose can easily render the design and, subsequently, the Rubric unproductive and unfit. Further, another important issue related to design is that there are high chances of inconsistencies in the language associated with the description of the various dimensions related to a given Rubric, and this is an aspect that affects the designers, especially those who have little or no experience in designing Rubrics. The absence of standard guides acts as a bottleneck. Moreover, the language for lesser grade description can discourage students as well, and vague language usage can result in confusion related to the interpretation in their minds.

10.6.2 Rubrics in Education: The Emerging Trends

Rubrics used by instructors to evaluate the performance of students and give them feedback for a specific assignment is gaining in popularity in recent times. They are also referred to as instructional Rubrics, and they contain a description of about one or two pages that clarifies the evaluation procedure and logic to the students in advance. Instructional Rubrics are easy to use and explain to the students if the Rubric is designed carefully. The feedback helps the students to improve upon their performance in the future. Instructional Rubrics also help in making the teacher expectations very clear and specific, and if designed well, they help in communicating the same to the students effectively. The feedback that instructional Rubrics provide to the students can be potentially very informative and motivate the students to address their weak areas and perform better in the future. In some cases, especially for experienced instructors who have been using instructional Rubrics for a long time, they may also be used to facilitate self-evaluation by the students as well. They enhance critical thinking and analytical skills and promote original thinking among the students.

It can clearly be seen that when compared to traditional assessment, instructional Rubrics have emerged as a more powerful and flexible tool that is beneficial both for the teachers as well as the students, and subsequently, their use is likely to increase further in the future.

10.7 DISCUSSIONS

'Rubrics' as a tool for assessment purposes have been around for the last few decades. This article tried to develop a theoretical framework aimed toward a proper understanding of 'Rubrics' and their usage in education. In this section of the article, the authors try to assess and look into the potential of Rubrics in education and also throw light upon the possible areas that need addressing from researchers, educationists and policy-makers in the near future. Rubrics as a tool is in the present times is very widely used with reasonably successful outcomes as well. However, the lack of formal definition is something that acts as a 'gap', and a standard definition of the term will facilitate better understanding and usage.

Further, the design elements associated with Rubrics constitute the most important aspect that determines the degree to which a Rubric is effective. In this context, it is felt that collaboration among instructors will be helpful in ensuring that the design considerations associated with a given Rubric are addressed properly and are subsequently used more productively. In the recent past, there has been a great deal of emphasis upon certain key areas in education like skill-based learning, development of analytical and critical thinking skills, use of tools to enhance the practical knowledge of students and so on. Traditional education and traditional modes of assessment have exhibited limitations in properly addressing the same. 'Rubrics' in education has emerged as a viable and potent tool to address the needs of students.

10.8 CONCLUSION

It is to be noted that the present paper has been very limited in its scope, and it has focused mainly upon Rubrics and their usage in assessment; but it should be clearly understood that the tool is very flexible and can be effectively employed in other areas such as teaching review by gathering student feedback related to the teacher's performance, evaluation of activities performed by peers and colleagues and so on. The authors sincerely hope that this article succeeds in enhancing the clarity of the term Rubrics and its usage in education and also encourages fellow researchers to carry out more elaborate studies related to Rubrics and their applications across diverse areas.

REFERENCES

Blyman, K. K., K. M. Arney, B. Adams, and T. A. Hudson. 2020. "Does your course effectively promote creativity? Introducing the mathematical problem solving creativity rubric." *Journal of Humanistic Mathematics* 10(2): 157–193.

Brookhart, Susan M. 2018. "Appropriate criteria: key to effective rubrics." *Frontiers in Education* 3: 22. Frontiers.

Byerley, Cameron, and Patrick W. Thompson. 2017. "Secondary mathematics teachers' meanings for measure, slope, and rate of change." *The Journal of Mathematical Behavior* 48: 168–193.

Cheng, Liying, Christopher DeLuca, Heather Braund, Wei Yan, and Amir Rasooli. 2020. "Teachers' grading decisions and practices across cultures: Exploring the value, consistency, and construction of grades across Canadian and Chinese secondary schools." *Studies in Educational Evaluation* 67: 100928.

Clabough, Erin B. D., and Seth W. Clabough. 2016. "Using rubrics as a scientific writing instructional method in early stage undergraduate neuroscience study." *Journal of Undergraduate Neuroscience Education* 15(1): A85.

Cockett, Andrea, and Carole Jackson. 2018. "The use of assessment rubrics to enhance feedback in higher education: An integrative literature review." *Nurse Education Today* 69: 8–13.

Cooper, B. S., and A. Gargan. 2009. "Rubrics in education." *Phi Delta Kappan* 91(1): 54–55.

Debattista, Martin. 2018. "A comprehensive rubric for instructional design in e-learning." The International Journal of Information and Learning Technology.

Dickinson, P., and J. Adams. 2017. "Values in evaluation–The use of rubrics." *Evaluation and Program Planning* 65: 113–116.

Elsheikh, Aymen. 2018. "Rubrics." The TESOL Encyclopedia of English Language Teaching, 1–8.

Ford, Channing R., Kimberly Garza, Jan Kavookjian, and Erika L. Kleppinger. 2019. "Assessing student pharmacist communication skills: Development and implementation of a communication rubric." *Currents in Pharmacy Teaching and Learning* 11(11): 1123–1131.

Fraile, Juan, Ernesto Panadero, and Rodrigo Pardo. 2017. "Co-creating rubrics: The effects on self-regulated learning, self-efficacy and performance of establishing assessment criteria with students." *Studies in Educational Evaluation* 53: 69–76.

Fu, Qing-Ke, Chi-Jen Lin, and Gwo-Jen Hwang. 2019. "Research trends and applications of technology-supported peer assessment: A review of selected journal publications from 2007 to 2016." *Journal of Computers in Education* 6(2): 191–213.

Gallardo, Katherina. 2020. "Competency-based assessment and the use of performance-based evaluation rubrics in higher education: Challenges towards the next decades." *Competency* 78(1): 61.

Jönsson, Anders, and Ernesto Panadero. 2017. "The use and design of rubrics to support assessment for learning." In *Scaling up Assessment for Learning in Higher Education*, pp. 99–111. Singapore: Springer.

Junisbai, Barbara, M. Sara Lowe, and Natalie Tagge. 2016. "A pragmatic and flexible approach to information literacy: Findings from a three-year study of faculty-librarian collaboration." *The Journal of Academic Librarianship* 42(5): 604–611.

Lin, Qiao, Yue Yin, Xiaodan Tang, Roxana Hadad, and Xiaoming Zhai. 2020. "Assessing learning in technology-rich maker activities: A systematic review of empirical research." *Computers & Education* 157: 103944.

Luft, J. A.1999. "Rubrics: Design and use in science teacher education." *Journal of Science Teacher Education* 10(2): 107–121.

Menéndez-Varela, J. L., and E. Gregori-Giralt. 2018a. "The reliability and sources of error of using rubrics-based assessment for student projects." *Assessment & Evaluation in Higher Education*43(3): 488–499.

Menéndez-Varela, José-Luis, and Eva Gregori-Giralt. 2018b. "Rubrics for developing students' professional judgement: A study of sustainable assessment in arts education." *Studies in Educational Evaluation* 58: 70–79.

Minnich, Margo, Amanda J. Kirkpatrick, Joely T. Goodman, Ali Whittaker, Helen Stanton Chapple, Anne M. Schoening, and Maya M. Khanna. 2018. "Writing across the curriculum: reliability testing of a standardized rubric." *Journal of Nursing Education* 57(6): 366–370.

Panadero, Ernesto, and Anders Jonsson. 2020. "A critical review of the arguments against the use of rubrics." *Educational Research Review* 100329.

Papadakis, S., M. Kalogiannakis, and N. Zaranis. 2017. "Designing and creating an educational app rubric for preschool teachers." *Education and Information Technologies* 22(6): 3147–3165.

Riddle, Emma Jane, Marilyn Smith, and Steven A. Frankforter. 2016. "A rubric for evaluating student analyses of business cases." *Journal of Management Education* 40(5): 595–618.

Samir, Aynaz, and Mona Tabatabaee-Yazdi. 2020. "Translation quality assessment rubric: A Rasch model-based validation." *International Journal of Language Testing* 10(2): 101–128.

Savic, Milos, Gulden Karakok, Gail Tang, Houssein El Turkey, and Emilie Naccarato. 2017. "Formative assessment of creativity in undergraduate mathematics: Using a creativity-in-progress rubric (CPR) on proving." In *Creativity and giftedness*, pp. 23–46. Cham: Springer.

Sharma, V. 2019. "Teacher perspicacity to using rubrics in students' EFL learning and assessment." *Journal of English Language Teaching and Applied Linguistics* 1(1): 16–31.

Suhardiyati, Y., M. Sukirlan, and A. Nurweni. 2018. "The influence of what am I? Game toward students' speaking achievement." *U-JET* 7(4).

Tam, Cheung On. 2018. "Evaluating students' performance in responding to art: the development and validation of an art criticism assessment rubric." *International Journal of Art & Design Education* 37(3): 519–529.

Tenam Zemach, M., and J. E. Flynn (Eds.). 2015. *Rubric Nation: Critical Inquiries on the Impact of Rubrics in Education.* IAP.

Tobón, Sergio, Luis Gibran Juárez-Hernández, Sergio Raúl Herrera-Meza, and Cesar Núñez. 2020. "Assessing school principal leadership practices. Validity and reliability of a rubric." *Educación, XX1* 23(2).

Trace, Jonathan, Valerie Meier, and Gerriet Janssen. 2016. "'I can see that': Developing shared rubric category interpretations through score negotiation." *Assessing Writing* 30: 32–43.

Wang, W. 2017. "Using rubrics in student self-assessment: Student perceptions in the English as a foreign language writing context." *Assessment & Evaluation in Higher Education* 42(8): 1280–1292.

Whetstone, Devon, and Heather Moulaison-Sandy. 2020. "Quantifying authorship: A comparison of authorship rubrics from five disciplines." *Proceedings of the Association for Information Science and Technology* 57(1): e277.

11 A Systematic Review of Barriers to Crowdsourcing in Science in Higher Education

Regina Lenart-Gansiniec

CONTENTS

DOI: 10.1201/9781003132097-11

11.1 INTRODUCTION

Non-academic and public participation in scientific research is an emerging, nascent and growing practice. In recent years, crowdsourcing in science has been presented as inbound open innovation in science practices Beck et al. (2020), a new way of conducting contemporary scientific research activities, example of opening of science, new collaborative forms of knowledge creation, an alternative to research projects, a strategy for organizing work of researchers (Lukyanenko et al. 2019), a tool for research Law et al. (2017) and in some ways, it is the digital version of citizen science Eklund et al. (2019).

The benefits of crowdsourcing in science are well examined in the academic literature and include the following: improvement of knowledge building processes and optimization of teaching and administrative processes Llorente and Morant (2015). In addition, crowdsourcing can be applied to possible examination questions, micro courses for other students, adaptive learning, and in research, create joint projects between universities and industry Berbegal-Mirabent et al. (2020), in the field of MOOCs, and giving rise to Collaborative Open Online Courses Zhou et al. (2018).

Crowdsourcing promises to scale up research to previously unreachable magnitude by its access to large crowds. Researchers employ crowdsourcing for a number of tasks, such as content creation (Doan et al. 2011; Kalev et al. 2013), communicating, collecting, classifying data (Beck et al. 2019), processing and analyzing research data (Law et al. 2017), enlisting participants for surveys, research, experiments, panels, focus groups, statistical analyses, transcriptions (Schlagwein and Daneshgar 2014), generating innovative research questions, hypotheses, research proposals, testing research at an early stage, establishing cooperation and seeking collaborators for joint research, obtaining an assessment and opinion on the concept of a research project or article (Ipeirotis et al. 2010; Uhlmann et al. 2019), solving problems arising in the course of writing an article or conducting research, determining the reliability and generalization of the results (Pan et al. 2017) and dissemination of the results (Beck et al. 2019).

Research on crowdsourcing in science is extensive and has provided important insights into the potential of crowdsourcing in science (Riesch and Potter 2014), requirements of the crowdsourcing platform (Schlagwein and Daneshgar 2014), barriers to crowdsourcing projects conducted by researchers (Burgess et al. 2017; Law et al. 2017), identification of the type of crowdsourcing suitable for purposes of research (Sauermann et al. 2019), the potential of the virtual community for crowdsourcing, ways of motivating the virtual community, the characteristics of the virtual community involved in the endeavors of researchers (Franzoni and Sauermann 2014).

However, some significant gaps remain Beck et al. (2020). In particular, it is widely argued that employees, especially academic workers play a role in influencing the effectiveness of these practices (Hui and Gerber 2015; Beck et al. 2019). It is important to involve academic teachers, and to convince them of their potential and effectiveness in crowdsourcing in science (Riesch and Potter 2014), because "the teachers are an important and key element in school, and therefore their professional attributes are a significant factor in the processes of school change" (Avidov-Ungar and Magen-Nagar 2014, p. 229).

Despite these arguments, need to be diffused into higher education institutions and inviting non-academic participations is necessary Beck et al. (2020), mostly scientists are skeptical or demonstrate a negative attitude towards crowdsourcing in science (Kim and Adler 2015; Poliakoff and Webb 2007; Schlagwein and Daneshgar 2014; Riesch and Potter 2014). Research on crowdsourcing in science is extensive and has provided important insights into the potential of crowdsourcing in science (Riesch and Potter 2014), requirements of the crowdsourcing platform (Schlagwein and Daneshgar 2014), identification of the type of crowdsourcing suitable for research purposes (Sauermann et al. 2019), the potential of the virtual community for crowdsourcing, ways of motivating the virtual community, the characteristics of the virtual community involved in the endeavors of researchers (Franzoni and Sauermann 2014). A limited number of studies (Burgess et al. 2017; Law et al. 2017) utilized the responses to crowdsourcing in science. However, none of these studies referred to the nature of barriers exclusively, especially in management science.

Therefore, the purpose of this paper is to identify barriers that contain the adoption and use of crowdsourcing in science. This study followed a systematic review process focused on summarization of knowledge, reviewing 27 empirical researches on crowdsourcing in science. An extensive search of the literature was conducted. The selected references were analyzed into a number of categories: macro-level (organizational, technical, characteristics, context) and micro-level (individual and social factors). The systematic literature review was chosen because it is repeatable and clear and includes several clear steps. It also enables the identification and synthesis of the results of all major research and theoretical approaches, which in turn enables the identification of existing cognitive and research gaps (Tranfield et al. 2003).

11.2 THEORETICAL FRAMEWORK

11.2.1 CROWDSOURCING

The increasing interest in crowdsourcing among researchers was initiated in 2006 by the editor of "Wired" magazine, Howe. Howe (2006) claimed that "the act of a company or institution taking a function once performed by employees and outsourcing it to an undefined (and generally large) network of people in the form of an open call." Crowdsourcing is to apply principles of open-source software development, a form of peer production, to other contexts of social life, a collective effort in attaining knowledge, material or financial resources to, again, produce knowledge, materials and values. Crowdsourcing is understood to be a tool that enables the inclusion of virtual communities in diversified activities that for the time being have been undertaken by organizations' personnel only. A tool may be understood to be platforms that provide IT infrastructure and their crowds' participations to the potential pool for organizations. Crowdsourcing is used when the dispersed knowledge of individuals and groups is leveraged to take advantage of bottom-up crowd-derived inputs and processes with efficient top-down engagement from organizations through IT, to solve problems, complete tasks or generate ideas.

Crowdsourcing has gained on popularity in management sciences owing to its potential, among others: business process improvement, creating open innovations, building of competitive advantage, access to experience, innovativeness, information, crowd skills and work, which are located outside the organization. It started to be linked to initiating collaboration and relations with virtual communities, further on using their wisdom to solve problems, participation management, increasing transparency and openness of public organizations. Crowdsourcing also enables crisis management, expands existing activity and offer of the organization, creates the organization's image, improves communication with the surroundings, and optimizes costs of the organization's activity. It also enables access to knowledge and creativity resources, facilitates acquiring of new contents and data (Majchrzak and Malhotra 2013).

11.2.2 Crowdsourcing in Science

Crowdsourcing in science is at an emergent stage, there are discrepancies of what is considered to be academic crowdsourcing. While a common term for crowdsourcing in science, it is variously referred to as: "online citizen science," "crowdsourcing citizen science," "crowdsourced science," "crowdsourcing science," "citizen cyberscience," "virtual citizen science," "crowd science," "crowd research," "scientific crowdsourcing," "Science 2.0," "crowdsourcing in the science," "crowdsourcing research," "crowdsourcing for science" and "academic crowdsourcing." Some authors suggested that there are several features that differentiate crowdsourcing in science from citizen science. Firstly, in crowdsourcing, the process and outcomes are only legitimized by the initiator and the crowd has no other role than providing input with no decision-making power. Secondly, in citizen science, the result is an exclusively scientific outcome, but in crowdsourcing can take a variety of other forms. Thirdly, in citizen science access to the result of the activity belongs to the knowledge commons, whereas in crowdsourcing, the result belongs to the initiator of the activity.

Crowdsourcing in science is based on online collaborative, massively collaborative science, public participation, volunteer engagement in knowledge creation. In this context, crowdsourcing in science could be conceived as a set of workflow tasks, assets, processes and outputs. The literature stresses that crowdsourcing in science is as one of the types of civic science is an alternative to research projects, digitized version of citizen science.

11.3 METHODOLOGY

11.3.1 Search Strategy

The term "crowdsourcing in science" has been defined and used differently (Beck et al. 2020). In order to embrace different facets of crowdsourcing in science, eight search terms were used to identify and select studies for this literature review: citizen science, online citizen science, open innovation in science, crowd science, participatory science, massively collaborative science and crowdsourcing science (Table 11.1).

TABLE 11.1
Different Facets of Crowdsourcing in Science

Term	Definition	References
Citizen science	Partnerships between scientists and the public in scientific research. The participants can participate in resource gathering, research question defining, data collecting and analyzing, disseminating results and evaluating success of a project	Mäkipää et al. (2020)
Online citizen science	An extension of citizen science, where the tasks to be completed are aided, or completely mediated, through the Internet	Doyle et al. (2018)
Open innovation in science	A process of purposively enabling, initiating and managing inbound, outbound, and coupled knowledge flows and (inter/transdisciplinary) collaboration across organizational and disciplinary boundaries and along all stages of the scientific research process, from the formulation of research questions and the obtainment of funding or development of methods (i.e. conceptualization) to data collection, data processing, and data analyses (exploration and/or testing) and the dissemination of results through writing, translation into innovation, or other forms of codifying scientific insight (i.e. documentation)	Beck et al. (2020)
Crowd science	Research that is characterized by two features: participation in a project is open to a wide base of potential contributors and intermediate inputs such as data or problem solving algorithms are made openly available	Franzoni and Sauermann (2014)
Participatory science	Engagement of non-professionals in scientific investigation, whether by contributing resources, asking questions, collecting data, or interpreting results	Heaton et al. (2016)
Massively collaborative science	Computing paradigm intended to bridge the gap between machine and human computation	Correia et al. (2018)
Crowdsourcing science	Coming up with research ideas, assembling the research team, designing the study, collecting and analyzing the data, replicating the results, writing the article, obtaining reviewer feedback, and deciding next steps for the program of research	Uhlmann et al. (2019)

In this context, two methods were used to carry out this systematic literature review. Firstly, Scopus and Web of Science restricted to Social Science Citation Index in title, abstract and keywords of articles. The terms [citizen science*, online citizen science*, open innovation in science*, crowd science*, networked science*, massively collaborative science*, scientific crowdsourcing*, and crowdsourcing

TABLE 11.2
Research Articles Using Keywords

Databases/Keyword (Publication Title)	Scopus	Web of Science
Citizen science	95	140
Online citizen science	7	14
Open innovation in science	15	36
Crowd science	8	20
Participatory science	58	461
Massively collaborative science	1	0
Crowdsourcing science	6	10
Crowdsourcing in science	5	2

science*] were searched along with the term [barriers*]. This strategy resulted in a total of 885 articles. Table 11.2 shows the number of articles identified.

Secondly, stage of the literature search involved search through articles on crowdsourcing published in the following ten leading higher education journals: *Review of Educational Research*, *Review of Higher Education*, *Journal of Higher Education*, *Higher Education Quarterly*, *Higher Education Research & Development*, *Studies in Higher Education*, *Higher Education Policy*, *Higher Education*, *Research in Higher Education*, *Internet and Higher Education*. When selecting them, they were guided by the indications (Tight 2018) and results of Scimago Journal & Country Rank 2018. Same term strings were selected in title, abstract, keywords and full text. This strategy did not produce studies.

The "Preferred Reporting Items for Systematic Review and Meta-Analyses" (PRISMA) (Moher et al. 2009) to identify eligible studies was employed.

1. Field: articles should study crowdsourcing in research in higher education. I defined higher education as "all universities, colleges of technology and other institutions of post-secondary education, whatever their source of finance or legal status; all research institutes, experimental stations and clinics operating under the direct control of or administered by or associated with higher education institutions" (OECD 2002).
2. Topic: only empirical studies were eligible for the literature review (literature reviews and conceptual works were not included) because we are interested in empirical evidence concerning crowdsourcing in science.
3. Language: publications written in English.
4. Year of publication: articles that were published between 2006 and September 2020 were searched and included. The starting date includes the first appearance of the concept of crowdsourcing in literature (Howe 2006).
5. Publication types: international, peer-reviewed, full-text publications.

Firstly, I identified 878 articles (Scopus: 195; Web of Science: 683). Secondly, I reduced the sample by 729 articles (redundancies between the databases). Next, the abstracts of the remaining 125 articles were checked regarding the inclusion criteria. Next, leading to the removal of a further 90 articles and left 35 articles. Finally, the compiled database of publications, containing 27 articles, was subjected to an in-depth analysis of the content.

11.4 RESULTS

11.4.1 Number of Publications

Within the methodology of the systematic literature review, an analysis of the number of publications on crowdsourcing (Figure 11.1) and an analysis of the number of publications qualified for contents analysis were carried out. The synthesized results of all the publications identified for review show that the number of studies has increased sharply in recent years: 76% of all papers were published between 2014 and 2019. Other publications appeared in 2009–2013.

The conducted analysis of the number of publications on crowdsourcing enables stating that this issue enjoys interest among researchers. The trend value of publications found in English language databases equal to $R^2 = 0.796$ proves a growing tendency of publications. Such conclusions enable stating the validity of considerations on crowdsourcing in science. Based on the above, the author draws the conclusion about a current, although young, present research area.

All publications qualified for analysis were of empirical character (100%). In most of the analyzed studies, a qualitative approach dominates (98%). These were multiple (91%) or single (7%) approaches. In one case, it was a focus study (2%). Quantitative research was less common, only in one case researchers reached for an online survey to investigate how members of the public view the process of crowdfunding in science and the lack of a peer review process in Japan. Respondents in the survey were usually members of the virtual community (83%), in other cases academic teachers (16%). Moreover, none of the authors undertook the problem of recognizing behavioral antecedents of crowdsourcing in science importance.

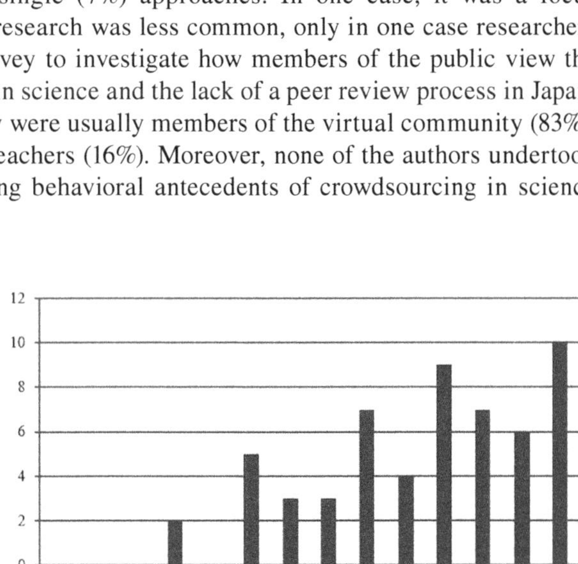

	2006	2007	2008	2009	2010	2011	2012	2013	2014	2015	2016	2017	2018	2019
■ Number of publications in English language databases	0	0	0	2	0	5	3	3	7	4	9	7	6	10

FIGURE 11.1 The number of papers that were published in each year (2006–2019). (*Source:* Own elaboration. As for: 15.10.2019.)

FIGURE 11.2 Word cloud – an analysis of keywords frequency. (*Source:* Visualization prepared through Wordle (http://www.edwordle.net).)

11.4.2 Keyword Frequency

The obtained database of publications was analyzed with bibliometric techniques. First of all, the analysis of keyword frequency, geographical distribution of research, applied types of crowdsourcing in science, research methods and problems were carried out. Keywords used by the authors in publications not only make it easier to find papers in databases but also reflect the specificity of the paper. The keywords identified in publications have been visualized by "word clouds." It must be emphasized that 9 out of 56 papers qualified for the analysis did not have any keywords. In a graphical presentation of keywords, the frequency of presence is determined by the font's size and bold feature (Figure 11.2). Keywords most frequently used in publications selected from databases are science (35), citizen (15), research (14), crowdsourcing (13).

11.4.3 Geographical Distribution of Research

The crowdsourcing in science research teams identified by the affiliate of the first author are limited to the United States (36%) and the United Kingdom (16%). This means that the American-English perspective is of key importance when it comes to researching crowdsourcing in science. This may affect the external credibility of the findings, the ability to draw conclusions and their usefulness in a different cultural context. In addition, most of the analyzed studies were conducted in one country (50; 90%), which may lead to a lack of comparability between individual states.

11.4.4 Most Cited Authors, Works and Journals

The most cited authors and the most cited works identified in articles about crowdsourcing in science were analyzed (Table 11.3). Authors leading in the number of published papers on a topic are recognized as active researchers in that particular discipline. In Table 11.1, the top ten most-cited authors are based on their affiliation or contribution to an article. Crowdsourcing in science literature is influenced by Zhao and Zhu (2014), Franzoni and Sauermann (2014). These authors are the most cited, with 218 and 175 citations, respectively. The overview of leading authors and affiliations indicates that European authors are most actively engaged in crowdsourcing in

TABLE 11.3
Top Ten Most-Cited Authors

Rank	Authors (Year)	Title	Total Citations	Average per Year
1.	Zhao and Zhu (2014)	Evaluation on crowdsourcing research: Current status and future direction	218	31.14
2.	Franzoni and Sauermann (2014)	Crowd science: The organization of scientific research in open collaborative projects	175	25.00
3.	See (20196)	Crowdsourcing, citizen science or volunteered geographic information? The current state of crowdsourced geographic information	136	27.20
4.	Sauermann and Franzoni (2015)	Crowd science user contribution patterns and their implications	102	17.00
5.	Nov et al. (2014)	Scientists@Home: What drives the quantity and quality of online citizen science participation?	95	13.57
6.	Sheehan (2018)	Crowdsourcing research: Data collection with Amazon's Mechanical Turk	69	23.00
7.	Waldrop (2008)	Information technology – Science 2.0	60	4.62
8.	Fausto et al. (2012)	Research Blogging: Indexing and registering the change in Science 2.0	49	5.44
9.	Cox et al. (2015)	Defining and measuring success in online citizen science: A case study of Zooniverse projects	47	7.83
10.	Curtis (2015)	Motivation to participate in an online citizen science game: A study of Foldit	35	5.83

science research suggesting their dominance and contribution to crowdsourcing in science literature Zhao and Zhu (2014).

The span of citations of the most referenced 10 articles ranged from 35 to 216 times. The year in which the most number from among the top 10 articles were published was 2008–2016. These publications are regarded as highly influential and they build the foundation and further develop crowdsourcing in science. Six of these cited publications (rank: two to five, nine to ten) focus on collaboration, participation of non-researchers. The other four of the most cited publication (in table: rank one and six to eight) deal with that evaluation on crowdsourcing research, measurement of online citizen science and data collection with a special platform for crowdsourcing in science. To investigate the results further, the average number of citations received within the 1,211 citing articles per year up until 2008 is measured.

Overall, the most cited journals indicate the multidisciplinary scope of crowdsourcing in science, especially in the areas of communication, information science (e.g. *JCOM Journal of Science Communication*, *Communication Monographs*, *ASLIB Journal of Information Management*), multidisciplinary, interdisciplinary across science, engineering, medicine, social science and humanities (*PLOS ONE*), biology (*American Biology Teacher*), chemistry (*Chemical Engineering News*) and medicine (*American Journal of Preventive Medicine*).

11.4.5 Types of Crowdsourcing in Science

Based on the literature review, the following five types of crowdsourcing in science can be distinguished as follows:

- citizen-science initiatives – include anyone willing to collect data involve a high degree of independence between actors and thus fall into the bottom-left quadrant,
- posing a research question to specialists (e.g. moral-judgment researchers) and asking them to independently design studies to test the same idea,
- iterative contests in which topic experts work together to improve experimental interventions and the collective development of open-source software,
- open peer review, in which anyone can publicly comment on a scientific manuscript or article,
- crowdfunding for scientific research – an open call for money from the general public.

In the analyzed crowdsourcing in science studies, by far the largest category included posing a research question to specialists (67%). Much less attention was paid to citizen-science initiatives (14%) and crowdfunding for scientific research (13%) in the literature. Definitely, the smallest category included open peer review (2%). In two cases, the researchers reached for all five types of crowdsourcing in science identified in literature (4%).

11.4.6 Topic and Highlighting Significant Areas of Research

The content analysis showed that we divided our crowdsourcing in science research into six different categories: significance (45%), types of crowdsourcing in science

(18%), crowdsourcing in science from the perspective of virtual communities (16%), success factors (12%), crowdsourcing in science from the perspective of researchers (7%), reasons for the failure of crowdsourcing in science initiatives (2%). In the case of crowdsourcing in science from the perspective of virtual communities, the research focused on mechanisms of cooperation between virtual community and researchers (34%), ways of motivating virtual community (22%), characteristics of virtual community participating in research projects (11%), opinions of virtual community about crowdsourcing in science (11%), analysis of potential (11%) and needs of virtual community (11%). Studies focused on crowdsourcing in science from the perspective of researchers are as follows: challenges for researchers (25%), collaboration of researchers with the virtual community (25%), involvement of researchers in crowdsourcing in science (25%) and the needs of researchers (25%).

11.4.7 Macro-Level Barriers

There is a consensus in the literature that academic teachers are not required to improve their academic research or increase its reliability (LeBel et al. 2018), to involve society members in their research projects, to establish cooperation with scientists from other research centers or to increase the quality of the research conducted. It is highlighted that individual research or small team research is likely to generate mistakes. In this context, crowdsourcing in science is becoming an important element of the changing scientific landscape and a major issue in ongoing school reforms (Law et al. 2017).

11.4.8 Supporting Infrastructures

Firstly, a supportive infrastructure is a factor of crowdsourcing in science adoption, technical infrastructure like platforms understood as mediators between researchers and crowd. According to Hirth et al. (2017), these platforms are dedicated to data collection (Sheehan 2018), analyses and survey construction, measurement, experiments, longitudinal research (Stritch et al. 2017). In this respect, a consensus has emerged and it is suggested that current crowdsourcing platforms are some barriers because typically they concentrate on microtasks and do not meet the needs of academic research well, time-consuming studies are required (Beck et al. 2020; Hirth et al. 2017). For example, Schlagwein and Daneshgar (2014) suggested that it is difficult for academics to use existing crowdsourcing platforms for doing research issues. There is also another barrier, i.e. crowdsourcing platforms create distance (Eklund et al. 2019). The literature suggests that conditions facilitating the use of crowdsourcing, such as connection to the Internet, technical support, appropriate technical infrastructure may increase the intention to use crowdsourcing in science. When there are no favorable conditions, the barriers may be too high, and as a consequence, the intentions of potential crowdsourcing in science users to use the open data and open data technologies should be lower.

11.4.9 Peer Effects

Thirdly, the adoption of a crowdsourcing in science is inherently dependent upon an environment of trust, creativity, openness and collaboration. Poliakoff and Webb (2007) suggest that "although scientists may have a positive attitude towards participating in public engagement activities and believe that significant others would approve of their taking part, they may not feel confident about their ability to engage with the public." For example, Shirk et al. (2012), based on narrative research with nine crowdsourcing researchers, show that some researchers could do cutting research but at the cost of derision of their peers and little or no recognition. According to Guinan, Boudreau, and Lakhani (2013), scientists' funding decisions are the veracity of the academics and feasibility of the proposed task and research.

11.4.10 Characteristics of Crowdsourcing in Science

Fourthly, crowdsourcing in science based on a large number of crowd, potential contributors. This feature may be a barrier of crowdsourcing in science use and adoption. In this context, the open call nature of crowdsourcing in science leaves control over those who participate mostly outside the researcher's control, which inherently leads to uncertainty about the composition of the crowd. In this context, the loss of control over who participates in a crowdsourced research project poses problems for studies. Academic workers are not sure who will participate in crowdsourcing initiatives (Eklund et al. 2019). Next, the data quality and transparency of the project would be problematic. In addition, crowdsourcing in science is not good for all research (Behrend et al. 2011). What is more, scientists do not meet research participants, which means it is difficult to answer some research questions. Wexler (2011) suggests that crowdsourcing ignores the disruptive potential of crowds, it glosses over potential interpretations, arguments, and problems of the method.

11.4.11 Micro-Level Barriers

11.4.11.1 Attitude

The academics' perspectives on crowdsourcing in science are an important factor of crowdsourcing in science. Firstly, papers mostly suggested that scientists were skeptical about crowdsourcing. For example, Poliakoff and Webb (2007) have examined 1,000 scientific workers. They stated that "scientists who decide not to participate in public engagement activities do so because (a) they did not participate in the past, (b) they have a negative attitude towards participation, (c) they feel that they lack the skills to take part, and (d) they do not believe that their colleagues participate in public engagement activities" (Poliakoff and Webb 2007, p. 259). Other authors, Riesch and Potter (2014) interviewed 30 British scientists. Their survey showed negative beliefs in the respondents towards the potential and effectiveness of crowdsourcing. In turn, Schlagwein and Daneshgar (2014) interviewed 28 researchers from Asia and the Pacific.

Most of the respondents had negative views on the usefulness of crowdsourcing for research purposes. Kim and Adler (2015) found that personal motivations such as perceived career benefit and risk, perceived effort and attitude towards data sharing are important factors of crowdsourcing in science. Law et al. (2017) interviewed 18 researchers. Their research demonstrates that the feasibility, desirability and utility are important factors of absorptive capacity. Beck et al. (2019) analyzed the crowdsourcing platform, on which they invited individuals with personal or professional experiential knowledge on accidental injuries to participate for a small monetary incentive. A total of 722 people were examined. Secondly, some scientists are afraid of using new technology like crowdsourcing in science (Schlagwein and Daneshgar, 2014). Also, it is highly probable that the academic teachers who have a high level of anxiety related to the use of new technology in conducting the research will more often prefer traditional forms of conducting research when compared to the persons with lower fears related to the use of new technology in research. Negative emotions related to new technology may be sourced in imagination or experience of an individual.

11.4.12 Performance Expectancy

Performance expectancy is defined here as the degree to which an individual believes that using the system will help him or her to attain gains in job performance (Venkatesh et al. 2003). The expected performance means that using crowdsourcing in science may increase the academic teachers' expectations on better results when they reach for crowdsourcing in science. Results of research conducted so far suggest that expected performance is the strongest determinant of the intention of a given behavior. The literature stresses that performance expectancy is linked to behavioral intentions and motivation. Previous research has shown that motivation is an important antecedent to crowdsourcing. In the case of crowdsourcing in science, it may mean that academic teachers are more likely to use traditional working methods if they believe that crowdsourcing in science will not help them in achieving better results or higher salary. It is thought that the lack of user-friendly interfaces in crowdsourcing in science may discourage academic teachers from crowdsourcing in science.

11.4.13 Effort Expectancy

Effort expectancy is defined as the degree of ease associated with the use of a given technology, which influences the intention to use that technology (Venkatesh et al. 2003). In the context of crowdsourcing in science, we believe that academic teachers can analyze their expectations as to the extent to which crowdsourcing in science is easy or difficult to use, thus the perceived ease of use will affect their intentions to adopt this solution. It should be stressed that the use of crowdsourcing requires a specific effort, including coordination of a very large group of people, promotional activities, motivating the virtual community. Furthermore, various types of crowdsourcing in science, created in another context, may require other legal, cultural or technical conditions. Therefore, if academic teachers encounter a certain level of difficulty in using crowdsourcing in science, they may not apply it.

11.4.14 Social Influence

Social influence is defined as "the degree to which an individual perceives that important others believe he or she should use the new system" (Venkatesh et al. 2003). Previous research has shown that social impact is important for the behavioral intention to use the technology. This means that the closest surroundings, especially the coworkers, can influence whether someone reaches for crowdsourcing in science or not. Efforts to promote crowdsourcing in science among potential users may have a positive impact on the intention to use crowdsourcing in science, although it should be noted that research in this area is limited. Social impact can also come from management, friends, family and others whose opinion is important to academic teachers. Positive feedback from these individuals may encourage academic teachers to use crowdsourcing in science, thereby increasing their intention to use crowdsourcing in science.

11.4.14.1 Demographic Factors and Personality Traits

Several studies suggest that demographic factors are very important in crowdsourcing in science. For example, Uhlmann et al. (2019) claim that "early-career researchers from less well-known institutions, underrepresented demographic groups, and countries that lack economic resources may never have a fair chance to compete." Beck et al. (2020) suggested that older researchers are more skeptical about crowdsourcing in science. Linek et al. (2017) suggested that females are less open to crowdsourcing in science compared to males. Other studies suggest a relationship between personality traits and application of crowdsourcing in science (Linek et al. 2017; Matzler et al. 2008). For example, Metzler et al. (2008) suggested that there are positive relations between openness to share knowledge, agreeableness, conscientiousness and openness to crowdsourcing in science.

11.4.15 Trust

Trust has proven to be a key factor in the behavioral intention to adopt the technology. The literature points to correlations between trust and the performance and efficiency of the whole crowdsourcing process. Crowdsourcing in science needs compromise between maintaining high standards of data quality and the simplicity, commitment and pleasure of the platform's design for potential participants. Hence, the crowdsourcing success may depend on the initiator's trust towards the knowledge, skills or experience of the virtual community involved in a given undertaking, and the faith that the virtual community member will perform given activities.

11.5 CONCLUSION

Based on a systematic literature review, reviewing 27 empirical research on crowdsourcing in science is performed. The majority of the reviewed studies focused on the micro level of crowdsourcing in science barriers. Papers mostly suggested that the most important barrier is seen in skeptical or negative attitudes towards crowdsourcing in science, especially in views on the usefulness and qualitative of crowdsourcing for research purposes.

Firstly, a systematic literature review shows some macro-level factors. For instance, technical infrastructure is an important factor of crowdsourcing in science used. Another barrier is found in peer effects. In this context, the lack of approval from colleagues discourages to crowdsourcing in science. In addition, research skeptical of projects and belief that crowdsourcing in science data are of lower quality are barriers.

Secondly, demographic dimensions, such as gender and age, are perceived to be important factors that influence whether academics use crowdsourcing for research. Also, the literature points to correlations between trust and the performance and efficiency of the whole crowdsourcing process.

As with any emerging area of study and practice, there are a few research questions that require exploration. Based on the analysis of the content of the publication, the recommended directions of research field development were identified (Table 11.4). Three directions for further research can be selected on the basis of the results of a systematic literature review: (1) Acceptance of new crowdsourcing in science technologies by academics, (2) Academic concerns about the adoption of crowdsourcing in science, (3) Characteristics of crowdsourcing in science initiators.

In conclusion, on the basis of a systematic literature review, many authors suggest the need for research into the adoption of new crowdsourcing in science technologies by academic teachers. The second direction of research aims to identify the concerns of academic teachers regarding the adoption of crowdsourcing in science. Whereas the third direction of research indicates the necessity of characterizing crowdsourcing in science initiators. The above directions of further research are described in the literature in a fragmentary and selective way.

Furthermore, future research efforts should focus on developing a conceptual framework for crowdsourcing in science, including academic motivation for crowdsourcing in science, taking into account different types of crowdsourcing in science. Currently, such a broad and comprehensive approach to crowdsourcing in science remains in the field of theoretical interest of researchers. There is a need for future studies to better understand scientists' individual and characteristics, especially relationship between personality traits and application of crowdsourcing in science. Future research is still needed to develop a more of institutional support of crowdsourcing in science use. However, the motivations of scientists can depend on collaborative research practices and organizational support.

TABLE 11.4
Recommended Directions of Research over Crowdsourcing in Science

No.	Author/Authors	Directions of Research
1.	Dwivedi et al. (2015)	Adoption of new crowdsourcing in science technologies by academic teachers
2.	Dowthwaite and Sprinks (2019); Lewandowski and Specht (2015); Scheliga et al. (2018)	Academic staff's concerns about adopting crowdsourcing in science
3.	Hevner et al. (2014); Peffers et al. (2017)	Requirements for a dedicated crowdsourcing in science platform
4.	Franzoni and Sauermann (2014)	Crowdsourcing performance in science
5.	Wiggins and Crowston (2011)	Crowdsourcing classification in science
6.	Schlagwein and Daneshgar (2014)	Ethical aspects of crowdsourcing in science
7.	Law et al. (2017); Sauermann et al. (2019); Skarlatidou et al. (2019)	Virtual community motivation to participate in crowdsourcing in science
8.	Friesike et al. (2015); Rotman et al. (2012); Sturm and Tscholl (2019)	Motivation of academic teachers for crowdsourcing in science
9.	Singh and Fleming (2010)	Interactions between people in crowdsourcing in science project teams
10.	Conley (2014)	Conducting research on crowdsourcing in science by management scientists
11.	Erdt et al. (2017); Woolley et al. (2016)	Taking into account demographic data and its importance in the participation of academic teachers in crowdsourcing in science
12.	Riesch and Potter (2013)	Limitations of crowdsourcing in science
13.	Sauermann et al. (2019)	Characteristics of crowdsourcing in science initiators
14.	Schäfer et al. (2016)	Factors of crowdsourcing in science
15.	Schlagwein and Daneshgar (2014)	Conducting research on crowdsourcing in science using qualitative and quantitative methods

ACKNOWLEDGMENT

This project was financed from the funds provided by the National Science Centre, Poland, awarded on the basis of decision number DEC-2019/35/B/HS4/01446.

REFERENCES

Avidov-Ungar, Orit, and Noga Magen-Nagar. 2014. "Teachers in a changing world: Attitudes toward organizational change." *Journal of Computers in Education* 1 (4): 227–249.

Beck, Susanne, Carsten Bergenholtz, Marcel Bogers, Tiare-Maria Brasseur, Marie Louise Conradsen, Diletta Di Marco, Andreas P. Distel, Leonhard Dobusch, Daniel Dörler, Agnes Effert, Benedikt Fecher, Despoina Filiou, Lars Frederiksen, Thomas Gillier, Christoph Grimpe, Marc Gruber, Carolin Haeussler, Florian Heigl, Karin Hoisl, Katie Hyslop, Olga Kokshagina, Marcel LaFlamme, Cornelia Lawson, Hila Lifshitz-Assaf, Wolfgang Lukas, Markus Nordberg, Maria Theresa Norn, Marion Poetz, Marisa Ponti, Gernot Pruschak, Laia Pujol Priego, Agnieszka Radziwon, Janet Rafner, Gergana Romanova, Alexander Ruser, Henry Sauermann, Sonali K. Shah, Jacob F. Sherson, Julia Suess-Reyes, Christopher L. Tucci, Philipp Tuertscher, Jane Bjørn Vedel, Theresa Velden, Roberto Verganti, Jonathan Wareham, Andrea Wiggins, and Sunny Mosangzi Xu. 2020. "The open innovation in science research field: A collaborative conceptualisation approach." *Industry and Innovation* 1–50. https://www.tandfonline.com/doi/full/10.1080/13662716.2020.1792274.

Beck, Susanne, Tiare-Maria Brasseur, Marion Kristin Poetz, and Henry Sauermann. 2019. "What's the problem? How crowdsourcing contributes to identifying scientific research questions." *Academy of Management Proceedings* 2019 (1): 15282.

Behrend, Tara S., David J. Sharek, Adam W. Meade, and Eric N. Wiebe. 2011. "The viability of crowdsourcing for survey research." *Behavior Research Methods* 43 (3): 800.

Berbegal-Mirabent, Jasmina, Dolors Gil-Doménech, and Doming E. Ribeiro-Soriano. 2020. "Fostering university-industry collaborations through university teaching." *Knowledge Management Research & Practice* 18 (3): 263–275.

Burgess, H. K., L. B. DeBey, H. E. Froehlich, N. Schmidt, E. J. Theobald, A. K. Ettinger, J. HilleRisLambers, J. Tewksbury, and J. K. Parrish. 2017. "The science of citizen science: Exploring barriers to use as a primary research tool." *Biological Conservation* 208: 113–120.

Conley, C., and J. Tosti-Kharas. 2014. "Crowdsourcing Content Analysis for Managerial Research." *Management Decision* 52 (4).

Correia, António, Daniel Schneider, Benjamim Fonseca, and Hugo Paredes. 2018. "Crowdsourcing and Massively Collaborative Science: A Systematic Literature Review and Mapping Study." In: A. Rodrigues, B. Fonseca, and N. Preguiça (eds), *Collaboration and Technology*. (CRIWG 2018. Lecture Notes in Computer Science, vol. 11001.) Springer, Cham.

Cox, Joe, E. Y. Oh, Brooke Simmons, Chris Lintott, Karen Masters, Gary Graham, Anita Greenhill, and Kate Holmes. 2015. "Defining and measuring success in online citizen science: A case study of zooniverse projects." *Computing in Science & Engineering* 17: 1.

Curtis V. 2015. "Motivation to participate in an online citizen science game: A study of Foldit." *Science Communication* 37 (6):723–746.

Dwivedi, Y. K., D. Wastell, S. Laumer, H. Z. Henriksen, M. D. Myers, D. Bunker, A. Elbanna, M. Ravishankar, and S. C. Srivastava. 2015. "Research on information systems failures and successes: Status update and future directions." *Information Systems Frontiers* 17 (1): 143–157.

Doan, Anhai, Raghu Ramakrishnan, and Alon Y. Halevy. 2011. "Crowdsourcing systems on the World-Wide Web." *Communications of the ACM* 54 (4): 86–96.

Doyle, Cathal, Yevgeniya Li, Markus Luczak-Roesch, Dayle Anderson, Brigitte Glasson, Matthew Boucher, Carol Brieseman, Dianne Christenson, and Melissa Coton. 2018. What is online citizen science anyway? An educational perspective. https://arxiv.org/ftp/arxiv/papers/1805/1805.00441.pdf

Dowthwaite, Liz, & James Sprinks. 2019. "Citizen science and the professional-amateur divide: Lessons from differing online practices." *Journal of Science Communication* 18: 1–18.

Eklund, Lina, Isabell Stamm, and Wanda Katja Liebermann. 2019. "The crowd in crowdsourcing: Crowdsourcing as a pragmatic research method." *First Monday* 24 (10).

Erdt, M., H. H. Aung, A. S. Aw, C. Rapple, Y.-L. Theng. 2017. "Analysing researchers' outreach efforts and the association with publication metrics: A case study of Kudos." *PLoS ONE* 12 (8): e0183217.

Fausto, S., F. A. Machado, L. F. J. Bento, A. Iamarino, T. R. Nahas, et al. 2012. "Research blogging: Indexing and registering the change in Science 2.0." *PLOS ONE 7* (12): e50109.

Franzoni, Chiara, and Henry Sauermann. 2014. "Crowd science: The organization of scientific research in open collaborative projects." *Research Policy* 43 (1): 1–20.

Friesike, S., B. Widenmayer, O. Gassmann, et al. 2015. "Opening science: Towards an agenda of open science in academia and industry." *Journal of Technolofy Transfer* 40: 581–601.

Guinan, E., K. J. Boudreau, and K. R. Lakhani. 2013. "Experiments in Open Innovation at Harvard Medical School: What happens when an elite academic institution starts to rethink how research gets done?" *MIT Sloan Management Review* 54(3): 45–52.

Heaton, Lorna, Florence Millerand, Xiao Liu, and Élodie Crespel. 2016. "Participatory science: Encouraging public engagement in ONEM." *International Journal of Science Education, Part B* 6 (1): 1–22.

Hevner, Alan R., Salvatore T. March, Jinsoo Park, and Sudha Ram. 2004. "Design science in information systems research." *Management Information Systems Quarterly* 28 (1): 75–105.

Hirth, Matthias, Jason Jacques, Peter Rodgers, Ognjen Scekic, and Michael Wybrow. 2017. "Crowdsourcing Technology to Support Academic Research." In D. Archambault, H. Purchase, and T. Hoßfeld (eds), Evaluation in the Crowd: Crowdsourcing and Human-Centered Experiments (pp. 70–95). (Lecture Notes in Computer Science; Vol. 10264.) Springer, Cham.

Howe, Jeff. 2006. "The rise of crowdsourcing." *Wired* 14.

Hui, Julie S., and Elizabeth M. Gerber. 2015. "Crowdfunding Science: Sharing Research with an Extended Audience." Proceedings of the 18th ACM Conference on Computer Supported Cooperative Work & Social Computing, Vancouver, BC, Canada.

Ipeirotis, Panagiotis G., Foster Provost, and Jing Wang. 2010. "Quality management on Amazon Mechanical Turk." Proceedings of the ACM SIGKDD Workshop on Human Computation, Washington DC.

Kalev, Leetaru, Wang Shaowen, Cao Guofeng, Padmanabhan Anand, and Eric Shook. 2013. "Mapping the global Twitter heartbeat: The geography of Twitter." *First Monday* 18 (5): 64–67. https://doi.org/10.5210/fm.v18i5.4366

Kim, Youngseek, and Melissa Adler. 2015. "Social scientists' data sharing behaviors: Investigating the roles of individual motivations, institutional pressures, and data repositories." *International Journal of Information Management* 35 (4): 408–418.

Law, Edith, Krzysztof Z. Gajos, Andrea Wiggins, Mary L. Gray, and Alex Williams. 2017. "Crowdsourcing as a Tool for Research: Implications of Uncertainty." Proceedings of the 2017 ACM Conference on Computer Supported Cooperative Work and Social Computing, Portland, Oregon, USA.

LeBel, E. P., R. J. McCarthy, B. D. Earp, M. Elson, and W. Vanpaemel. 2018. "A unified framework to quantify the credibility of scientific findings." *Advances in Methods and Practices in Psychological Science* 1 (3): 389–402.

Lewandowski, E., and H. Specht. 2015. "Influence of volunteer and project characteristics on data quality of biological surveys." *Conservation Biology* 29: 713–723.

Linek, Stephanie B., Benedikt Fecher, Sascha Friesike, and Marcel Hebing. 2017. "Data sharing as social dilemma: Influence of the researcher's personality." *PLOS One* 12 (8): e0183216.

Llorente, Roberto, and Maria Morant. 2015. "Crowdsourcing in Higher Education." In F. Garrigos Simon, I. Gil Pechuán, and S. Estelles-Miguel (eds), *Advances in Crowdsourcing* (pp. 87–95). Springer, Switzerland.

Lukyanenko, Roman, Jeffrey Parsons, Yolanda F. Wiersma, and Mahed Maddah. 2019. "Expecting the unexpected: Effects of data collection design choices on the quality of crowdsourced user-generated content." *MIS Quarterly* 43 (2): 623–648.

Majchrzak, A., and A. Malhotra. 2013. "Towards an information systems perspective and research agenda on crowdsourcing for innovation." *The Journal of Strategic Information Systems* 22 (4): 257–268.

Mäkipää, Juho-Pekka, Duong Dang, Teemu Mäenpää, and Tomi Pasanen. 2020. *Citizen Science in Information Systems Research: Evidence From a Systematic Literature Review*. https://osuva.uwasa.fi/handle/10024/10401

Matzler, K., B. Renzi, J. Muller, S. Herting, and T. A. Mooradian. 2008. "Personality traits and knowledge sharing." *Journal of Economic Psychology* 29 (3): 301–313.

Matzler, Kurt, Birgit Renzl, Julia Müller, Stephan Herting, and Todd A. Mooradian. 2008. "Personality traits and knowledge sharing." *Journal of Economic Psychology* 29 (3): 301–313.

Moher, David, Alessandro Liberati, Jennifer Tetzlaff, Douglas G. Altman, and Prisma Group The. 2009. "Preferred reporting items for systematic reviews and meta-analyses: The PRISMA statement." *PLOS Medicine* 6 (7): e1000097.

Nov, O, O Arazy, and D Anderson. 2014. Scientists@Home: What Drives the Quantity and Quality of Online Citizen Science Participation? *PLoS ONE* 9 (4): e90375.

OECD. 2002. *Frascati Manual 2002*. https://www.oecd-ilibrary.org/science-and-technology/frascati-manual-2002_9789264199040-en

Pan, Stephen W., Gabriella Stein, Barry Bayus, Weiming Tang, Allison Mathews, Cheng Wang, Chongyi Wei, and Joseph D. Tucker. 2017. "Systematic review of innovation design contests for health: Spurring innovation and mass engagement." *BMJ Innovations* 3: 227–237.

Poliakoff, Ellen, and Thomas Webb. 2007. "What factors predict scientists' intentions to participate in public engagement of science activities?" *Science Communication* 29 (2): 242–263.

Riesch, Hauke, and Clive Potter. 2014. "Citizen science as seen by scientists: Methodological, epistemological and ethical dimensions." *Public Understanding of Science* 23 (1): 107–120.

Rotman, Dana, Jennifer Hammock, Jenny Preece, Derek Hansen, Carol Boston, Anne Bowser, and Yurong He. 2014. "Motivations affecting initial and long-term participation in citizen science projects in three countries." 10.9776/14054.

Sauermann, Henry, and Chiara Franzoni. 2015. "Crowd science user contribution patterns and their implications." *PNAS* 112 (3): 679–684.

Sauermann, Henry, Chiara Franzoni, and Kourosh Shafi. 2019. "Crowdfunding scientific research: Descriptive insights and correlates of funding success." *PLOS ONE* 14 (1): e0208384.

Schäfer, M. S., J. Metag, J. Feustle, and L. Herzog. 2016. "Selling science 2.0: What scientific projects receive crowd funding online?" *Public Understanding of Science* 27(5): 496–514.

Scheliga, K., S. Friesike, C. Puschmann, and B. Fecher. 2018. "Setting up crowd science projects." *Public Understanding of Science* 27(5): 515–534.

Schlagwein, Daniel, and Farhad Daneshgar. 2014. "User Requirements of a Crowdsourcing Platform for Researchers: Findings from a Series of Focus Groups." PACIS. https://aisel.aisnet.org/pacis2014/195/

See, Linda. 2019. "A review of citizen science and crowdsourcing in applications of pluvial flooding." *Frontiers in Earth Science* 7 (44): 1–7.

Sheehan, K. B. (2018). "Crowdsourcing research: Data collection with Amazon's Mechanical Turk." *Communication Monographs* 85 (1): 140–156. https://doi.org/10.1080/03637751.20 17.1342043

Shirk, J. L., H. L. Ballard, C. C. Wilderman, T. Phillips, A. Wiggins, R. Jordan, E. McCallie, M. Minarchek, B. V. Lewenstein, M. E. Krasny, and R. Bonney. 2012. Public participation in scientific research: A framework for deliberate design. *Ecology and Society* 17(2): 29.

Singh, Jasjit, and Lee Fleming. 2010. "Lone inventors as sources of breakthroughs: Myth or reality?" *Management Science* 56 (1): 41–56.

Skarlatidou, A., M. Suškevičs, C. Göbel, B. Prūse, L. Tauginienė, A. Mascarenhas, et al. 2019. "The value of stakeholder mapping to enhance co-creation in citizen science initiatives." *Citizen Science: Theory and Practice* 4 (1): 24.

Stritch, Justin M., Mogens Jin Pedersen, and Gabel Taggart. 2017. "The opportunities and limitations of using mechanical Turk (MTURK) in public administration and management scholarship." *International Public Management Journal* 20 (3): 489–511.

Sturm, U., and M. Tscholl. 2019. "The role of digital user feedback in a user-centred development process in citizen science." *Journal of Science Communication* 18 (01): A03.

Tight, Malcolm. 2018. "Higher education journals: Their characteristics and contribution." *Higher Education Research & Development* 37 (3): 607–619.

Tranfield, David, David Denyer, and Palminder Smart. 2003. "Towards a methodology for developing evidence-informed management knowledge by means of systematic review." *British Journal of Management* 14 (3): 207–222.

Uhlmann, Eric Luis, Charles R. Ebersole, Christopher R. Chartier, Timothy M. Errington, Mallory C. Kidwell, Calvin K. Lai, Randy J. McCarthy, Amy Riegelman, Raphael Silberzahn, and Brian A. Nosek. 2019. "Scientific utopia III: Crowdsourcing science." *Perspectives on Psychological Science* 14 (5): 711–733.

Venkatesh, Viswanath, Michael Morris, Gordon Davis, and Fred Davis. 2003. "User acceptance of information technology: Toward a unified view." *MIS Quarterly* 27: 425–478.

Waldrop, M. M. 2008. "Science 2.0." *Scientific American* 298: 69–73.

Wexler, Mark. 2011. "Reconfiguring the sociology of the crowd: Exploring crowdsourcing." *International Journal of Sociology and Social Policy* 31: 6–20.

Wiggins, Andrea, and Kevin Crowston. 2011. "From conservation to crowdsourcing: A typology of citizen science." Proceedings of the Annual Hawaii International Conference on System Sciences, 1–10.

Woolley, J. P., M. L. McGowan, H. J. Teare, V. Coathup, J. R. Fishman, R. A. Settersten Jr., S. Sterckx, J. Kaye, and E. T. Juengst. 2016. "Citizen science or scientific citizenship? Disentangling the uses of public engagement rhetoric in national research initiatives." *BMC Medical Ethics* 17 (1): 33.

Zhao, Yuxiang, and Qinghua Zhu. 2014. "Evaluation on crowdsourcing research: Current status and future direction." *Information Systems Frontiers* 16 (3): 417–434.

Zhou, Mi Jamie, Baozhou Lu, Weiguo Patrick Fan, and G. Alan Wang. 2018. "Project description and crowdfunding success: An exploratory study." *Information Systems Frontiers* 20 (2): 259–274.

12 Effects of Technology-Based Feedback on Learning

Irum Alvi

CONTENTS

12.1 INTRODUCTION

The Webster's II new college dictionary(2001) defines Feedback as a method in which the elements producing the result are modified, amended, reinforced by the result itself; it is comprehended as a reaction, which starts the process. In the context of learning, feedback may be described as a form of communication or process used to enlighten the learners of the correctness of an answer, usually to a query (Carter 1984; Cohen 1985; Kulhavy 1977). In the context of digitalized teaching, Technology-based feedback may be given to a learner after input with a single-mindedness of influencing the perceptions of a learner. However, many learners find feedback process problematic (Hopfenbeck 2020) and don't know how to act upon it (Carless 2019) or depend on the student teachers' competence (DeLuca et al. 2019), as such many feedback strategies fail to produce the desired effect.

DOI: 10.1201/9781003132097-12

Feedback plays a main role in improving the learners' awareness and abilities (Azevedo and Bernard 1995; Bangert-Drowns et al. 1991; Corbett and Anderson 1989; Epstein et al. 2002; Moreno 2004; Pridemore and Klein 1995). It includes guidance, reminders, and hints. Feedback may be multifaceted and include multidimensional, helpful, timely, unambiguous, reliable, sporadic, conditional, and candid information given to the learners (Brophy 1981; Schwartz and White 2000). Technology-based feedback may include information, in the form of a message being displayed on the computer screens received after the completion of any task with the intent to enhance the learners' insight, reasoning, and subsequently the responses of the learners (Moreno 2004; Wager and Wager 1985), but perspectives of both perspectives between teachers and students are always important (Prastiyani et al. 2020) for both formal and informal feedback (Trimmer and Guest 2020).

12.2 LITERATURE REVIEW

The influence of feedback on the learners has been dealt with in detail by many researchers (Bangert-Drowns et al. 1991; Kluger and DeNisi 1996; Kulhavy and Wager 1993; Mory 2004). Feedback, especially good feedback, can play a major role in helping the learners (Albertson 1986; Azevedo and Bernard 1995; Narciss and Huth 2004), although different types of feedback may elicit different responses from them, as such the enormous research work in this field has led to a conglomeration of conflicting and contradicting results.

With the advent of new technology that offers more options for giving feedback to the learners, the matter has become all the more complex (Dempsey et al. 1993). Technology-based feedback has grown increasingly into a widespread form of evaluation since the advent of new technologies. It has considerably transformed the practice of evaluation (Tseng and Tsai 2007). Technology-based feedback has become possible due to online functions, such as submission of assignments, storing data, communicating information, and managing reviews (Liu et al. 2001). Technology-based feedback has numerous benefits over other forms of face-to-face feedback (Tsai and Liang 2009; Yang and Tsai 2010), as it permits the teachers to communicate more freely with the students, who in turn may be able to ponder on and incessantly review the work-based on it (Yang 2011). Online systems permit the instructors to screen students' undertakings and improvement carefully (Lin et al. 2001) and empower the collection of facts possible by saving information on tasks, projects, involvement, and communicating it. Finally, the feedback in the form of praise and appreciation given by the teacher has a substantial affirmative influence on students.

A plethora of research work focused on the effect of information within the feedback for improving the precision of the learners' responses to a problem and for rectifying the errors and misconceptions in their minds (Azevedo and Bernard 1995; Birenbaum and Tatsuoka 1987; Cheng et al. 2005; Cohen, 1985; Kulhavy 1977; Sleeman et al. 1989) and for motivating the learners to learn (Lepper and Chabay 1985; Narciss and Huth 2004). Research has also shown that detailed feedback has a more pronounced effect on the learners than vague feedback or hints (Albertson 1986; Grant et al. 1982; Hannafin 1983; Moreno 2004; Pridemore and

Klein 1995; Roper 1977). Feedback may be used for facilitating learning and presentation (Bandura 1991; Bandura and Cervone 1983; Fedor 1991; Ilgen et al. 1979), for abridging the gap between presentation and the preferred level of presentation (Locke and Latham 1990; Song and Keller 2001), for reducing uncertainty about task performance (Ashford 1986; Ashford et al. 2003), for lessening the mental load of the learners, particularly the beginners (Sweller et al. 1998; Paas et al. 2003), for rectifying wrong strategies, practical mistakes, or misunderstandings (Ilgen et al. 1979; Mory 2004).

The idea of immediate response was dealt with by Skinner, who found that feedback must follow an answer as immediately as probable to be effective. The effect the Delayed Feedback was initially observed by Kulhavy (1977), who reported that if the feedback is delayed for a day or more it helps in retention (Sturges 1969, 1972) due to the delay-retention influence (Brackbill et al. 1962; Brackbill and Kappy 1962), especially in the context of written instructions (Kulhavy and Stock 1989). Some researchers maintain that initial errors are overlooked after a period of time (Bardwell 1981; Kulhavy and Anderson 1972; Surber and Anderson 1975); others maintain the delay does not have an effect (Peeck et al. 1985; Phye et al. 1976) early replies are disremembered (Peeck and Tillema 1979). Kulik and Kulik (1988) found that for actual classroom quizzes Immediate Feedback was more useful than delayed, which challenges the use of Delayed Feedback for learning purposes.

Immediate Feedback is informative educative feedback sent to the learners as speedily as digitalized resources permit during teaching or testing, while Delayed Feedback is informative, educative feedback sent to the learners after a definite delay interlude all through teaching or testing. Dempsey et al. (1993) propose that delayed response is equivalent to suppression of required information from the learner. Many researchers agree that Delayed Feedback may hamper the acquirement of desirable information (Dempsey et al. 1993; Kulhavy 1977). However, the findings by Richards (1989) favor temporarily Delayed Feedback.

Anderson et al. (1989) found no significant difference between subjects presented with Immediate Feedback and those subjects presents with Delayed Feedback. The Delayed Feedback may foster the improvement of skills including error recognition and self-improvement. Contradicting findings show feedback has an insignificant influence on learning (Corbett and Anderson 1989, 1990; Gilman 1969; Hodes 1985; Kulhavy et al. 1985; Merrill 1987).

12.2.1 Purpose and Originality

Modern digitized learning has opened the door to interactivity, the ability to keep accurate records of learners' responses, and the possibility to acclimatize feedback according to the requirements of the learners. The purpose of the current study is to assess the significance of Technology-based feedback for English language learners in the Indian context. Further, recent developments in the use of language learning using multimedia have opened vast areas for researchers to consider.

The present study scrutinizes the effects of Technology-based feedback, delayed and immediate to find which helps the language learners to learn better. It focuses on only two forms of feedback and the effects on the learners' acquisition of the English Tenses.

Providing no feedback or feedback, immediate or delayed, was incorporated in the learning mode using an English Language Learning Software Program. For observing the effects of feedback on the learners, information was communicated to the learners of the experimental groups, with the intent to modify their understanding of the use of the English Tenses. The feedback given to the learners allowed them the chance to better comprehend the established standards and rules for one grammatical item, the Tenses. The first experimental group received the feedback immediately upon submission of their responses, while for the second experimental group, the same information was delayed. The purpose was to verify the findings of Butler et al. (2007) and Butler and Roediger (2008), who found feedback delays of ten minutes through one day are effective for learners.

This present study is based on sound theoretical knowledge and backs the current body of knowledge about feedback as well as technology for teaching and assessment, two entirely diverse fields of research, using empirical insights in the perspective of digitalization and use of technology for providing learners with quality feedback. The paper also reveals the possibility of using diverse forms of feedback for language learners. Its novelty also lies in the systemic approach adopted for evaluating the influence of feedback on learners from two different Mediums of instruction and in offering a comprehensive evaluation of feedback given; in exposing the possibility of using diverse forms of feedback, in new arguments about their effectiveness for technology-based applications, and in time efficiency of feedback. The study is also novel as entirely different statistical tools are used for the evaluation of the effectiveness and influence of feedback on the learners. The present study is significant as it addresses a gap in extant literature due to the paucity of investigational research design in field of feedback, as sometimes scholars fail to find effects of the experiments and the results are not published due to null results (Frans et al. 2019; Herrington and Maynard 2019). The study focuses on the Medium of instruction which has not been dealt with by other researchers. The research will help teachers in giving effect and constructive feedback (Whitney and Ackerman 2020) which is required for acquisition of the English Tenses.

12.2.2 Research Questions

The study addressed the following questions:

1. Are there any statistically substantial differences between the learners' total test mean scores in Pre and Post-tests of the control and experimental groups, attributed to whether they received feedback or not?
2. Are there any statistically substantial differences between the Hindi and English Medium learners' mean scores in Pre and Post-tests attributed to whether they received feedback or not?
3. Are there any statistically substantial differences between the learners' attainment mean scores, ascribed to the type of Technology-based feedback they received, delayed or immediate?

4. Are there any statistically substantial differences between the Hindi and English Medium learners' mean scores in Pre and Post-tests, attributed to the feedback they received, both delayed and immediate?

12.2.3 Hypotheses for ANCOVA

Hypothesis (H_o): There is no substantial difference between treatment conditions in the Independent Factor (Type of feedback) on the Dependent Factors (Post-test of the three groups)

$$H_o: M_1 = M_2 = M_3 \ldots\ldots = M_K$$

Alternate Hypothesis (Ha): There is a substantial difference between treatment conditions in the Independent Factor (Type of feedback) on the Dependent Factors (Post-test of the three groups)

H_a: Means are not equivalent
$M_1 = M_2 = M_3 \ldots\ldots = M_K$ = Means of the different groups

12.3 Methodology

12.3.1 Research Sample

The subjects of the study were learners from two different Mediums of education, Hindi and English. The Hindi Medium learners started learning English late in comparison to the English Medium learners. Moreover, while the English Medium learners studied most subjects in primary, secondary, and higher secondary schools in English language, Hindi Medium learners often studied these subjects in Hindi. The sample of the study comprised of 60 learners who had sought admission in the First Year in a Technical College. They had previous experience of using computers for educational and instructional purposes. Besides, they had access to well-equipped computerized digital labs. They had easy access to the English Language Learning Software. They were equally assigned three groups i.e. one control and two experimental groups. All groups were taught the Tenses using English Language Learning Software which provided multiple options for giving feedback, no feedback, Immediate Feedback, and Delayed Feedback.

12.3.2 Research Instruments

To device the study, three types of instruments were used: English Language Learning Software Program and Pre and Post-tests and oral interviews.

A Pre-test was administered on all the groups to check the initial group knowledge with regards to the use of Tenses and to help control non-randomization effect, which was considered as a potential threat to the internal cogency with the study design. The quasi-experiment lasted for eight weeks. The content covered for all the groups was the same and covered the grammatical aspects under study. During the study, the same material/content was covered and all the learners used the same materials and texts for learning the English Tenses. The drills and practice exercises were also identical but the experimental groups received Technology-based feedback, while the control group received no feedback.

The Pre-test and Post-test each comprised 20 multiple-choice items with 4 options for each item. At the start, the guidelines were given. The learners were to select the right answer for each item. The time allocated for the online test was 20 minutes. One mark was allocated for each item; the maximum score was 20. The learners' acquaintance with regards to the Tenses was measured by the Pre-test given to the control and experimental groups at the onset of the study.

The Post-test was developed for evaluating the differential effects of the type of Technology-based feedback the groups received. The Post-test consisted of an equal number of MCQ items. The Post-test was given to all the learners to calculate the influence of the Technology-based feedback method used.

12.3.3 English Language Learning Software Program

For the present study, an English Language Learning Software Program was used to teach the Tenses to the learners and find out its effect on the learners' acquisition of Tenses. The English Language Learning Software Program comprised of the following:

- Introduction to Tenses
- Construction of the Tenses
- Use of the Tenses
- Explanation of the rules for Tenses
- Examples
- Practice Exercises
- Tests
- Feedback – immediate or delayed or no feedback.

No feedback shows the learners were given a question and were told to answer, but no suggestion was given as to the accuracy of the learners' answers. Immediate Feedback was given to the learners as soon as they submitted their responses. Delayed Feedback was given to the learners the next day.

12.3.4 ANCOVA Assumptions

The following assumptions were considered:

Random Sample, Normality which assumes that there are no outliers, Homogeneity of variance, which states the dependent factors must have equal variances, when Levene's test is not noteworthy, homogeneity of variances is met. Other assumptions were the linear association between the DV and covariates, and the covariate is consistent and calculated without error.

The data recovered from the subjects was analyzed for the above assumptions. For normality, the null hypothesis is that the data has a normal distribution. As the

TABLE 12.1
Tests of Normality

		Kolmogorov-Smirnov[a]			Shapiro–Wilk		
	Group	**Statistic**	**df**	**Sig.**	**Statistic**	**df**	**Sig.**
Pre-test	Control	0.126	20	0.200*	0.942	20	0.263
	Immediate Feedback	0.110	20	0.200*	0.959	20	0.531
	Delayed Feedback	0.234	20	0.006	0.921	20	0.103
Post-test	Control	0.176	20	0.105	0.911	20	0.067
	Immediate Feedback	0.140	20	0.200*	0.953	20	0.407
	Delayed Feedback	0.164	20	0.167	0.921	20	0.105

Notes: *This is a lower bound of the true significance.
[a] Lilliefors significance correction.

dataset was small, the Shapiro–Wilk test was conducted. The p-value for the Pre-test for the three groups, Control, Immediate and Delayed Feedback were 0.263, 0.531, and 0.103, and for the Post-test the p-value were 0.067, 0.407, and 0.105, respectively. Thus, the null hypothesis was accepted and it was established that the data came from a distribution which has normality, as p-values were larger than the alpha level. Table 12.1 shows results for the Normality tests.

However, since the experiment may be affected by sample size and normality, Q–Q plots were made for additional authentication of the normality of the dataset. Histograms were created for the variables and variables in each group to scrutinize the normality of the data. For the univariate outcomes, utilizing Levene's test, the homogeneity of variance hypothesis was met for the dependent variable ($p > 0.05$) with and without adding the covariate, the Pre-test, as shown above. Next, the box's test showed that the assumption of homogeneity of covariance was fulfilled. The summary plots of the dataset explicitly depicted the quartiles, median, and highest values. The box signified the inter quartile score array, the middle 50% estimation. The whiskers extended from the higher and lower verge of the box to the uppermost and lowermost estimations for each variable. A line across the box indicated the median. The vertical axis showed the test scores of the test variables and the horizontal axis showed Medium scores of the learners. The graph showed that the range of scores for Post-test was greater than for the Pre-tests scores and the range (middle 50% of records), was higher on the scale in Post-test than in Pre-test for two of the three groups. The graph suggested that Post-test scores were more than Pre-test scores and there might be some degree of discrepancy between Post-test scores and Pre-test scores for these groups.

Moreover, no outliers were present in the data that is there was no value that did not fall in the inner limits. There was no extreme value. No stars/asterisks were observed. There were no outliers for the variables for Hindi and English Medium learners. Next, the scatterplot for Pre-test and Post-test were also examined and the data was found to be suitable for further analysis.

Analysis of COVAriance – ANCOVA was performed to find the differences between the experimental group, which was given Feedback, Immediate, and Delayed using English Language Learning Software Program and the control group that was not given any feedback.

12.4 FINDINGS

The first question was concerned with the presence of statistically significant differences, if any, among the learners' total test mean scores in use of Tenses ascribed to the process of giving or not giving feedback.

Table 12.2 displays the mean and standard deviation (St. D) estimations of the scores in the Post-test. They do not contain any modifications attributable to the use of a covariate in the exploration made later.

As shown in Table 12.2, there are statistically substantial dissimilarities among the mean scores in the test result of the experimental groups which were given feedback. The experimental groups' mean scores were 16.05 and 15.30 for Immediate Feedback and Delayed Feedback experimental groups, respectively while for the control group given no feedback the mean score was 12.45. It can be seen that there are substantial dissimilarities between the mean score of the learners who were given feedback that is the experimental groups and the mean scores of the learners who were not given any feedback statistically significant difference at $\alpha < 0.05$. This clearly demonstrates a noteworthy effect of Feedback on the learners' learning of the Tenses. This influence was in favor of the experimental groups.

The second question sought to find statistically substantial differences, if any, between the Hindi and English learners' mean scores in use of Tenses attributed to the type of feedback, delayed or Immediate Feedback received. The learners' mean scores and St. D. in the Post-test for all the groups, the two control and experimental were calculated.

Table 12.3 shows the mean and St. D. estimations of the learners' scores in the Post-test according to their Medium of education. The values show no changes attributable to the use of a covariate in the exploration made later.

As shown in Table 12.3, there are substantial differences between the mean score in the result of Hindi and English Medium learners. The Hindi Medium experimental groups' mean scores were 16.05 and 14.30 for Immediate and Delayed Feedback, respectively while for the English Medium experimental groups mean scores were

TABLE 12.2

Means and Standard Deviations According to Feedback Method

Group	Total Mean	Std. Deviation	N
Control	12.45	3.395	20
Immediate Feedback	16.05	1.986	20
Delayed Feedback	15.30	1.658	20
Total	14.60	2.883	60

TABLE 12.3
Means and Standard Deviations According to Learners' Medium of Education

Group	Medium	Mean	Std. Deviation	N
Control	Hindi	11.20	3.393	10
	English	13.70	3.057	10
Immediate Feedback	Hindi	16.70	1.494	10
	English	15.40	2.271	10
Delayed Feedback	Hindi	14.30	1.337	10
	English	16.30	1.337	10
Total	Hindi	14.07	3.172	30
	English	15.13	2.501	30

15.40 and 16.30 for Immediate Feedback and Delayed Feedback, respectively. Thus, there exist statistically substantial differences between the mean score of Hindi Medium learners and the mean score of English Medium learners who were given feedback. From the above table, it is clear that they were more in favor of Immediate Feedback for the Hindi Medium group and for Delayed Feedback for the English Medium group.

To discover the statistical implication of the differences, ANCOVA, Analysis of Covariance was applied to the outcomes of the Post-test. The variables taken for the test were method of feedback, and medium of education – as evidenced by $F(5, 54) = 1.076$, $p = 0.348$ and $F(5, 54) = 1.445$, $p = 0.223$. That is, $p\ (0.348) > \alpha\ (0.05)$ with the covariate and $p\ (0.223) > \alpha\ (0.05)$ without the covariate, Pre-test. The covariate was involved in the exploration to regulate the differences in the independent variable. Although it was not the fundamental motivation of the exploration, the results were part of the output. The main aim of the covariate test was to help estimate the connection between the covariate and the dependent variable, controlling for the variable. The difference amid the dependent variable (attainment in the Post-test) was estimated, at the significance level ($\alpha < 0.05$), to evaluates the differences between the adjusted means for groups.

12.5 RESULTS

This table shows if the different interventions were statistically substantially different when adjusted for the covariate or if there exists a statistically substantial difference in Post-test scores between the different groups when their means had been adjusted for the Pre-test, first without the covariate and next with the addition of the covariate i.e. Pre-test. Table 12.4 Tests of Between-Subjects Effects (without the covariate).

Table 12.5 shows the Between-Subjects Effects with the addition of the covariate, i.e. Pre-test.

TABLE 12.4

Tests of Between-Subjects Effects (Without the Covariate Pre-test)

Dependent Variable: Post-Test

Source	Type III Sum of Squares	df	Mean Square	F	Sig.	Partial Eta Squared	Noncent. Parameter	Observed Power[b]
Corrected model	324.402[a]	11	29.491	8.528	0.000	0.662	93.804	1.000
Intercept	426.325	1	426.325	123.276	0.000	0.720	123.276	1.000
Group MH and E	68.717	5	13.743	3.974	0.004	0.293	19.870	0.923
Group M.H and E * Pre-test	120.402	6	20.067	5.803	0.000	0.420	34.815	0.995
Error	165.998	48	3.458					
Total	13280.000	60						
Corrected total	490.400	59						

Notes: [a]R Squared = 0.662 (Adjusted R Squared = 0.584).
[b] Computed using alpha = 0.05; M = Medium; H = Hindi; E = English.

To interpret these results, the statistical significance value (i.e. p-value) was focused upon to find if there were statistically substantial differences in Post-test score, taken as dependent variable, and between the groups, taken as independent variable when adjusted for Pre-test, which is the covariate. It was found that there was a statistically substantial difference among the adjusted means values ($p < 0.0005$).

TABLE 12.5

Tests of Between-Subjects Effects with Covariate (Pre-test)

Tests of Between-Subjects Effects

Dependent Variable: Post-Test

Source	Type III Sum of Squares	df	Mean Square	F	Sig.	Partial Eta Squared	Noncent. Parameter	Observed Power[b]
Corrected model	297.739[a]	6	49.623	13.651	0.000	0.607	81.906	1.000
Intercept	449.171	1	449.171	123.564	0.000	0.700	123.564	1.000
Pre-test	93.739	1	93.739	25.787	0.000	0.327	25.787	0.999
Group	143.247	2	71.624	19.703	0.000	0.426	39.406	1.000
M.H and E	10.718	1	10.718	2.948	0.092	0.053	2.948	0.392
Group MH and E	27.558	2	13.779	3.791	0.029	0.125	7.581	0.666
Error	192.661	53	3.635					
Total	13280.000	60						
Corrected total	490.400	59						

Notes: [a]R Squared = 0.607 (Adjusted R Squared = 0.563).
[b] Computed using alpha = 0.05; M = Medium; H = Hindi; E = English.

TABLE 12.6

Evaluating Covariates in the Study at the Value: Pre-test

Group * Medium of Education Hindi = 1, English = 2

Dependent Variable: Post-test

				95% Confidence Interval	
Group	**Medium**	**Mean**	**Std. Error**	***Lower Bound***	***Upper Bound***
Control	Hindi	11.729[a]	0.601	10.521	12.937
	English	13.261[a]	0.607	12.040	14.482
Immediate Feedback	Hindi	16.627[a]	0.592	15.437	17.817
	English	15.486[a]	0.589	14.302	16.671
Delayed Feedback	Hindi	14.316[a]	0.592	13.125	15.506
	English	16.216[a]	0.593	15.024	17.409

Note: [a]Covariates appearing in the model are evaluated at the following values: Pre-test = 12.27.

It was noticed that the values for the mean underwent a change compared to the previous values as indicated in Table 12.6. The new values represented the adjusted mean values, indicating the original means adjusted for the Covariate. Figure 12.1 shows the Profile Plot, in which the Covariance appears in the Model.

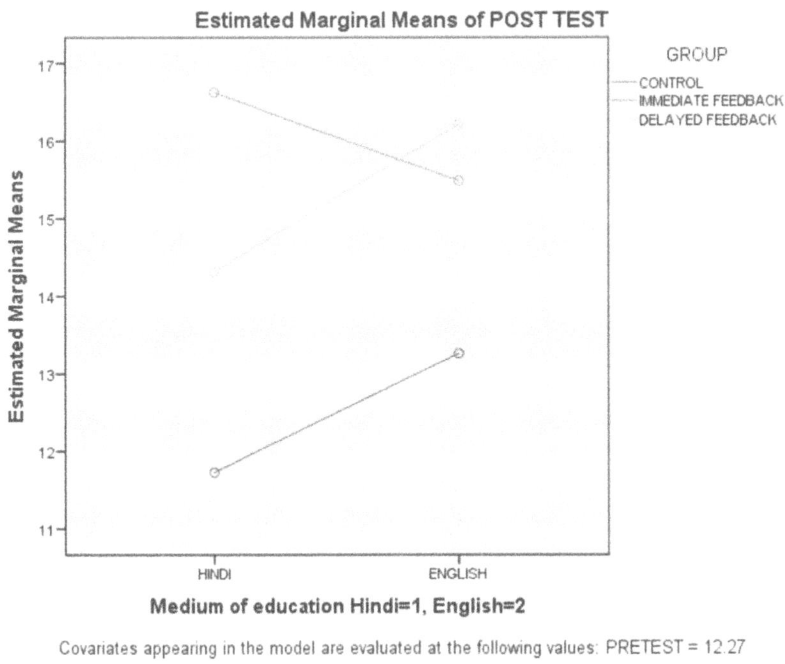

FIGURE 12.1 Profile plot indicating covariance appearing in the model.

The Profile Plot gives a graphic depiction of the findings of the study. The lines represent the assessed mean for the test score for both groups. There is some interaction, which is clearly discernible.

12.6 DISCUSSION

The purpose of the present study was to scrutinize the effects of Technology-based feedback, delayed and immediate, and to find their influence on language learners. At the outset of the study, learners' mean score of control and experimental groups in the pretest showed no statistically substantial differences among the three groups, which indicates that the knowledge of the subjects with regards to the use of the Tenses before implementing the experiment was approximately similar. This also reveals that any improvement in the test scores in the use of the English Tenses may be attributed to the feedback technique used. By the conclusion of the experiment, the total mean scores of the experimental groups were more than those of the other two groups. This increase may thus be attributed to the technique used. This indicates that the use of feedback, both immediate and delayed enhanced learners' ability to produce more grammatically correct sentences within the available time more than the control group. The feedback given to the experimental group markedly improved the learners' ability to grasp and use the English Tenses.

The outcomes of the study are constant with other studies piloted by researchers such as Metcalfe et al. (2009) who found Delayed Feedback was quite effective for language learners, and Mehta et al. (2013), who stressed the effectiveness of feedback for trainees. The findings are different from those of Chaudhary (cited by Devi and Revathi 2016), who found conventional methods worked better than Technology-based feedback for language learners. However, not many studies have focused on the influence of Medium on the language learners. As such, the present study's findings disclose that there are statistically substantial differences attributed to the variable – the learners' Medium of education. This means that Hindi Medium learners can gain more from Immediate Feedback as compared with Delayed Feedback.

12.6.1 Conclusion and Recommendations

This study develops the present appreciation and comprehension of the relationship between Technology-based feedback and language learning. Moreover, it recommends a conceivable elucidation for the advantages of feedback for enhancing the students' performances. Delving into how the Medium of the instruction and feedback influences the students' acquisition of Tenses, the study is original and presents how thoughtful feedback on specific problems influence the students' learning.

Studies on Technology-based feedback are very limited. The findings have numerous consequences for teachers and researchers involved in designing and implementation of Technology-based feedback for language learners. First, the teachers need to be thoughtful and aware of the fact that Technology-based feedback influence is different for English and Hindi Medium students. Different types of Technology-based feedback lead to different learning outcomes on the part of English and Hindi Medium students. Second, training may play an important role as such the teachers should be trained. Teachers should also get suggestions from the students. The findings may help teachers in planning strategies to intensify the efficiency of Technology-based feedback for the students. The teachers should provide thoughtful and meaningful feedback to the students. The results suggest Technology-based feedback is effective and may assist in boosting the interest, and self-efficacy of students, to further enhance their performances using motivation (Gnepp 2020).

12.6.2 Practical and Social Implications

Numerous practical and social recommendations stem out of the study, including how to provide quality feedback to the learners and its implementation; how to increase the effectiveness of Technology-based feedback in education, particularly digital learning, which is the new normal in the post-pandemic times; how to scrutinize the efficiency of Technology-based feedback on the learners. Technology-based feedback in education is deliberated upon, along with the influence, within the context of an educational system with two Mediums of instruction, demonstrating the interrelations and interdependencies for the learners. Raising the quality and measure of Technology-based feedback will positively affect the learners and benefit society itself and raise the standard of education.

12.6.3 Limitations of the Study and Scope for Further Studies

The study has several limitations of its own. The study was limited to the First Year B. Tech learners in the current session and the outcomes of the study are based on the effect of feedback on these learners only. Moreover, the study is limited to one aspect of the English language, the Tenses and results are based on tests, the Pre and Post, conducted after a period of two months. There was a constraint of time also. Further research may be conducted on the effect of Formative Feedback, both immediate and delayed on the learners, concentrating on other topics to create a more inclusive notion about the influence of Technology-based feedback on the learners, from Hindi and English Medium. Augmenting the effects of feedback on students is an intricate issue. Further studies are required to inform teachers and researchers on ways to nurture "mindful reception" to feedback (Bangert-Drowns et al. 1991). As such, further research may be carried out to study the effect of detailed feedback in similar experiments.

REFERENCES

Albertson, L. M. 1986. "Personalized feedback and cognitive achievement in computer assisted instruction." *Journal of Instructional Psychology*, 13 (2): 55–57.

Anderson, J. R., Conrad, F. G., and Corbett, A. T. 1989. "Skill acquisition and the LISP tutor." *Cognitive Science*, 13 (4): 467–505. https://doi.org/10.1207/s15516709cog1304_1

Ashford, S. J. 1986. "Feedback-seeking in individual adaptation: A resource perspective." *Academy of Management Journal*, 29: 465–487.

Ashford, S. J., Blatt, R., and Vande Walle, D. 2003. "Reflections on the looking glass: A review of research on feedback-seeking behavior in organizations." *Journal of Management*, 29: 773–799.

Azevedo, R., and Bernard, R. M. 1995. "A meta-analysis of the effects of feedback in computer based instruction." *Journal of Educational Computing Research*, 13 (2): 111–127.
Bandura, A. 1991. "Social theory of self-regulation." *Organizational Behavior and Human Decision Processes*, 50: 248–287.
Bandura, A., and Cervone, D. 1983. "Self-evaluation and self-efficacy mechanisms governing the motivational effects of goal systems." *Journal of Personality and Social Psychology*, 45 (5): 1017–1028.
Bangert-Drowns, R. L., Kulik, C. C., Kulik, J. A., and Morgan, M. T. 1991. "The instructional effect of feedback in test-like events." *Review of Educational Research*, 61 (2): 213–238.
Bardwell, R. 1981. "Feedback: How does it function?" *Journal of Experimental Education*, 50: 4–9.
Birenbaum, M., and Tatsuoka, K. K. 1987. "Effects of 'on-line' test feedback on the seriousness of subsequent errors." *Journal of Educational Measurement*, 24 (2): 145–155.
Brackbill, Y., Bravos, A., and Starr, R. H. 1962. "Delayed-improved retention of a difficult task." *Journal of Comparative and Physiological Psychology*, 55 (6): 947–952.
Brackbill, Y., and Kappy, M. S. 1962. "Delay of reinforcement and retention." *Journal of Comparative and Physiological Psychology*, 55 (1): 14–18.
Brophy, J. E. 1981. "Teacher praise: A functional analysis." *Review of Educational Research*, 51 (1): 5–32.
Butler, A. C., Karpicke, J. D., and Roediger, H. L. 2007. "The effect of type and timing of feedback on learning from multiple choice tests." *Journal of Experimental Psychology: Applied*, 13: 273–281. doi: 10.1037/1076-898X.13.4.273
Butler, A. C., and Roediger, H. L. 2008. "Feedback enhances the positive effects and reduces the negative effects of multiple-choice testing." *Memory & Cognition*, 36: 604–616. doi: 10.3758/MC.36.3.604
Carter, J. 1984. "Instructional learner feedback: A literature review with implications for software development." *The Computing Teacher*, 12 (2): 53–55.
Carless, D. 2019. "Longitudinal perspectives on students' experiences of feedback: A need for teacher–student partnerships." *Higher Education Research and Development*, 39 (3): 425–438. https://doi.org/10.1080/07294360.2019.1684455
Cheng, S. Y., Lin, C. S., Chen, H. S., and Heh, J. S. 2005. "Learning and diagnosis of individual and class conceptual perspectives: An intelligent systems approach using clustering techniques." *Computers and Education*, 44 (3): 257–283.
Cohen, V. B. 1985. "A reexamination of feedback in computer-based instruction: Implications for instructional design." *Educational Technology*, 25 (1): 33–37.
Corbett, A. T., and Anderson, J. R. 1989. "Feedback timing and student control in the LISP intelligent tutoring system." In D. Bierman, J. Brueker, and J. Sandberg (Eds.), *Proceedings of the Fourth International Conference on Artificial Intelligence and Education*: 64–72. Springfield, VA: IOS.
Corbett, A. T., and Anderson, J. R. 1990. "The effect of feedback control on learning to program with the LISP tutor." In *Proceedings: Twelfth Annual Conference of the Cognitive Science Society*: 796–803. Hillsdale, NJ: Lawrence Erlbaum.
DeLuca, C., Schneider, C., Coombs, A., Pozas, M., and Rasooli, A. 2019. "A cross-cultural comparison of German and Canadian student teachers' assessment competence." *Assessment in Education: Principles, Policy and Practice*, 27 (4): 1–20. doi:10.1080/0969594X.2019.1703171.
Dempsey, J., Driscoll, M., and Swindell, L. 1993. "Text-based feedback." In J. Dempsey and G. Sales (Eds.), *Interactive instruction and feedback*: 21–54. Englewood Cliffs: Educational Technology Publications.
Devi, A. V., and Revathi, K. 2016. "A Brief Review of Literature on Immediate Feedback Studies in CALL." *Rupkatha Journal on Interdisciplinary Studies in Humanities*, 8 (3): 262–268.

Epstein, M. L., Lazarus, A. D., Calvano, T. B., Matthews, K. A., Hendel, R. A., Epstein, B. B., et al. 2002. "Immediate feedback assessment technique promotes learning and corrects inaccurate first responses." *The Psychological Record*, 52: 187–201.
Fedor, D. B. 1991. "Recipient responses to performance feedback: A proposed model and its implications." *Research in Personnel and Human Resources Management*, 9: 73–120.
Frans, N., Post, W. J., Oenema-Mostert, C. E., and Minnaert, A. E. M. G. 2019. "Preschool/Kindergarten teachers' conceptions of standardised testing." *Assessment in Education: Principles, Policy and Practice*, 27 (1): 87–108. doi:10.1080/0969594X.2019.1688763.
Gilman, D. A. 1969. "Comparison of several feedback methods for correcting errors by computer-assisted instruction." *Journal of Educational Psychology*, 60 (6): 503–508.
Gnepp, J. 2020. "The future of feedback: Motivating performance improvement through future-focused feedback." *PLoS One*, 15 (6): e0234444. doi:10.1371/journal.pone.0234444.
Grant, L., McAvoy, R., and Keenan, J. B. 1982. "Prompting and feedback variables in concept programming." *Teaching of Psychology*, 9 (3): 173–177.
Hannafin, M. J. 1983. "The effects of systemized feedback on learning in natural classroom setting." *Educational Research Quarterly*, 7: 22–29.
Herrington, C. D., and Maynard, R. 2019. "Editors' introduction: Randomized controlled trials meet the real world: The nature and consequences of null findings." *Educational Researcher*, 48 (9): 577–579.
Hodes, C. L. 1985. "Relative effectiveness of corrective and non-corrective feedback in computer assisted instruction on learning and achievement." *Journal of Educational Technology Systems*, 13 (4): 249–254.
Hopfenbeck, T. N. 2020. "Making feedback effective?" *Assessment in Education: Principles, Policy and Practice*, 27 (1): 1–5. doi: 10.1080/0969594X.2020.1728908.
Ilgen, D. R., Fisher, C. D., and Taylor, M. S. 1979. "Consequences of individual feedback on behavior in organizations." *Journal of Applied Psychology*, 64: 349–371.
Kluger, A. N., and DeNisi, A. 1996. "The effects of feedback interventions on performance: A historical review, a meta-analysis, and a preliminary feedback intervention theory." *Psychological Bulletin*, 119 (2): 254–284.
Kulhavy, R. W. 1977. "Feedback in written instruction." *Review of Educational Research*, 47 (1): 211–232.
Kulhavy, R. W., and Anderson, R. C. 1972. "Delay-retention effect with multiple-choice tests." *Journal of Educational Psychology*, 63 (5): 505–512.
Kulhavy, R. W., and Stock, W. 1989. Feedback in written instruction: The place of response certitude. *Educational Psychology Review*, 1 (4): 279–308.
Kulhavy, R. W., and Wager, W. 1993. "Feedback in programmed instruction: Historical context and implications for practice." In J.V. Dempsey and G.C. Sales (Eds.), *Interactive instruction and feedback*: 3–20. Englewood Cliffs, NJ: Educational Technology Publications.
Kulhavy, R. W., White, M. T., Topp, B. W., Chan, A. L., and Adams, J. 1985. "Feedback complexity and corrective efficiency." *Contemporary Educational Psychology*, 10 (3): 285–291.
Kulik, J. A., and Kulik, C. C. 1988. "Timing of feedback and verbal learning." *Review of Educational Research*, 58 (1): 79–97.
Lepper, M. R., and Chabay, R. W. 1985. "Intrinsic motivation and instruction: Conflicting views on the role of motivational processes in computer-based education." *Educational Psychologist*, 20 (4): 217–230.
Liu, E. Z., Lin, S. S., Chiu, C. H., and Yuan, S. M. 2001. "Web-based peer review: The learner as both adapter and reviewer". *IEEE Transactions on Education*, 44(3): 246–251.
Locke, E. A., and Latham, G. P. 1990. *A theory of goal setting and task performance.* Englewood Cliffs, NJ: Prentice Hall.

Mehta, F., Brown, J., and Shaw, N.J. 2013. "Do trainees value feedback in case-based discussion assessments?" *Medical Teacher*, 35(5): 1166–1172. doi: 10.3109/0142159X.2012.731100.

Metcalfe, J., Kornell, N., and Finn, B. 2009. "Delayed versus immediate feedback in children's and adults' vocabulary learning." *Memory & Cognition*, 37 (8): 1077–1087. doi:10.3758/MC.37.8.1077.

Merrill, J. 1987. "Levels of questioning and forms of feedback: Instructional factors in courseware design". *Journal of Computer-Based Instruction*, 14(1): 18–22.

Moreno, R. 2004. "Decreasing cognitive load for novice learners: Effects of explanatory versus corrective feedback in discovery-based multimedia." *Instructional Science*, 32: 99–113.

Mory, E. H. 2004. "Feedback research review." In D. Jonassen (Ed.), *Handbook of research on educational communications and technology*: 745–783. Mahwah, NJ: Erlbaum Associates.

Narciss, S., and Huth, K. 2004. "How to design informative tutoring feedback for multimedia learning." In H. M. Niegemann, D. Leutner, and R. Brunken (Eds.), *Instructional design for multimedia learning*: 181–195. Munster, New York: Waxmann.

Paas, F., Renkl, A., and Sweller, J. 2003. "Cognitive load theory and instructional design: Recent developments." *Educational Psychologist*, 38: 1–4.

Peeck, J., and Tillema, H. H. 1979. "Delay of feedback and retention of correct and incorrect responses." *Journal of Experimental Education*, 47: 171–178.

Peeck, J., van den Bosch, A. B., and Kreupeling, W. J. 1985. "Effects of informative feedback in relation to retention of initial responses." *Contemporary Educational Psychology*, 10: 303–313.

Phye, G. D., Gugliamella, J., and Sola, J. 1976. "Effects of delayed retention on multiple-choice test performance." *Contemporary Educational Psychology*, 1: 26–36.

Prastiyani, N. H. N., Felaza, E., and Findyartini, A. 2020. "Exploration of constructive feedback practices in chairside teaching: A case study." *European Journal Dental Education*, 24 (3): 580–589. doi. https://doi.org/10.1111/eje.12539.

Pridemore, D. R., and Klein, J. D. 1995. "Control of practice and level of feedback in computer based instruction." *Contemporary Educational Psychology*, 20: 444–450.

Richards, D. R. 1989. "A comparison of three computer-generated feedback strategies." In *Proceedings of Selected Research Papers, Association for Educational Communications and Technology, Research and Theory Division*, 11: 357–368.

Roper, W. J. 1977. "Feedback in computer assisted instruction." *Programmed Learning and Educational Technology*, 14 (1): 43–49.

Schwartz, F., and White, K. 2000. "Making sense of it all: Giving and getting online course feedback." In K. W. White and B. H. Weight (Eds.), *The online teaching guide: A handbook of attitudes, strategies, and techniques for the virtual classroom*: 57–72. Boston: Allyn and Bacon.

Sleeman, D. H., Kelly, A. E., Martinak, R., Ward, R. D., and Moore, J. L. 1989. "Studies of diagnosis and remediation with high school algebra learners." *Cognitive Science*, 13: 551–568.

Song, S. H., and Keller, J. M. 2001. "Effectiveness of motivationally adaptive computer-assisted instruction on the dynamic aspects of motivation." *Educational Technology Research and Development*, 49 (2): 5–22.

Sturges, P. T. 1969. "Verbal retention as a function of the informativeness and delay of information feedback." *Journal of Educational Psychology*, 60: 11–14.

Sturges, P. T. 1972. "Information delay and retention: Effect of information in feedback and tests." *Journal of Educational Psychology*, 63: 32–43.

Surber, J. R., and Anderson, R. C. 1975. "Delay-retention effect in natural classroom settings." *Journal of Educational Psychology*, 67 (2): 170–173.

Sweller, J., Van Merriënboer, J., and Paas, F. 1998. "Cognitive architecture and instructional design." *Educational Psychology Review*, 10: 251–296.

Trimmer, J. T., and Guest, J. S. 2020. "A framework of assumptions for constructive review feedback." *Environmental Science and Technology*, 54 (19): 11648–11650. doi: 10.1021/acs.est.0c04119.

Tsai, C. C., and Liang, J. C. 2009. "The development of science activities via on-line peer assessment: The role of scientific epistemological views." *Instructional Science*, 37 (3): 293–310.

Tseng, S. C., and Tsai, C. C. 2007. "On-line peer assessment and the role of the peer feedback: A study of high school computer course." *Computers and Education*, 49 (4): 1161–1174.

Wager, W., and Wager, S. 1985. "Presenting questions, processing responses, and providing feedback in CAI." *Journal of Instructional Development*, 8 (4): 2–8.

Webster's II new college dictionary. 2001. Boston: Houghton Mifflin.

Whitney, T., and Ackerman, K. B. 2020. "Acknowledging student behavior: A review of methods promoting positive and constructive feedback." *Beyond Behavior*, 29 (2): 86–94. doi: https://doi.org/10.1177/1074295620902474.

Yang, Y.-F. 2011. "A reciprocal peer review system to support college students' writing." *British Journal of Educational Technology*, 42(4):687–700. doi:10.1111/j.1467-8535.2010.01059.x.

Yang, Y. F., and Tsai, C. C. 2010. "Conceptions of and approaches to learning through online peer assessment." *Learning and Instruction*, 20 (1): 72–83.

13 Storyboarding

A Pedagogical Tool for Digital Learning

Prajna Pani

CONTENTS

13.1 INTRODUCTION

This chapter introduces storyboarding as an alternative pedagogical tool. The first phase of the chapter outlines storyboarding concepts in teaching and learning, implications, phases of storytelling and digital storytelling. The second phase is an introduction to the storyboard tools, such as Storyboader, Boords, Canva, StudioBinder. These are simple guidelines to create engaging stories and teach lessons relevant to the current times. The concluding phase of the chapter supports the instructional designers and content developers with storyboard tools to make an engaging and

DOI: 10.1201/9781003132097-13

visual presentation of the learning scenarios, learning objectives, learning outcomes, methodology and flow of an eLearning course. This chapter illustrates a course "Contemporary Development Communication" to elaborate the pedagogical applications of storyboard tools in eLearning. The capabilities of storyboard tools that allow communication, collaboration, creativity, critical thinking, reflection and engaging stories play a crucial role in 21st-century teaching and learning. This chapter highlights how storyboarding can be used in eLearning from instructional designer's perspectives and makes recommendations for future pedagogical practices to create engaging and visually enriching experiences. It lays out the framework for digital learning. Steve Penfold (2015) believes storyboarding can enhance the efficiency of an eLearning design and development process.

13.2 STORYBOARD CONCEPTS AND DIGITAL LEARNING

The storyboard is a useful concept in teaching and learning, with the motto to improve the design and avoid anything that might be difficult to achieve or understand. A storyboard is a precise visual or graphical presentation of a narrative. The storyboard concept is based upon standard ideas that appreciate (1) clarity by giving a high-level modeling approach, (2) simplicity which empowers everyone to turn into a storyboard creator, (3) visual appearance as diagrams, (4) the opportunity to be executed by basic standard tools. Storyboard inspires students to think innovatively and critically about the story that they will compose. It is on the grounds that in making a storyboard, the students are expected to envision their story in a series (Jantke and Knauf 2005). Lottier (1986), as cited by (Hasan and Wijaya 2016), explains that storyboarding is a process to pull out a learner's creativity by using the creative right brain and the critical left brain to create thoughts and ideas and afterward to look at those ideas critically. Students are more active in the technology-enabled class than the teacher, furthermore, it changes the role of the facilitator and learner, the classroom situation becomes learner-centered. Learners can actively become involved in their learning process through digital storyboard tools, the learners' creativity can be fostered, and collaborative learning and teaching practices can be enhanced significantly. Student engagement in learning activities occurs in different forms and levels of learning. However, there is a significant point that students do not always recognize this engagement, completely welcome it or discover it to be adequately well instructed to think of as a meaningful or pleasurable learning experience (Zamorski 2002).

It is the educational rationale and not the technical tool that is of essential concern behind the integrative moves (Salomon 2000). Probably the best strategies to practice powerful learning, with content that learners restore, is achieved by using stories and narratives. Storytelling and narratives have established themselves as the most powerful practices which people possess. Since there is a lesson behind every story with some morale or guidance, it implies that it is an essential tool (Kolagani 2019).

Effective learning and student engagement depend on the activity type in which students take charge of learning. A storyboard is a useful tool for content developers and instructional designers. The storyboard should have contributions from all stakeholders in the content. A storyboard goes through several revisions, discussed with people with different perspectives and different skills. This helps in addressing any gaps in the course design and avoidance of errors. It provides satisfactory answers to questions that the designers and developers have in the content creation process. There is nothing right or wrong in storyboarding. But, what is all-important is the designer should focus on target achievement. A storyboard might include the overall structure and instructional strategy, and underlying pedagogical ideas, audience analysis, ensure learning objectives the courseware/material addresses, define any pre-requisite material. It provides a clear statement of how the objectives will be met i.e., what the learner will do with the material the facilitator produces and how it will take place. It also helps to measure learning, a navigation system, the overall layout of the pages, outlines of the different page types used, description of the overall graphical style and details of graphics/sounds/video required for each page, details of other interactivity.

In designing courses, storyboards often illustrate the learning journey. Storyboards help teachers, instructional designers, content creators and learners to collaborate and translate their ideas into relevant visual experiences. This chapter is a visual presentation of a course titled contemporary development communication offered to management students. The instructional designer or the course facilitator lays out the framework for the course that needs to be created through a storyboard. Storyboards are visual representations of illustrations or sequential presentations of images shown to pre-visualize images, concepts, animation, graphic or interactive media. An increased sharing of power between the instructor and the student is considered to be an emerging trend. This is manifest as a changing instructional role, toward more support and negotiation over content and methods, and a focus on developing and supporting student autonomy. On the student side, this can mean "an emphasis on students supporting each other through new social media, peer assessment, discussion groups, even online study groups but with guidance, support, and feedback from learning and content experts" (teachonline.ca 2020).

An eLearning storyboard is an arrangement of boards which the instructional designer spreads out the eLearning module framework. Instructional design principles guide the process, solidifying and organizing course content to engage the learner in the material (think dialogue and interactivity). Beyond written content, the instructional designer also considers how to encourage the learners to think metaphorically. What images, graphs, icons, charts, screenshots or animations will best support the points being made? (Narum 2017). Insights and experiences come from seeing (Arnheim 2004), and here the visual presentation of the story becomes a tool to visually outline the thoughts and ideas of the learners, which prompts new insights. Digital moviemaking offers a chance to harmonize the use of technology to support learner-centered pedagogy and inter-disciplinary approaches rooted in discipline-specific instructional methods (Hofer and Swan 2005). This process can help create critical thinking through questioning, probing and sequential thinking (Fornis and Peden-McAlpine 2007), thus providing context-specific learning. Storyboards are a useful tool for effectively presenting collaborative work to a variety of audiences, especially in an e-classroom setting. Collaborative critical reflection on practice is a significant process, as it provides space for the thinking necessary to enable deeper reflection (Regan 2008). The storyboarding approach aims to create critical and sequential thinking, linking theory to practice in the lived experiences of learners themselves, as opposed to a more didactic input from the teaching team (Lillyman et al. 2011). Online and hybrid courses provide opportunities for faculty to use technology and digital applications, such as storyboard tools, to enhance student learning and engagement. Despite the fact that storyboarding is relatively new, the methodology can improve learner's comprehension of ideas and applications by developing a story in a progression of visual scenes with or without critique (Dexter 2016). In reflecting critically on the use of storyboarding, it is highlighted how illustrative structures become points of departure for creative thinking and creating the possibility of multiple stories and perspectives in the production of creative knowledge (Pahl 2017).

13.3 IMPLICATIONS

The innovation has opened up numerous pathways for instructional designers. With regards to eLearning, the content is the king. It is significant to invest a great deal of energy thinking on alternative approaches to get the message across. For this reason, storyboards are wonderful and can elevate the eLearning experience for both the students and the instructional designers. A storyboard gets everybody engaged with the task on the same page by showing precisely what content will be created and how. A digital storyboard stimulates the course creator, also called the instructional designer, to brainstorm the course's structure and flow. Why are you designing it this manner? Is there any other better method to present the information? What value will the course bring to the lives of students? What more can be done to improve the thinking abilities of students? Will it meet the expectations of the learners? Will it lead to a good understanding, innovation or problem-solving? Will the course raise the curiosity of the learners? Along these lines, a storyboard is a collaborative assignment. A digital storyboard welcomes coordinated effort on the course design and content.

TABLE 13.1
Top Four Skills of 21st Century Derived from the Literature Review (Laar et al 2020)

Critical Thinking	Creativity	Communication	Collaboration
• Researching, and brainstorming to solve problems. • Learners use research tools to investigate and share findings.	• Demonstrate learning through patent, production, and innovation. • Learners are motivated to use technology/digital tools to create and innovate.	• Use digital content to express ideas and communicate with others locally and at a distance. • Learners should communicate through digital tools to build interpersonal skills.	• Digital platforms empower personal learning and collaborative learning. • Learners are encouraged to use technology to work collaboratively on projects, products and for conferencing.

Critical thinking, creativity, communication and collaboration are fundamental life skills (Maneen 2016). In a true self-learning and learn-by-doing environment, course participants will gain knowledge about 21st-century skills and competencies, and learn to innovate, create, think out-of-box and communicate with others. This chapter aims to provide learners with the knowledge, technical know-how and the confidence to incorporate these life skills into the curriculum and University environment through the storyboarding tools. Thus, storyboarding is an effective technique to engage students in critical thinking, creativity, constant communication and collaboration.

13.4 PHASES OF STORYTELLING

A storyboard is an integral part of an interactive eLearning course. A storyboard will help the facilitator to organize both the course content and thoughts on student engagement and experience. It includes additional development/design notes. Each module of the course is a scene. Organizing ideas and course planning are the two essential phases of storytelling.

13.5 DIGITAL STORYTELLING

The compelling digital storytelling process should include the following features.

13.5.1 Logical and Coherent Presentation

Storyboarding requires a clear, logical and coherent progression of ideas or concepts.

13.5.2 Interesting and Captivating

The beginning should be catchy to set the tone of the story and captivate the attention of the audience. It can begin with a thought-provoking question, an exciting scenario, a meaningful image, an engaging game or a used case to grab the attention of the audience. The story should be moved by the show, not tell mechanism.

13.5.3 Concept

When developing a storyboard, keep the visuals simple. The focus should be on the story, not the template or tool. The images in each frame should describe the action and explain the concepts. The developer should use easy-to-understand sketches and images from own resources. Images can also be uploaded from the web.

13.6 STORYBOARD TOOLS

Storyboard templates are available in various formats, such as printable PDF, Photoshop, PowerPoint, Microsoft Word, Animation, Video.

13.6.1 Storyboarder

Storyboarder is one of the powerful open-source storyboarding tools with features like drawing tools, boards, frames to add metadata for the video. It can be opened in Photoshop to edit and change the layout.

13.6.2 Boords

Boords is an easy-to-use video storytelling tool. It is a storyboard creator for team collaboration. Some of its features include drag and drop images, animation, a built-in editor to create ready-to-share storyboards, a feedback system and team or client interaction in real-time. The storyboard can be exported with multiple options.

13.6.3 Canva

Digital stories are designed with Canva, a free storyboard creator. Storyboard creator has a library of images, illustrations, graphics to choose from, grids, frames and links to share and collaborate.

13.6.4 StudioBinder

StudioBinder is a production management platform. It offers project management solutions. Content can be produced collaboratively. It has call sheets, workflow, script breakdowns, shot lists and storyboards, shooting schedules, project contacts, tasks, calendars and files, media library.

13.6.5 Plot

Plot App can be used to see the script in the storyboard format. It is considered to be a game-changer for storyboarding. The instructional designer can choose the frames that include various sections like comment, script and action. Simple illustrations with included brushes, shapes and clipart can be used while designing the storyline. The course designer can use pictures or copy and paste from other websites. Course facilitators can invite collaborators in the design process and export.

13.7 COURSE DESCRIPTION USING STORYBOARD TOOL

13.7.1 Learning Scenarios

A communication course from a development management program is described using a storyboard tool. Learners engage with community people, identify problems and conduct focus group discussions to find solutions to improve the quality of life. They listen to the voices of community people, create blogsite and share their experiences and learnings. Students publish presentations, video documentaries and other materials on their blogsites. The course facilitators conduct review presentations. Students are stimulated to think and reflect. Each student makes a presentation on stakeholder analysis, problem identification, situation analysis and proposed solution to the identified community problems.

The course "contemporary development communication" begins with the stakeholder analysis, problem identification, situation analysis and solution to identified problems. Thus, the course begins with the needs analysis. And here you are asking the most relevant questions like, what issues or problems the learners are trying to address through this course? What are the best ways? Who do they approach to solve the issues? How do they approach to solve the issues? And, what do we want the learner to know or be able to do after the course is done? This last question helps to define the learning objectives.

13.7.2 Learning Objectives

This course aims to enable the students to

- understand the processes and approaches to contemporary development communication;
- learn situation analysis, problem tree analysis and participatory communication appraisal in the field;
- understand strategies for awareness-raising and communication campaigns relevant to current times (e.g., Education in schools of rural areas).

13.7.3 Learning Outcomes

After completion of the course, students will be able to

- demonstrate an understanding of the approaches to contemporary development communication (e.g., Education in schools of rural areas);
- perform situation analysis and problem-tree analysis to address the development issues
- design and implement development communication strategies combining participatory methods with communication processes, social media and digital tools best suited for a specific situation (e.g., Education in schools of rural areas).

13.7.4 Methodology

The methodology of the contemporary development communication course is articulated in three phases (1) doing and experiencing, (2) planning and applying and (3) reflecting and discussing. These three components of experiential learning are illustrated using the Canva education template.

How to design a course storyboard?

- Assess the learner needs
- Decide on design approach or instructional approach

- Design content
- Choose a storyboard template
- Present content in a sequential manner
- Choose an authoring tool

13.8 DIGITAL STORYBOARD TEMPLATE

Name: Contemporary Development Communication **Page: XX**

Storyboard Scene 1: Image Description/Drawing	*Storyboard Scene 2: Image Description/Drawing*
Image Credit: Class on Storyboard Tools https://www.udemy.com/course/contemporary-development-communication	*Image Credit: Class on Problem Tree Analysis* https://www.udemy.com/course/contemporary-development-communication
Spoken Text: Students will use storyboard tools (e.g., Canva, Broods), concepts and frameworks for giving a voice to their story/sharing the unheard voice during (or post) COVID-19 from the field. Students will also convert their storyboards into visuals and upload them to their Blogsites.	*Spoken Text:* Students will brainstorm the entire day to develop a problem tree on the challenges of Education in rural areas representing cause-effect relationships. Problem Tree Analysis will include: identify root causes of the focal problem, effects of the problem, design a problem tree showing the cause and effect relationships between the problems.
Written Text: To start seeing and exploring various methods of digital storytelling – and applying it to work – students will be responsible for bringing to the class's attention an example of a good/bad, effective/ineffective or ethical/ unethical digitally told communication.	*Written Text:* Students will present the problem tree and communication plan on online education during COVID-19 in the review meeting. Students will also scan and upload the problem tree image and communication plan to the blogsite.
Online Resources https://www.canva.com/ https://theplot.io/ https://boords.com/best-storyboard-software https://www.youtube.com/watch?v=miJ_k_xeUyA https://www.youtube.com/watch?v=Kfpe5RewJwM	*Online Resources* https://www.youtube.com/watch?v=-j-_Y7D35H4 https://sswm.info/taxonomy/term/2647/problem-tree-analysis https://www.youtube.com/watch?v=-j-_Y7D35H4 https://sswm.info/taxonomy/term/2647/problem-tree-analysis

Storyboard Scene 3: Image Description/Drawing
Image Credit: Students of School of Management and School of Agriculture
Spoken Text: Students run and manage awareness campaigns on literacy in the adopted villages and social media platforms throughout the course.
Written Text: Students distribute messages in the form of presentations, sponsored events and/or promotional items at regular intervals to reinforce the campaign message.
Online Resource https://www.culturepartnership.eu/en/publishing/digital-communication

This chapter uses some selected features of the Canva education template. Canva education template includes the lesson plan, worksheet, certificate, storyboard, bookmark, class schedule. The learning process is described into scenes for storyboard, for example, stakeholder analysis, surveys, focus group discussion and problem identification. Students use the story tree template from Canva to identify the problems and the root causes of the problems. This tree narrates the development of the story.

Participation in community engagement programs immerses students in learning through reflection on doing. Reflection takes place during the experience and after the experience to derive meaning from it and to identify what has been learned. It is through this reflective process that learners acquire new skills, develop new attitudes and new ways of thinking. Learners discover about themselves, other people, the world, the challenges and opportunities during the reflective learning journey.

13.9 CONCLUSION

Storyboarding is a powerful pedagogical tool for eLearning course design. The chapter explains digital storyboarding, storyboard tools, such as Storyboarder, Boords, Canva, StoryBinder and Plot. The design is inclusive of the topmost skills of the 21st century, i.e., critical thinking, creativity, communication and collaboration. A communication course "contemporary development communication" is designed with storyboarding tool with learning objectives and outcomes. Storyboarding can be used in higher education to stimulate thinking and create engaging stories. Instructional designers can use storyboard tools to convey complex ideas to the students. Teachers can use the storyboarding concepts to validate the course before content delivery. In case of budgetary constraints, free storyboard tools can be used by instructional designers and students. This chapter addresses instructional designers and subject matter experts engaged with content development as well as all those who need to find out about storyboarding as an alternative pedagogy to develop an eLearning course.

REFERENCES

Arhneim, R. 2004. "*Visual Thinking*", London, University of California Press.

Dexter, Y. 2016. "Storyboarding as an aid to learning about death in children's nursing". *Nursing Children & Young People*, 28, 16–21. doi:10.7748/ncyp.28.5.16.s21

Fornis, S., Peden-McAlpine, C. 2007. "Evaluation of a reflective learning intervention to improve critical thinking in novice nurses". *Journal of Advanced Nursing*, 57, no. 4, 410–421.

Hasan, D. N., and Wijaya, M. S. 2016. "Storyboard in teaching writing narrative text". *English Education: Jurnal Tadris Bahasa Inggris*, 9, no. 2, 262–275.

Hofer, M., and Swan, K. O. 2005. "Digital moviemaking – the harmonization of technology, pedagogy and content". *International Journal of Technology in Teaching and Learning*, 1, no. 2, 102–110.

Kolagani, Sushmitha. 2019. "4 Smart Compelling Story Narratives to Make Your Learning". Learning Evangelist – Learning in the 21st Century/Lernen im 21. Jahrhundert.

Laar et al. 2020. Determinants of 21st-Century Skills and 21st-Century Digital Skills for Workers: *A Systematic Literature Review.* https://doi.org/10.1177/2158244019900176

Lillyman, S., Gutteridge, R., and Berridge, R. 2011. "Using a storyboarding technique in the classroom to address end of life experiences in practice and engage student nurses in deeper reflection". *Nurse Education in Practice*, 11, no. 3.

Lottier, L. F. 1986. "Storyboarding your way to successful training". *Public Personal Management*, 15, no. 4, 421–427.

Maneen, C. 2016. "A Case Study of Arts Integration Practices in Developing the 21st Century Skills of Critical". https://digitalcommons.gardner-webb.edu/

Narum, Claire. 2017. "Elevate Your eLearning: How to Storyboard Like a Boss". https://www.dashe.com/blog/how-to-storyboard-what-is-a-storyboard

Jantke P., and Knauf, R. 2005. "Didactic design through storyboarding: Standard concepts for standard tools". In Baltes, B. R., Edwards, L., Galindo, F., Hvorecky, J., Jantke, K. P., Jololian, L., Leith, P., van der Merwe, A., Morison, J., Nejdl, W., Ramamoorthy, C. V., Seker, R., Shaffer, B., Skliarova, I., Sklyarov, V., and Waldron, J., editors. "First International Workshop on Dissemination of E-Learning Technologies and Applications". DELTA2005: Proceedings of the 4th International Symposium on Information and Communication Technologies, 20–25.

Pahl, K. 2017. "Dialogic objects: Material knowledge as a challenge to educational practice". In D. Pillay, K. Pithouse-Morgan & I. Naicker (Eds.), Object medleys: Interpretive possibilities for educational research (pp. 29–44). Sense Publishers.

Penfold, Steve. 2015. "Why storyboards for eLearning are important (4 reasons)". https://www.elucidat.com/blog/why-storyboards-for-elearning-are-important

Pillay, K. Pithouse-Morgan, & I. Naicker (Eds.), *Object Medleys: Interpretive Possibilities for Educational Research* (pp. 29–44). Sense Publishers.

Regan, P. 2008. "Reflective practice: How far, how deep?" *Reflective Practice*, 9, no. 2, 219–229.

Salomon, G. 2000. "It's not just the tool, but the educational rationale that counts (keynote address)". Proceedings of ED-MEDIA Montreal. http://www.aace.org/conf/edmedia/00/salomonkeynote.htm

Zamorski, B. 2002. "Research-led teaching and learning in higher education: A case". *Teaching in Higher Education*, 7, no. 4, 411–427.

Section III

Enhancing Teaching Quality in Digital Age

14 Online Social Capital and Its Role in Students' Career Development

Najmul Hoda

CONTENTS

14.1 INTRODUCTION

The developments in computer-mediated communication technology have altered the way individuals communicate with each other. They also resulted in the proliferation of online networking sites (ONSs) that revolutionized the networking behaviors of individuals (Bano et al. 2019). ONSs offer innovative services such as likes, shares, updating profiles, etc. (Dhir and Tsai 2017). Research shows that the young use these sites more than the old, and these ONSs have affected the lives of users in multiple

DOI: 10.1201/9781003132097-14

ways. One of the most noticeable impacts of these ONS is that they accrue social capital similar to offline social networks (Ellison, Steinfield, and Lampe 2007). The advantages of social capital accrue both in terms of physical resources as well as emotional support (Putnam 2001; Coleman 2009; Portes 1998). The social capital accrued through ONSs is termed online social capital (Williams 2006), and it aids students in various stages of their career development, starting from their entry into the higher education system (Sikolia and Mberia 2019). Extant literature has dealt with the benefits of ONS usage on students ranging from academic performance to psychosocial outcomes (Rios, Wohn, and Lee 2019; Wohn et al. 2013). This chapter aims to review and synthesize extant research on online social capital focusing on students' career development. A theoretical model derived from the identified constructs in various studies is also included in the following sections.

14.1.1 Social Capital Theory

Häuberer (2011) posits that the concept of social capital reportedly finds its origin in the pioneering works of Bourdieu (1986) and Coleman (1988). Putnam (2001) extended the theoretical framework provided by Coleman. The definitions of social capital present in various studies (Spottswood and Wohn 2020; Abbas and Mesch 2018; Coleman 1988; Ellison, Steinfield, and Lampe 2007) all point to the following:

i. *it is a potential pool of tangible and intangible resources*
ii. *it results from social networking and social relationships*
iii. *it results in resources that may be capitalized.*

Social capital has been categorized into different types. Putnam (1995) categorized social capital into: "bonding social capital and bridging social capital." Some researchers (like Ellison, Steinfield, and Lampe 2007) added new categories, namely *maintained social capital*, to Putnam's classification. Nahapiet and Goshal (1998) classified it differently as "structural, cognitive and relational social capital." The nature of social capital differs from other forms of capital (Chen et al. 2020) and may be developed by the investment in network relationships (Lounsbury et al. 2002). Social capital is characterized by "interactive relationships and reciprocal values" (Ellison, Steinfield, and Lampe 2007) and "weak or strong relationships" (Na 2015). These networks are a source of both tangible and intangible benefits to the users.

14.1.2 Social Media and Online Networking Sites

Social media and technology-based applications have dominated the communications landscape in the recent past. Pan et al. (2019) quoted Kaplan and Hanlein (2010), who defined social media as "internet-based applications that allow the creation and exchange of content which is user-generated." Social media includes ONSs (e.g. Facebook, LinkedIn) as well as other forms of online communication services such as blogs (e.g. Twitter), instant messaging (e.g. WhatsApp), online gaming (e.g. World of Warcraft), video streaming (e.g. YouTube) and discussion forums (e.g. Reddit) (Ryan et al. 2017; Johnston et al. 2013; Dhir and Tsai 2017; Ruparel et al. 2020).

ONSs are the most effective medium for networking activities. Boyd and Ellison defined ONSs as, "web-based services that enable users to 1) create a public or semi-public profile within a bounded system, 2) specify a contact log of direct or indirect friends, and 3) see and check the list of contacts is created by other friends within the system" (Boyd and Ellison 2007, p. 211). Among the various social media, ONSs like Facebook have seen a sharp rise in their usage (Asterhan and Rosenberg 2015). This is one of the reasons why Facebook has received greater attention in the research on social media (Pang 2018). ONSs are used by people from across the demographic and social classes (Mahmood, Zakar, and Zakar 2018). One of the highlighted features of ONS is that they allow visibility of social ties (Naseri 2017).

Research findings support the theory that online networking facilitates building and maintaining social relationships, eventually leading to the accrual of social capital (Boyd and Ellison 2007; Greenhow and Askari 2017; Liu et al. 2013; Badoer, Hollings, and Chester 2020). The next section discusses "online social capital," which is the social capital accrued through online networking.

14.2 ONLINE SOCIAL CAPITAL

The online social capital is embedded in online social networks, similar to the offline networks (Boyd and Ellison 2007; Ellison, Steinfield, and Lampe 2011; Ellison et al. 2014; Williams 2006). Ahn (2012) noted that the focus of earlier studies was mainly on offline social capital. Among the two main categories of social capital, it was initially hypothesized and tested that online networking leads to the creation of "bridging social capital" only (Abbas and Mesch 2018). However, this notion too got revised soon and studies confirmed that online networking forms both "bridging and bonding social capital" (Ellison, Steinfield, and Lampe 2007; Jang and Dworkin 2014; Lin and Lu 2011; Phua, Jin, and Kim 2017b).

Ellison et al. (2014) explained how online social capital is formed. This was further extended to students' online social capital (Chen and Starobin 2019; You and Hon 2019). They highlighted that the users of these online networks must actively and intentionally engage in activities that attract responses from others; for example, actively using the features of liking, sharing, commenting, posting on others' timelines, etc., available on Facebook. Research shows that both types of social capital formation take place online. Online networking allows connections with unknown people and also develops ties that result in the creation of "bridging social capital" (Best and Krueger 2006). Further, more intense actions like "online self-disclosure" helps in more intense and close relationships with network members, thereby resulting in the formation of strong ties or bonding social capital (Shane-Simpson et al. 2018; Kang, Jiang, and Tan 2017). Other factors such as time spent, interactions and usage frequency also affect the formation of online social capital Kim and Kim (2017).

14.2.1 Factors Affecting Online Social Capital

Based on the literature review, factors that affect the formation of online social capital are presented in Table 14.1. The description of each factor has been included in

TABLE 14.1
Factors Affecting Online Social Capital

Factors	Details	Studies
Intensity of usage	This is a measure of different items included in the scale created by Ellison, Steinfield, and Lampe (2007).	Mahmood, Zakar, and Zakar (2018); Steinfield, Ellison, and Lampe (2008); Ellison, Steinfield, and Lampe (2011)
Usage time	Time spent daily on the SNS.	Bano et al. (2019); Mahmood, Zakar, and Zakar (2018); Li et al. (2019); Ahmad, Mustafa, and Ullah (2016)
Purpose and motivations	The specific reasons for which the users register and log in to their SNS accounts.	Kasperski and Blau (2020); Dorethy, Fiebert, and Warren (2014); Ahmad, Mustafa, and Ullah (2016); Papacharissi (2009); Lee et al. (2016)
Preference for SNS	The preference for a particular SNS among the various sites available.	Shane-Simpson et al. (2018)
Online self-disclosure and information seeking	The particulars about oneself a user wants to share and the information about others look for in the ONS members.	Shane-Simpson et al. (2018); Ellison, Steinfield, and Lampe (2011); Rykov, Koltsova, and Sinyavskaya (2020)
Privacy settings	Control settings in an ONS.	Phua, Jin, and Kim (2017a); Stutzman (2006); Ellison, Steinfield, and Lampe (2011)
Number of connections/friends	This includes mainly the first connections of a user.	Pang (2018); Mahmood, Zakar, and Zakar (2018); Ellison et al. (2014); Steinfield, Ellison, and Lampe (2008); Ellison and Vitak (2015)

the table. These factors are affecting the formation of "online bonding and online bridging social capital." There is a need for assessing the individual and combined impacts of these factors.

Figure 14.1 summarizes the various factors that influence "online social capital." Each of the seven factors may be having varying degrees of influence on online social capital formation. Some of the factors like the number of connections and time spent are quantitative factors, whereas the other factors depict the behavioral dimensions.

14.3 ONLINE SOCIAL CAPITAL IN STUDENTS' CAREERS

The use of online networking brings significant advantages to college students. The benefits of social capital have been reported in several studies (Peng 2019; Warren, Sulaiman, and Jaafar 2015; Wen, Geng, and Ye 2016; Dika and Singh 2002). These benefits may be realized at various stages of students' life cycle (Benson and Morgan

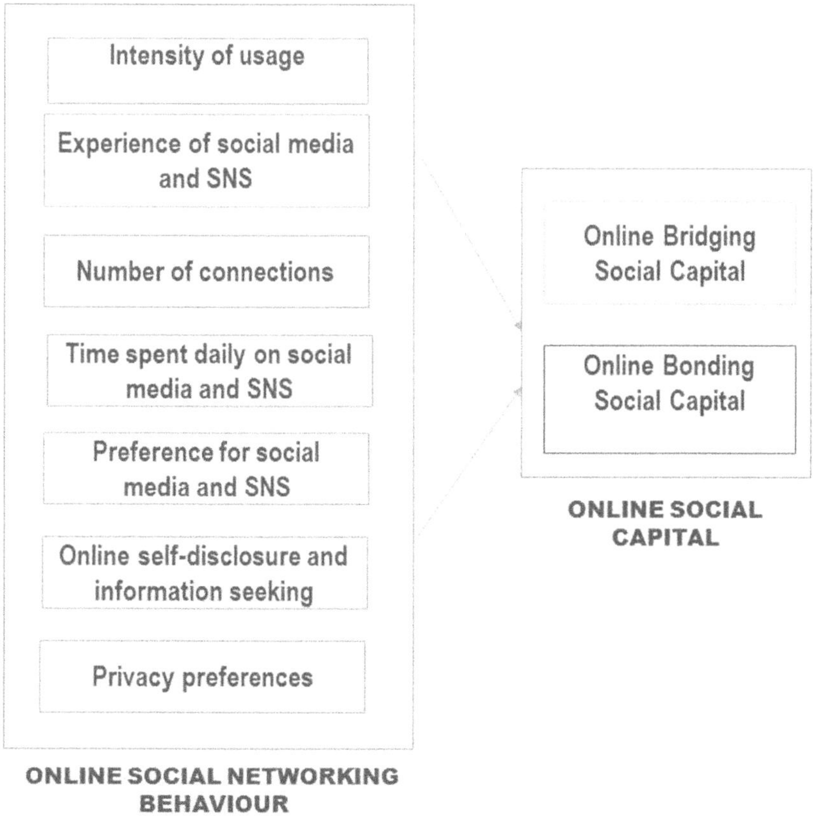

FIGURE 14.1 OSN behavior and online social capital. (Created by author.)

2013; Benson, Filippaios, and Morgan 2010). Greenhow and Burton (2011) cited a review-based study by Dika and Singh (2002) in which the benefits of students social capital have been classified into various categories, namely "educational achievement (i.e., grades and test scores), educational attainment (i.e. high school graduation and college enrollment) and psychosocial factors (i.e. school engagement, self-concept, motivation, homework effort, and perceived importance of school)." This section describes the factors affecting online social capital and the benefits that accrue to college students in the various stages of their career and various forms.

14.3.1 Benefits in Education and Learning

Online networking has been reportedly playing a great role in improving the learning processes by encouraging "innovative and collaborative learning" (Kasperski and Blau 2020). They opine that the application of online networking in learning may be attributed to the Social Constructivist Theory postulated by Vygotsky. Venter (2019) quoted several studies to corroborate that online learning and collaboration

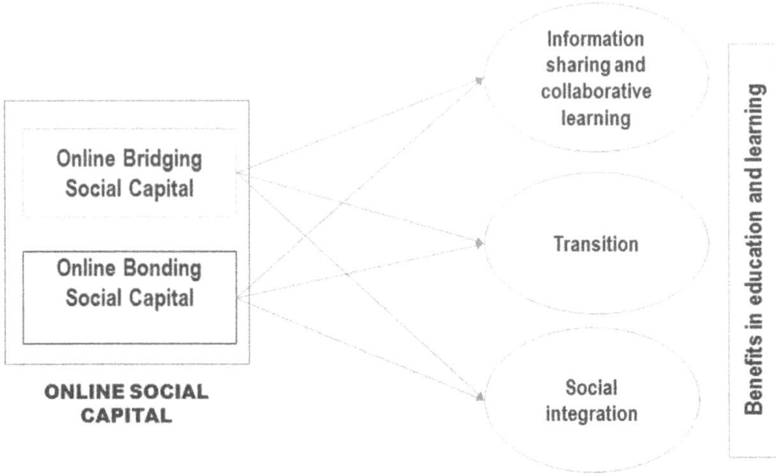

FIGURE 14.2 Online social capital and benefits in education and learning. (Created by author.)

among members result in unique learning benefits unavailable in offline mode. Both the teachers and students/learners are found to prefer SNS for enhanced learning outcomes (Heidari, Salimi, and Mehrvarz 2020). Such learning processes enhance students' learning experiences not just within the institutional premises but beyond the boundaries too. Northey et al. (2018) researched the students' preference for SNS for learning. They found that students prefer Facebook because of the ease of access, the experience of usage, and ease of integration in learning. Interestingly, Kasperski and Blau (2020) and Öztok et al. (2015) note that there is not much impact of cultural variations on the benefits of online networking in learning. Figure 14.2 presents the illustrative model to show the relationship between "online social capital" and its benefits in education and learning.

14.3.1.1 Information Sharing and Collaborative Learning

Social media and online networking allow easy and convenient access to information for the students (Wen, Geng, and Ye 2016; Ridings, Gefen, and Arinze 2002; Chen et al. 2020). Social capital is instrumental in information sharing among students (Venter 2019; Öztok et al. 2015; Valenzuela, Park, and Kee 2009). The information shared may be general classwork, competency-based information as well as information that provides social support (Wen, Geng, and Ye 2016; Forbush and Foucault-Welles 2016). Information sharing through social media is cheap, more trustworthy and rare (Mou and Lin 2017). Students feel more motivated and secured in seeking information through social media (Chang and Zhu 2012; Pan et al. 2019). Studies have pointed to the specific use of networking sites like blogs (Deng and Yuen 2011), Twitter (Carpenter, Tur, and Marín 2016) Facebook (Madge et al. 2009; Ellison, Steinfield, and Lampe 2011), WeChat (Pang 2018) and WhatsApp (Bano et al. 2019) for information sharing, among students.

14.3.1.2 Transition

The transition from school to college and college to work involves challenging phases that have considerable short-term and long-term impacts on a student's career. (Sohn et al. 2019) reported that students' experience of loss of ties during this transition and online social capital reduces the loneliness. Online networking offers great help to the students during this transition phase by providing a medium to preserve old relationships (Mahmood, Zakar, and Zakar 2018; Mikal et al. 2013). Both types of "online social capital" (bonding and bridging) positively influence students' transition.

14.3.1.3 Social Integration

The benefits of online networking also accrue in the form of social integration, which means a sense of belonging (Pan et al. 2019; Ewart and Snowden 2012; Moon and Shin 2019). Social integration affects students' satisfaction as well as retention within an institution (Bano et al. 2019; Ganesh, Haslinda, and Raghavan 2017). Severiens and Schmidt (2009) posited that social integration is important for the successful completion of a program by students. Pan et al. (2019) has cited empirical studies to prove that offline social capital reduces dropouts. Through an online medium, students can share their inner views more freely, thereby increasing their comfort with the institutional or organizational environment (Tian 2016; Chen et al. 2020; Forbush and Foucault-Welles 2016).

14.3.2 Psychosocial Benefits

The psychological and social benefits of online networking and online social capital have been reported in several studies (Spottswood and Wohn 2020; Lee, Chung, and Park 2018a). Psychological well-being has been expressed in different forms like "emotional support" (Vitak 2014); "reduced stress" (Kalaitzaki, Tsouvelas, and Koukouli 2020); "facing difficult situation" (Mou and Lin 2017); "experiencing pleasure" (Huang, Chen, and Kuo 2017); "overcoming the sense of loneliness and enhance life quality" (Forbush and Foucault-Welles 2016); "feelings of security and self-esteem" (Mou and Lin 2017); and "social connectedness" (Tibber, Zhao, and Butler 2020). The possible influences of "online social capital" on various psychosocial benefits are presented in Figure 14.3.

14.3.2.1 Well-Being

Pang (2018) quoted the definition of psychological well-being given by Harrington and Loffredo (2011). Psychological well-being is defined as, "how individuals self-assess and their ability to gratify fundamental needs of their lives, such as autonomy, positive relationships with others, as well as purpose in life." Based on empirical evidence, Lee, Chung, and Park (2018b) and Bano et al. (2019) proved that online networking influences students' well-being in positive ways. Madge et al. (2009) related online networking with well-being, specifying its role in supporting an individual in difficult situations and providing a pleasant experience. The unique networking tools present in online networking enhance users' well-being (Wang et al. 2014).

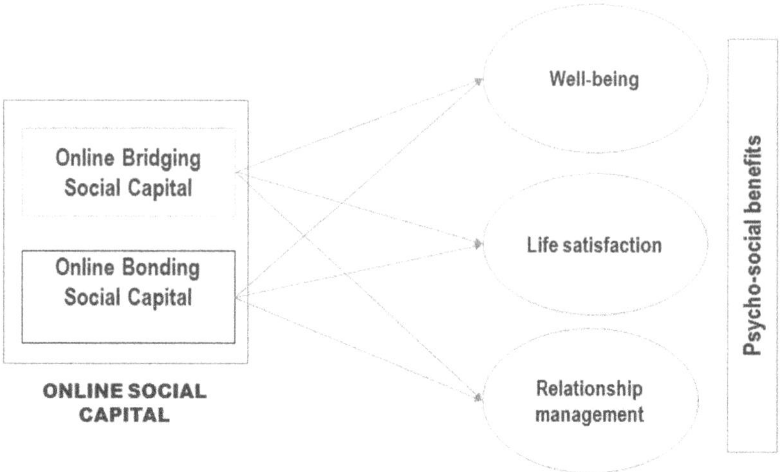

FIGURE 14.3 Online social capital and psychosocial benefits. (Created by author.)

14.3.2.2 Life Satisfaction

Researchers considered life satisfaction to be an important psychological construct (Ellison, Steinfield, and Lampe 2007; Williams 2006; Johnston et al. 2013). Pang (2018) summarized the definition of life satisfaction as ones' assessment of life and circumstances. Social capital and life satisfaction have been found to be positively associated in studies (Wen, Geng, and Ye 2016; Petersen and Johnston 2015).

14.3.2.3 Relationship Management

Online networking helps the students form and maintain relationships (Ellison and Boyd 2013). The process of maintaining relationships may include just routine social networking activities like status updates, liking and commenting (Bano et al. 2019) or more intense behaviors like contacting friends through personal messages, sharing personal information and even providing critique (Stutzman 2006; Annamdevula and Bellamkonda 2016).

14.3.3 Benefits in Career Selection and Professional Development

Online social capital positively affects the intention and behavior of students that eventually support them in their career selection and development. In the current times, entrepreneurship is being encouraged at policy levels in different countries. The intention to start an enterprise develops right at the college level. Further, online social capital benefits students in various specific ways that help them grow and prosper in career. These benefits are illustrated in Figure 14.4.

14.3.3.1 Entrepreneurial Intention

Entrepreneurship as a career choice is being considered by students. Hoda et al. (2020) have quoted two definitions of entrepreneurial intention that try to summarize the

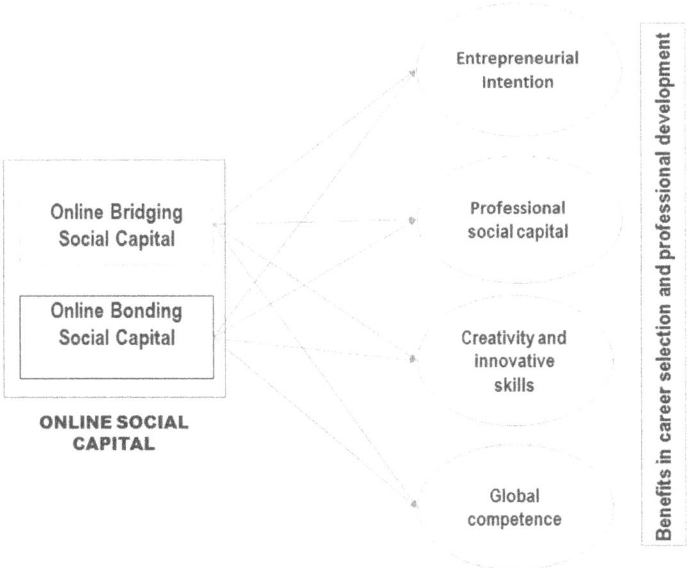

FIGURE 14.4 Online social capital and benefits in career selection and professional development. (Created by author.)

varying definitions of the term. They quoted the term defined by Gupta and Bhawe (2007) as "the planning and implementation of business ideas, which orientates by a mental process" and that defined by De Pillis and Reardon (2007) as "intention to venture into business." The intention to start an enterprise develops right at the college level, and social capital plays a significant role in positively increasing the intention (Liñán and Santos 2007; Kwon and Adler 2014; García-Villaverde, Parra-Requena, and Molina-Morales 2018; Tatarko and Schmidt 2016). One of the most popular explanations of Entrepreneurial intention is based on the theory of planned behavior (Ajzen 1991). It postulates that "entrepreneurial intention is a function of three antecedents namely personal attitude, subjective norms, and perceived behavioral control." Entrepreneurship research has proved that social capital positively influences these antecedents (Tatarko and Schmidt 2016; Malebana 2019; Sequeira, Mueller, and Mcgee 2007; Westlund and Gawell 2012). Pérez-Macías, Fernández-Fernández, and Rua Vieites (2019) have found significant relationship between online social capital and entrepreneurial intention.

14.3.3.2 Professional Social Capital

Social media usage and online networking are predictors of professional social capital (Spottswood and Wohn 2020; Smith, Smith, and Shaw 2017). Professional social capital increases the employment chances of users (Badoer, Hollings, and Chester 2020). Organizations from various sectors have started using social media for recruitment (Ruparel et al. 2020). This has resulted in the development of specialized professional social media platforms, thereby changing the recruitment process

(Heydenrych and Case 2018). Apart from this trend, online networking is positively related to career success (Bozionelos 2003) and access to plenty of resources that aid success (Benson and Morgan 2013). Adler and Kwon (2002) explained that the resources that an individual garners through the network may be in the form of information, influence and professional solidarity. Further, these professional networks provide evidence of the social skills of the potential employee (Benson, Morgan, and Filippaios 2014). Professional social networking skills are being included in the curriculum of various programs in different universities (Badoer, Hollings, and Chester 2020).

14.3.3.3 Creativity and Innovative Skills

Creativity or the ability to innovate is an important trait being considered in professional life. Research indicates that the social capital that individuals accrue through social networking is supportive in the acquisition of resources and skills that develop creativity (Nahapiet and Goshal 1998; Gu, Zhang, and Liu 2014). Interactions among students as well as student-teacher interactions serve as important ingredients in the development of creativity and innovative skills in students (Fischer et al. 2005). Social networking, including online, would facilitate the development of creativity in students (Gu, Zhang, and Liu 2014).

14.3.3.4 Global Competence

To succeed in the globalized world, students are required to be accommodative of diverse perspectives and to have an understanding of the varying contexts. Online networking provides the students an exposure to “heterogeneous perspectives” of different users in the same network (Meng, Zhu, and Cao 2017; Uchino et al. 2004). The diverse perspectives in online networks are referred to as “network heterogeneity” (Kim and Kim 2017; Kim and Lee 2016).

14.4 IMPLICATIONS

The chapter aimed to review the extant research on the role of online social capital in students’ careers. It was found that studies have focused on several dimensions of online social capital. The studies focusing on how online social capital is formed are discussed in Section 14.1. The benefits of online social capital in the students’ career development have been discussed in three separate sections explaining how online social capital may affect the students’ careers in terms of their learning, well-being, and career selection. Based on these discussions, the relationships between three main constructs, namely online social networking behavior, online social capital, and benefits to students, a theoretical model has been formulated (Figure 14.5). These relationships have been studied in parts in various studies described in the sections above. A comprehensive model, as presented in Figure 14.5, provides future directions for research dealing with the benefits of online networking and online social capital focusing on students’ benefits. The model also has policy implications. The model may be tested using various social media platforms and social networking sites (SNSs). A comparative analysis of the benefits considering different samples would help ascertain the uniformity or differences in the benefits through different

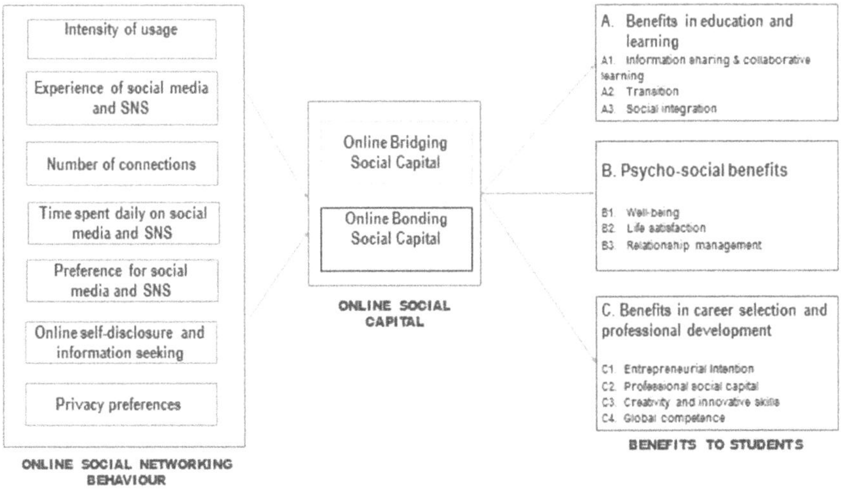

FIGURE 14.5 Conceptual model: Benefits of online social capital. (Created by author.)

platforms and sites. The model may also serve as a guide for making policy regarding the inclusion of professional social media skills in the curriculum. The different benefits might be treated as the different learning outcomes. A checklist of these learning outcomes based on this model may be used to measure the positive impact of "online social capital." Further, those social media platforms and sites that result in higher online social capital would be given more importance in the university learning system.

14.5 CONCLUSION

This chapter discussed the benefits of online social capital for students' careers and professional life. The benefits have been categorized into three broad categories in line with the findings of the extant literature. College life is an extremely important phase of an individual that has a long-lasting impact on the students' professional development. The students' use of online networking tools results in the formation of "online social capital." Online social capital differs from offline social capital due to the innovative tools offered by social media and networking sites. One of the direct benefits that accrue to students by the usage of social media is access to information that includes both general as well as rare information. As discussed, the most popular categorization of social capital is attributed to Putnam (1995). In terms of online social capital, both types of social capital, namely "bonding and bridging social capital," develop through online networking. Social capital brings resources to students that positively affect students' educational achievement, career success, as well as psychological well-being. These resources may be tangible (like information, money, jobs, etc.) or intangible (like well-being, confidence, creativity, etc.). Relationship management, traditionally a marketing term, has become a prominent tool for individual development too. Online networking facilitates the relationship management process by helping users manage both close and distant ties.

Social capital develops depending upon the investment by the individuals (Gerard 2012). The more the students can adequately invest in the development of online social capital, the more benefits would accrue to students in different forms and at different stages of career. Therefore, the institutions must incorporate courses in the curriculum that impart social media skills to students. The curriculum should include various dimensions of social media skills such as communication, security issues and professional social networking. The recruitment process and employee selection are now depending very highly on online social networks. Therefore, the students must possess these skills to succeed in securing their first jobs as well as further career development.

REFERENCES

Abbas, Rana, and Gustavo Mesch. 2018. "Do Rich Teens Get Richer? Facebook Use and the Link between Offline and Online Social Capital among Palestinian Youth in Israel." *Information Communication and Society* 21 (1): 63–79. doi:10.1080/1369118X.2016.1261168.

Adler, Paul S., and Seok Woo Kwon. 2002. "Social Capital: Prospects for a New Concept." Academy of Management Review. doi:10.5465/AMR.2002.5922314.

Ahmad, Saeed, Mudasir Mustafa, and Ahsan Ullah. 2016. "Association of Demographics, Motives and Intensity of Using Social Networking Sites with the Formation of Bonding and Bridging Social Capital in Pakistan." *Computers in Human Behavior* 57: 107–14. doi:10.1016/j.chb.2015.12.027.

Ahn, June. 2012. "Teenagers' Experiences with Social Network Sites: Relationships to Bridging and Bonding Social Capital." Information Society. doi:10.1080/01972243.2011.649394.

Ajzen, Icek. 1991. "The Theory of Planned Behavior." *Organizational Behavior and Human Decision Processes* 50 (2): 179–211. doi:10.1016/0749-5978(91)90020-T.

Annamdevula, Subrahmanyam, and Raja Shekhar Bellamkonda. 2016. "Effect of Student Perceived Service Quality on Student Satisfaction, Loyalty and Motivation in Indian Universities: Development of HiEduQual." *Journal of Modelling in Management* 11 (2): 488–517. doi:10.1108/JM2-01-2014-0010.

Asterhan, Christa S. C., and Hananel Rosenberg. 2015. "The Promise, Reality and Dilemmas of Secondary School Teacher-Student Interactions in Facebook: The Teacher Perspective." *Computers and Education* 85: 134–48. doi:10.1016/j.compedu.2015.02.003.

Badoer, Emilio, Yvette Hollings, and Andrea Chester. 2020. "Professional Networking for Undergraduate Students: A Scaffolded Approach." *Journal of Further and Higher Education* 00 (00): 1–14. doi:10.1080/0309877X.2020.1744543.

Bano, Shehar, Wu Cisheng, Ali Nawaz Khan, and Naseer Abbas Khan. 2019. "WhatsApp Use and Student's Psychological Well-Being: Role of Social Capital and Social Integration." *Children and Youth Services Review* 103 (June): 200–208. doi:10.1016/j.childyouth.2019.06.002.

Benson, Vladlena, and Stephanie Morgan. 2013. "Student Experience and Ubiquitous Learning in Higher Education: Impact of Wireless and Cloud Applications." *Creative Education* 04 (08): 1–5. doi:10.4236/ce.2013.48a001.

Benson, Vladlena, Stephanie Morgan, and Fragkiskos Filippaios. 2014. "Social Career Management: Social Media and Employability Skills Gap." *Computers in Human Behavior* 30: 519–25. doi:10.1016/j.chb.2013.06.015.

Benson, Vladlena, Fragkiskos Filippaios, and Stephanie Morgan. 2010. "Online Social Networks: Changing the Face of Business Education and Career Planning." *International Journal of E-Business Management* 4(1), 20–33. doi:10.3316/ijebm0401020.

Best, Samuel J., and Brian S. Krueger. 2006. "Online Interactions and Social Capital: Distinguishing between New and Existing Ties." Social Science Computer Review. doi:10.1177/0894439306286855.

Bourdieu, P. 1986. "The forms of social capital." In John G. Richardson (eds.), *Handbook of theory and research for the sociology of education*. Greenwood Press, Westport.

Boyd, Danah M., and Nicole B. Ellison. 2007. "Social Network Sites: Definition, History, and Scholarship." *Journal of Computer-Mediated Communication* 13 (1): 210–30. doi:10.1111/j.1083-6101.2007.00393.x.

Bozionelos, Nikos. 2003. "Intra-Organizational Network Resources: Relation To Career Success and Personality." *The International Journal of Organizational Analysis* 11 (1): 41–66. doi:10.1108/eb028962.

Carpenter, Jeffrey P., Gemma Tur, and Victoria I. Marín. 2016. "What Do U.S. and Spanish Pre-Service Teachers Think about Educational and Professional Use of Twitter? A Comparative Study." Teaching and Teacher Education. doi:10.1016/j.tate.2016.08.011.

Chang, Ya Ping, and Dong Hong Zhu. 2012. "The Role of Perceived Social Capital and Flow Experience in Building Users' Continuance Intention to Social Networking Sites in China." *Computers in Human Behavior* 28 (3): 995–1001. doi:10.1016/j.chb.2012.01.001.

Chen, X., J. Ma, J. Wei, and S. Yang. 2021. The role of perceived integration in WeChat usages for seeking information and sharing comments: A social capital perspective. *Information and Management* 58(1): 103280. doi:10.1016/j.im.2020.103280.

Chen, Yu, and Soko S. Starobin. 2019. "Formation of Social Capital for Community College Students: A Second-Order Confirmatory Factor Analysis." *Community College Review* 47 (1): 3–30. doi:10.1177/0091552118815758.

Coleman, James S. 1988. "Social Capital in the Creation of Human Capital." *American Journal of Sociology* 94: 95–120.

———. 2009. "Social Capital in the Creation of Human Capital." *Knowledge and Social Capital* 94: 17–42. doi:10.1086/228943.

Deng, Liping, and Allan H. K. Yuen. 2011. "Towards a Framework for Educational Affordances of Blogs." Computers and Education. doi:10.1016/j.compedu.2010.09.005.

Dhir, Amandeep, and Chin Chung Tsai. 2017. "Understanding the Relationship between Intensity and Gratifications of Facebook Use among Adolescents and Young Adults." Telematics and Informatics. doi:10.1016/j.tele.2016.08.017.

Dika, S. L., and K. Singh. 2002. Applications of social capital in educational literature: A critical synthesis. *Review of Educational Research* 72(1): 31–60. doi:10.3102/00346543072001031

Dorethy, Marcie D., Martin S. Fiebert, and Christopher R. Warren. 2014. "Examining Social Networking Site Behaviors: Photo Sharing and Impression Management on Facebook." *International Review of Social Sciences and Humanities*, 6(2), 111–116. https://s3.amazonaws.com/academia.edu.documents/34066691/10_IRSSH-723-V6N2.39115416_1.pdf?AWSAccessKeyId=AKIAIWOWYYGZ2Y53UL3A&Expires=1510653562&Signature=A33dcKTMzbNCNVR7GVbwL1NVt0U%3D&response-content-disposition=inline%3B filename%3DExamining_Social_Ne

Ellison, Nicole B., and Danah M. Boyd. 2013. "Sociality through Social Network Sites." *The Oxford Handbook of Internet Studies*, 151–72. doi:10.1093/oxfordhb/9780199589074.001.0001.

Ellison, Nicole B., Charles Steinfield, and Cliff Lampe. 2007. "The Benefits of Facebook 'Friends:' Social Capital and College Students' Use of Online Social Network Sites." *Journal of Computer-Mediated Communication* 12 (4): 1143–68. doi:10.1111/j.1083-6101.2007.00367.x.

———. 2011. "Connection Strategies: Social Capital Implications of Facebook-Enabled Communication Practices." *New Media and Society* 13 (6): 873–92. doi:10.1177/1461444810385389.

Ellison, N. B., and Vitak, J. 2015. Social network site affordances and their relationship to social capital processes. In S. S. Sundar (Ed.), *The handbook of the psychology of communication technology* (pp. 205–227). Wiley Blackwell. doi:10.1002/9781118426456.ch9.

Ellison, Nicole B., Jessica Vitak, Rebecca Gray, and Cliff Lampe. 2014. "Cultivating Social Resources on Social Network Sites: Facebook Relationship Maintenance Behaviors and Their Role in Social Capital Processes." *Journal of Computer-Mediated Communication* 19 (4): 855–70. doi:10.1111/jcc4.12078.

Ewart, Jacqui, and Collette Snowden. 2012. "The Media's Role in Social Inclusion and Exclusion." Media International Australia. doi:10.1177/1329878x1214200108.

Fischer, Gerhard, Elisa Giaccardi, Hal Eden, Masanori Sugimoto, and Yunwen Ye. 2005. "Beyond Binary Choices: Integrating Individual and Social Creativity." International Journal of Human Computer Studies. doi:10.1016/j.ijhcs.2005.04.014.

Forbush, Eric, and Brooke Foucault-Welles. 2016. "Social Media Use and Adaptation among Chinese Students Beginning to Study in the United States." International Journal of Intercultural Relations. doi:10.1016/j.ijintrel.2015.10.007.

Ganesh, R., Haslinda, A., & Raghavan, S. (2017). An investigation of Academic and Social integration in Private Higher Education Institution in Malaysia as a Moderating Variable in Relations to Satisfaction Toward Students' Retention. *International Review of Management and Business Research* 6 (4): 1543–1560.

García-Villaverde, Pedro M., Gloria Parra-Requena, and F. Xavier Molina-Morales. 2018. "Structural Social Capital and Knowledge Acquisition: Implications of Cluster Membership." *Entrepreneurship and Regional Development* 30 (5–6): 530–61. doi:10.1080/08985626.2017.1407366.

Gerard, Joseph G. 2012. "Linking in With LinkedIn®: Three Exercises That Enhance Professional Social Networking and Career Building." Journal of Management Education. doi:10.1177/1052562911413464.

Greenhow, Christine, and Emilia Askari. 2017. "Learning and Teaching with Social Network Sites: A Decade of Research in K-12 Related Education." Education and Information Technologies. doi:10.1007/s10639-015-9446-9.

Greenhow, Christine, and Lisa Burton. 2011. "Help from My Friends: Social Capital in the Social Network Sites of Low-Income Students." *Journal of Educational Computing Research* 45 (2): 223–45. doi:10.2190/EC.45.2.f.

Gu, Jibao, Yanbing Zhang, and Hefu Liu. 2014. "Importance of Social Capital to Student Creativity within Higher Education in China." *Thinking Skills and Creativity* 12 (96): 14–25. doi:10.1016/j.tsc.2013.12.001.

Gupta, Vishal K., and Nachiket M. Bhawe. 2007. "The Influence of Proactive Personality and Stereotype Threat on Women's Entrepreneurial Intentions." *Journal of Leadership & Organizational Studies* 13 (4): 73–85. doi:10.1177/10717919070130040901.

Harrington, Rick, and Donald A. Loffredo. 2011. "Insight, Rumination, and Self-Reflection as Predictors of Well-Being." Journal of Psychology: Interdisciplinary and Applied. doi:10.1080/00223980.2010.528072.

Häuberer, Julia. 2011. *Social Capital Theory: Towards a Methodological Foundation.* doi:10.1007/978-3-531-92646-9.

Heidari, Elham, Ghasem Salimi, and Mahboobe Mehrvarz. 2020. "The Influence of Online Social Networks and Online Social Capital on Constructing a New Graduate Students' Professional Identity." *Interactive Learning Environments* 0 (0): 1–18. doi:10.1080/10494820.2020.1769682.

Heydenrych, Hilton, and Jennifer M. Case. 2018. "Researching Graduate Destinations Using LinkedIn: An Exploratory Analysis of South African Chemical Engineering

Graduates." European Journal of Engineering Education. doi:10.1080/03043797.2017.1402865.

Hoda, Najmul, Naim Ahmad, Mobin Ahmad, Abdullah Kinsara, Afnan T. Mushtaq, Mohammad Hakeem, and Mwafaq Al-Hakami. 2020. "Validating the Entrepreneurial Intention Model on the University Students in Saudi Arabia." *Journal of Asian Finance, Economics and Business* 7 (11): 469–77.

Huang, Hsin Yi, Po Lin Chen, and Yu Chen Kuo. 2017. "Understanding the Facilitators and Inhibitors of Individuals' Social Network Site Usage." Online Information Review. doi:10.1108/OIR-10-2015-0319.

Jang, Juyoung, and Jodi Dworkin. 2014. "Does Social Network Site Use Matter for Mothers? Implications for Bonding and Bridging Capital." Computers in Human Behavior. doi:10.1016/j.chb.2014.02.049.

Johnston, Kevin, Maureen Tanner, Nishant Lalla, and Dori Kawalski. 2013. "Social Capital: The Benefit of Facebook Friends." Behaviour and Information Technology. doi:10.1080/0144929X.2010.550063.

Kalaitzaki, A., K. Tsouvelas, and S. Koukouli. 2020. "Social Capital, Social Support and Perceived Stress in College Students: The Role of Resilience and Life Satisfaction." *Stress & Health* 36 (4). doi:10.1002/smi.3008.

Kang, Lele, Qiqi Jiang, and Chuan Hoo Tan. 2017. "Remarkable Advocates: An Investigation of Geographic Distance and Social Capital for Crowdfunding." *Information and Management* 54 (3): 336–48. doi:10.1016/j.im.2016.09.001.

Kaplan, Andreas M., and Michael Haenlein. 2010. "Users of the World, Unite! The Challenges and Opportunities of Social Media." *Business Horizons* 53 (1): 59–68. doi:10.1016/j.bushor.2009.09.003.

Kasperski, Ronen, and Ina Blau. 2020. "Social Capital in High-Schools: Teacher-Student Relationships within an Online Social Network and Their Association with In-Class Interactions and Learning." *Interactive Learning Environments* 0 (0): 1–17. doi:10.1080/10494820.2020.1815220.

Kim, Bumsoo, and Yonghwan Kim. 2017. "College Students' Social Media Use and Communication Network Heterogeneity: Implications for Social Capital and Subjective Well-Being." *Computers in Human Behavior* 73: 620–28. doi:10.1016/j.chb.2017.03.033.

Kim, Cheonsoo, and Jae Kook Lee. 2016. "Social Media Type Matters: Investigating the Relationship Between Motivation and Online Social Network Heterogeneity." Journal of Broadcasting and Electronic Media. doi:10.1080/08838151.2016.1234481.

Kwon, Seok Woo, and Paul S. Adler. 2014. "Social Capital: Maturation of a Field of Research." *Academy of Management Review* 39 (4): 412–22. doi:10.5465/amr.2014.0210.

Lee, S., Chung, J. E., & Park, N. (2018a). Network Environments and Well-Being: An Examination of Personal Network Structure, Social Capital, and Perceived Social Support. *Health Communication* 33(1): 22–31. doi:10.1080/10410236.2016.1242032

———. 2018b. "Network Environments and Well-Being: An Examination of Personal Network Structure, Social Capital, and Perceived Social Support." *Health Communication* 33 (1): 22–31. doi:10.1080/10410236.2016.1242032.

Lee, Jee Young, Sora Park, Eun Yeong Na, and Eun Mee Kim. 2016. "A Comparative Study on the Relationship between Social Networking Site Use and Social Capital among Australian and Korean Youth." *Journal of Youth Studies*. doi:10.1080/13676261.2016.1145637.

Li, Stacey, Pratik Modi, Meng Shan (Sharon) Wu, Cheng Hao (Steve) Chen, and Bang Nguyen. 2019. "Conceptualising and Validating the Social Capital Construct in Consumer-Initiated Online Brand Communities (COBCs)." *Technological Forecasting and Social Change* 139 (October 2017): 303–10. doi:10.1016/j.techfore.2018.11.018.

Lin, Kuan Yu, and Hsi Peng Lu. 2011. "Intention to Continue Using Facebook Fan Pages from the Perspective of Social Capital Theory." *Cyberpsychology, Behavior, and Social Networking* 14 (10): 565–70. doi:10.1089/cyber.2010.0472.

Liñán, Francisco, and Francisco Javier Santos. 2007. "Does Social Capital Affect Entrepreneurial Intentions?" *International Advances in Economic Research* 13 (4): 443–53. doi:10.1007/s11294-007-9109-8.

Liu, Haihua, Junqi Shi, Yihao Liu, and Zitong Sheng. 2013. "The Moderating Role of Attachment Anxiety on Social Network Site Use Intensity and Social Capital." Psychological Reports. doi:10.2466/21.02.17.PR0.112.1.252-265.

Lounsbury, Michael, Nan Lin, Karen Cook, and Ronald S. Burt. 2002. "Social Capital: Theory and Research." *Contemporary Sociology* 31 (1): 28. doi:10.2307/3089402.

Madge, Clare, Julia Meek, Jane Wellens, and Tristram Hooley. 2009. "Facebook, Social Integration and Informal Learning at University: 'It Is More for Socialising and Talking to Friends about Work than for Actually Doing Work.'" *Learning, Media and Technology*. doi:10.1080/17439880902923606.

Mahmood, Qaisar Khalid, Rubeena Zakar, and Muhammad Zakria Zakar. 2018. "Role of Facebook Use in Predicting Bridging and Bonding Social Capital of Pakistani University Students." *Journal of Human Behavior in the Social Environment* 28 (7): 856–73. doi:10.1080/10911359.2018.1466750.

Malebana, M. J. 2019. "The Influencing Role of Social Capital in the Formation of Entrepreneurial Intention." *Southern African Business Review* 20 (1): 51–70. doi:10.25159/1998-8125/6043.

Meng, Qian, Chang Zhu, and Chun Cao. 2017. "The Role of Intergroup Contact and Acculturation Strategies in Developing Chinese International Students' Global Competence." Journal of Intercultural Communication Research. doi:10.1080/17475759.2017.1308423.

Mikal, Jude P., Ronald E. Rice, Audrey Abeyta, and Jenica Devilbiss. 2013. "Transition, Stress and Computer-Mediated Social Support." Computers in Human Behavior. doi:10.1016/j.chb.2012.12.012.

Moon, Rennie J., and Gi Wook Shin. 2019. "International Student Networks as Transnational Social Capital: Illustrations from Japan." Comparative Education. doi:10.1080/03050068.2019.1601919.

Mou, Yi, and Carolyn A. Lin. 2017. "The Impact of Online Social Capital on Social Trust and Risk Perception." *Asian Journal of Communication* 27 (6): 563–81. doi:10.1080/01292986.2017.1371198.

Na, Younkue. 2015. "The Study on Social Capital and Community Sense Formation for the Sustainability of Fashion Social Enterprises." *Fashion Business* 19 (5): 157–74. doi:10.12940/jfb.2015.19.5.157.

Nahapiet, J, and S Goshal. 1998. "Creating Organizational Capital through Intellectual and Social Capital." *Academy of Management Review* 23 (2): 242–66.

Naseri, Samaneh. 2017. "Online Social Network Sites and Social Capital: A Case of Facebook." *International Journal of Applied Sociology* 2017 (1): 13–19. doi:10.5923/j.ijas.20170701.02.

Northey, Gavin, Rahul Govind, Tania Bucic, Mathew Chylinski, Rebecca Dolan, and Patrick van Esch. 2018. "The Effect of 'Here and Now' Learning on Student Engagement and Academic Achievement." British Journal of Educational Technology. doi:10.1111/bjet.12589.

Öztok, Murat, Daniel Zingaro, Alexandra Makos, Clare Brett, and Jim Hewitt. 2015. "Capitalizing on Social Presence: The Relationship between Social Capital and Social Presence." Internet and Higher Education. doi:10.1016/j.iheduc.2015.04.002.

Pan, Youqin, Linda Coleman, Saverio Manago, and David Goodof. 2019. "Effects of Social Media Usage on Social Integration of University Students." *Competition Forum* 17 (2): 351.

Pang, Hua. 2018. "How Does Time Spent on WeChat Bolster Subjective Well-Being through Social Integration and Social Capital?" *Telematics and Informatics* 35 (8): 2147–56. doi:10.1016/j.tele.2018.07.015.

Papacharissi, Zizi. 2009. "The Virtual Geographies of Social Networks: A Comparative Analysis of Facebook, LinkedIn and ASmallWorld." New Media and Society. doi:10.1177/1461444808099577.

Peng, Michael Yao Ping. 2019. "Testing the Mediating Role of Student Learning Outcomes in the Relationship Among Students' Social Capital, International Mindsets, and Employability." Asia-Pacific Education Researcher. doi:10.1007/s40299-018-00431-3.

Pérez-Macías, Noemí, José Luis Fernández-Fernández, and Antonio Rua Vieites. 2019. "Entrepreneurial Intentions: Trust and Network Ties in Online and Face-to-Face Students." *Education and Training* 61 (4): 461–79. doi:10.1108/ET-05-2018-0126.

Petersen, C., and K. A. Johnston. 2015. "The Impact of Social Media Usage on the Cognitive Social Capital of University Students." *Informing Science* 18 (1): 1–31.

Phua, Joe, Seunga Venus Jin, and Jihoon (Jay) Kim. 2017a. "Gratifications of Using Facebook, Twitter, Instagram, or Snapchat to Follow Brands: The Moderating Effect of Social Comparison, Trust, Tie Strength, and Network Homophily on Brand Identification, Brand Engagement, Brand Commitment, and Membership Intention." *Telematics and Informatics* 34 (1): 412–24. doi:10.1016/j.tele.2016.06.004.

———. 2017b. "Uses and Gratifications of Social Networking Sites for Bridging and Bonding Social Capital: A Comparison of Facebook, Twitter, Instagram, and Snapchat." *Computers in Human Behavior* 72: 115–22. doi:10.1016/j.chb.2017.02.041.

Pillis, Emmeline De, and Kathleen K. Reardon. 2007. "The Influence of Personality Traits and Persuasive Messages on Entrepreneurial Intention: A Cross-Cultural Comparison." *Career Development International* 12 (4): 382–96. doi:10.1108/13620430710756762.

Portes, Alejandro. 1998. "Social Capital: Its Origins and Applications in Modern Sociology." In *Annual Review of Sociology*, 24: 1–24. doi:10.1146/annurev.soc.24.1.1.

Putnam, Robert D. 1995. "Bowling Alone: America's Declining Social Capital." Journal of Democracy. doi:10.1353/jod.1995.0002.

———. 2001. "Social Capital: Measurement and Consequences." *Canadian Journal of Policy Research* 2: 41–51.

Ridings, Catherine M., David Gefen, and BayArinze. 2002. "Some Antecedents and Effects of Trust in Virtual Communities." Journal of Strategic Information Systems. doi:10.1016/S0963-8687(02)00021-5.

Rios, Juan Sebastian, Donghee Yvette Wohn, and Yu Hao Lee. 2019. "Effect of Internet Literacy in Understanding Older Adults' Social Capital and Expected Internet Support." Communication Research Reports. doi:10.1080/08824096.2019.1586664.

Ruparel, Namita, Amandeep Dhir, Anushree Tandon, Puneet Kaur, and Jamid Ul Islam. 2020. "The Influence of Online Professional Social Media in Human Resource Management: A Systematic Literature Review." *Technology in Society* 63 (August): 101335. doi:10.1016/j.techsoc.2020.101335.

Ryan, Tracii, Kelly A. Allen, De Leon L. Gray, and Dennis M. McInerney. 2017. "How Social Are Social Media? A Review of Online Social Behaviour and Connectedness." *Journal of Relationships Research* 8: 1–8. doi:10.1017/jrr.2017.13.

Rykov, Yuri, Olessia Koltsova, and Yadviga Sinyavskaya. 2020. "Effects of User Behaviors on Accumulation of Social Capital in an Online Social Network." *PLoS ONE* 15 (4): 1–18. doi:10.1371/journal.pone.0231837.

Sequeira, Jennifer, Stephen L. Mueller, and Jeffrey E. Mcgee. 2007. "The Influence of Social Ties and Self-Efficacy in Forming Entrepreneurial Intentions and Motivating Nascent Behavior." *Journal of Developmental Entrepreneurship* 12 (03): 275–93. doi:10.1142/s108494670700068x.

Severiens, Sabine E., and Henk G. Schmidt. 2009. "Academic and Social Integration and Study Progress in Problem Based Learning." Higher Education. doi:10.1007/s10734-008-9181-x.

Shane-Simpson, Christina, Adriana Manago, Naomi Gaggi, and Kristen Gillespie-Lynch. 2018. "Why Do College Students Prefer Facebook, Twitter, or Instagram? Site Affordances, Tensions between Privacy and Self-Expression, and Implications for Social Capital." *Computers in Human Behavior* 86: 276–88. doi:10.1016/j.chb.2018.04.041.

Sikolia, G. S., & Mberia, H. K. (2019). 'Last seen now': Explaining teenage identities and social capital on social network sites in Kenya. *Journal of Development and Communication Studies* 6(1): 18–35. doi:10.4314/jdcs.v6i1.2.

Smith, Claudia, J. Brock Smith, and Eleanor Shaw. 2017. "Embracing Digital Networks: Entrepreneurs' Social Capital Online." *Journal of Business Venturing* 32 (1): 18–34. doi:10.1016/j.jbusvent.2016.10.003.

Sohn, Youngmi, Sungbum Woo, Duckhyun Jo, and Eunjoo Yang. 2019. "The Role of the Quality of College-Based Relationship on Social Media in College-to-Work Transition of Korean College Students: The Longitudinal Examination of Intimacy on Social Media, Social Capital, and Loneliness." *Japanese Psychological Research* 61 (4): 236–48. doi:10.1111/jpr.12234.

Spottswood, Erin L., and Donghee Yvette Wohn. 2020. "Online Social Capital: Recent Trends in Research." *Current Opinion in Psychology* 36: 147–52. doi:10.1016/j.copsyc.2020.07.031.

Steinfield, Charles, Nicole B. Ellison, and Cliff Lampe. 2008. "Social Capital, Self-Esteem, and Use of Online Social Network Sites: A Longitudinal Analysis." *Journal of Applied Developmental Psychology* 29 (6): 434–45. doi:10.1016/j.appdev.2008.07.002.

Stutzman, F. 2006. "An Evaluation of Identity-Sharing Behavior in Social Network Communities." International Digital and Media Arts Journal.

Tatarko, Alexander, and Peter Schmidt. 2016. "Individual Social Capital and the Implementation of Entrepreneurial Intentions: The Case of Russia." *Asian Journal of Social Psychology* 19 (1): 76–85. doi:10.1111/ajsp.12113.

Tian, Xiaoli. 2016. "Network Domains in Social Networking Sites: Expectations, Meanings, and Social Capital." *Information Communication and Society* 19 (2): 188–202. doi:10.1080/1369118X.2015.1050051.

Tibber, Marc S., Jiayuan Zhao, and Stephen Butler. 2020. "The Association between Self-Esteem and Dimensions and Classes of Cross-Platform Social Media Use in a Sample of Emerging Adults – Evidence from Regression and Latent Class Analyses." *Computers in Human Behavior* 109 (March): 106371. doi:10.1016/j.chb.2020.106371.

Uchino, Bert N., Julianne Holt-Lunstad, Timothy W. Smith, and Lindsey Bloor. 2004. "Heterogeneity in Social Networks: A Comparison of Different Models Linking Relationships to Psychological Outcomes." Journal of Social and Clinical Psychology. doi:10.1521/jscp.23.2.123.31014.

Valenzuela, Sebastián, Namsu Park, and Kerk F. Kee. 2009. "Is There Social Capital in a Social Network Site?: Facebook Use and College Student's Life Satisfaction, Trust, and Participation1." *Journal of Computer-Mediated Communication* 14 (4): 875–901. doi:10.1111/j.1083-6101.2009.01474.x.

Venter, A. 2019. "Social Media and Social Capital in Online Learning." *South African Journal of Higher Education* 33 (3): 241–58. doi:10.20853/33-3-3105.

Vitak, Jessica. 2014. "Unpacking Social Media's Role in Resource Provision: Variations across Relational and Communicative Properties." Societies. doi:10.3390/soc4040561.

Wang, Jin Liang, Linda A. Jackson, James Gaskin, and Hai Zhen Wang. 2014. "The Effects of Social Networking Site (SNS) Use on College Students' Friendship and Well-Being." Computers in Human Behavior. doi:10.1016/j.chb.2014.04.051.

Warren, Anne Marie, Ainin Sulaiman, and Noor Ismawati Jaafar. 2015. "Understanding Civic Engagement Behaviour on Facebook from a Social Capital Theory Perspective." Behaviour and Information Technology. doi:10.1080/0144929X.2014.934290.

Wen, Zhengbao, Xiaowei Geng, and Yinghua Ye. 2016. "Does the Use of WeChat Lead to Subjective Well-Being?: The Effect of Use Intensity and Motivations." Cyberpsychology, Behavior, and Social Networking. doi:10.1089/cyber.2016.0154.

Westlund, Hans, and Malin Gawell. 2012. "Building Social Capital For Social Entrepreneurship." *Annals of Public and Cooperative Economics* 83 (1): 101–16. doi:10.1111/j.1467-8292.2011.00456.x.

Williams, Dmitri. 2006. "On and Off the 'Net: Scales for Social Capital in an Online Era." *Journal of Computer-Mediated Communication* 11 (2): 593–628. doi:10.1111/j.1083-6101.2006.00029.x.

Wohn, Donghee Yvette, Nicole B. Ellison, M. Laeeq Khan, Ryan Fewins-Bliss, and Rebecca Gray. 2013. "The Role of Social Media in Shaping First-Generation High School Students' College Aspirations: A Social Capital Lens." Computers and Education. doi: 10.1016/j.compedu.2013.01.004.

You, Leping, and Linda Hon. 2019. "How Social Ties Contribute to Collective Actions on Social Media: A Social Capital Approach." *Public Relations Review* 45 (4): 101771. doi:10.1016/j.pubrev.2019.04.005.

15 Upskilling and Reskilling in the Digital Age
The Way Forward for Higher Educational Institutions

V. Padmaja and Kumar Mukul

CONTENTS

15.1 INTRODUCTION

'Education is not the learning of facts, but the training of minds to think.'

Albert Einstein

DOI: 10.1201/9781003132097-15

The evolution of technology and consequent changes in all the institutional and social constituents are omnipresent and having a paramount impact on our functioning. The evolutionary path guided by the technological disruptions are here to stay, are irreversible and touching our lives at each moment in diverse ways, sometimes by choice but most often without our control, knowledge and awareness.

In today's information society, emerging innovations have stimulated the development of disruptions (Selwyn and Facer 2014). Disruptive technologies are constantly changing and affecting people's lifestyles, working patterns, workplace practices and also influencing the global economy in major ways (Bolton et al. 2019). In such a scenario, the Higher Educational Institutions (HEIs) cannot remain relevant if they do not keep track of these technological advancements and respond appropriately. HEIs are impacted by the changes in technology which is the key driver for innovation and sustainable growth. Jarvis (2009, p. 210) stated, '*Education is one of the institutions most deserving of disruption.*' The domain of education, particularly higher education, requires and expects continuous technological interventions to stay relevant and respond to the needs of the stakeholders in the most appropriate ways. There needs to be a smooth interface between education and technological advancements.

The education sector is under major influence as a result of the recent trends brought in vogue by Industry 4.0. The constant but decisive drift toward universalization of technological outcomes like machine learning, AI (artificial intelligence), IoT (Internet of Things), robotics, virtual and remote work, cognitive and cloud computing, and so on, are transforming the landscape of the workplace and educational setups providing the skill-sets to develop and work on them. The nature of teaching-learning processes, pedagogical systems, teaching styles, skill requirements of students as well as tutors, the role of policymakers, administrators, as well as other stakeholders is going to change profoundly with the growing interventions of modern technologies. Education 4.0, the new nomenclature given to the new-age education system, profoundly depends on digital technologies to impart skill-sets to future workforces. The education sector owes a responsibility to come up with appropriate responses in the form of effective 'Education 4.0' revolution to contribute and guide the Industry 4.0 developments. In such a scenario, HEIs can hardly afford to ignore the technological changes transforming the industries, workplaces, businesses and teaching-learning practices. They need to be aware of, understand and proactively embrace the technological innovations and educate their students in such a way that they come out of the campuses as proficient leaders capable of driving the Industry 4.0 movements ahead. Strategies that envision the training of new generations to meet the needs of Industry 4.0 must integrate technology, human sensibilities and skill development approach prudently and holistically.

The recent policy initiatives across the globe, at the government and HEI leadership levels, illustrate that there is ample emphasis being given to the use of innovative teaching-learning approaches driven by technology to fill the skill gaps. For instance, in the Indian context, the National Education Policy (NEP) 2020 has been developed to embrace an approach to digital skills in schools and colleges. NEP is aimed at encouraging and incorporating vocational proficiency developments as one major outcome of education through a focus on job-skill enhancement among students. The impact of the latest trends and pertinent policy implementation can be seen globally too. Governments at central and local levels in various countries have invested heavily in the digitalization of the education system through the adoption of the latest offerings from information and communication technology (ICT) systems.

There seems to be an agreement that the skill development and upgradation challenges can be met by continuous adaptations in teaching pedagogies, such as experiential learning, flipped classrooms,[1] live online classes and MOOCs[2] (Massive Open Online Courses) combined with encouragement for vocational training. The focus in most of the initiatives is on developing skills that enable teachers/facilitators to implement them and students who are an outcome of these processes. Nevertheless, technology management's conceptual foundation is fragmented and poorly established from an empirical perspective. Hence, there is a need to delve and focus in this area on the part of research and academic fraternity.

15.2 OBJECTIVES

The following objectives have been framed for this chapter:

- To explore the interface between Industry 4.0 and Education 4.0 amid technological disruptions.
- To analyze the increasing need for HEIs to upskill and reskill to meet the needs and requirements of the future of work in the digital age.
- To focus on pertinent skills for the future by highlighting the replaceable and irreplaceable skills due to the growing impact of technology.
- To explore the possibilities of strategic revamp in teaching-learning processes facilitating the required upskilling and reskilling.
- To identify the drivers of the new age HEI ecosystem which will foster continuous innovation and skill-upgradation.

15.3 METHODOLOGY

This book chapter and study are focused on secondary data available in the form of existing literature and research papers published on higher education. A comprehensive survey of different works published in the areas of Educational Technology (Ed-Tech) in Education 4.0 with specific reference to HEIs was done. Research articles that were published during the last two decades in highly reputed and highly reputed journals indexed in Scopus, ABDC, Web of Science, EBSCO, JSTOR, Emerald, SAGE, ProQuest and Elsevier were referred. To understand the concepts encompassing upskilling and reskilling, Ed-Tech-related topics in higher education with specific reference to digital transformation in education, technological infusion

---

[1] A ***flipped classroom*** is an innovative pedagogical approach to learning where students are expected to read and work on the content/concepts at home and practice the application of those in schools and colleges.

[2] A **massive open online course (MOOC)** is a web-based pedagogical approach, aimed at unlimited participation of a diverse range of learners. It provides interactive, innovative, and focused course content with multiple options for learners to choose from.

and strategies for education management system leading to skill development were referred. Various websites which published articles on Industry 4.0 and Education 4.0 were referred to build and analyze the concepts in a systematic manner. For the Indian context, the NEP 2020 (see Table 15.1) was referred to draw inferences and relate to upskilling and reskilling.

15.4 DISCUSSION

Based on extensive secondary research by understanding the various concepts systematically as well as getting deep insights from various articles on the relevance of upskilling in the digital age, this chapter highlights various outlines and themes that converge on the vision of upskilling and reskilling. Also, various strategies that could be vital for HEIs to sustain the current technological disruptions have been discussed in the following segment.

15.4.1 Technological Disruptions, Education 4.0 and Recent Trends in Ed-Tech

New technology disruptions have transformed higher education and these disruptions have changed the face of education. Much depends on how administrators, faculty and students apply technologically enhanced learning. Technology-enhanced learning has taken on several different formats. Skog et al. (2018 p. 432) proclaimed digital disruption as 'digital technology-induced environmental turbulence.' They described the digital transformation as a 'systemic shock' caused by the cumulative impact of many technologies. According to research, integrating technology with successful classroom techniques improves academic success, increases student learning, increases student satisfaction and increases faculty acceptance (Missildine et al. 2013).

E-learning or web-based learning is the use of the internet to provide education in a versatile, flexible and simple way to promote individual learning, organizational success and performance objectives and goals (Clark and Mayer 2016; Maqableh et al. 2015). Ülker and Yılmaz (2016) mentioned that the use of a learning management system (LMS) is one of the approaches to e-learning. Jakobsone and Cakula (2012) proposed a new perspective on the future of e-learning systems by involving the use of innovative and new technological advancements and opportunities for educational institutions.

Recent trends have led to the emergence of some prominent models, such as virtual learning (or virtual classrooms), the flipped classrooms, blended classes, and the massive open online course commonly known as MOOC (worldscientific.com).

Virtual learning is a web-based learning form in which learning happens in remote ways in the absence of any face-to-face interaction. In virtual learning, the knowledge transfer is mediated through technology-based tools, such as online classes, webinars, web-conferencing, emails, chats, etc. Virtual learning is student-centered, wherein students set the pace of learning and this form is much more interactive than the traditional lecturing method. One big advantage of this form of learning is that

TABLE 15.1
NEP 2020 Policies Being Introduced in the Indian Context

National Education Policy 2020 being promulgated by the Government of India can act as a good blueprint for framing strategies of educational institutions. Given its focus on some transformational changes suggested and futuristic orientation, NEP 2020 may prove to be a timely and decisive intervention for revamp of prevailing ecosystems in HEIs and indicative for future evolutions in the education field. The proposals of NEP can form the bedrock for strategies of the future as far as educational ecosystem creation is concerned.
The NEP 2020 is seen as a major intervention toward upskilling teachers and students by ensuring a change in the pedagogical initiatives. It focuses on the development of core competencies, capacities and life skills, experimental learning at all stages, including 21st-century skills, a holistic progress card and promotion of critical and higher-order thinking to solve practical problems. As per the new policy, students should be motivated to collaborate and work in teams and achieve team targets rather than individual targets. HEIs should focus on experiential learning which will promote and develop curiosity or nurture innovation. Students should be encouraged to question constantly which will facilitate critical thinking about an experience. This, in turn, will allow students to hone their skills of communication and thereby encourage creativity. HEIs will provide a platform for students to discover inherent skills and present their ideas and opinions before others.
The NEP 2020 also focuses on introducing multidisciplinary courses, such as medical tourism, yoga, film-making, photography, to encourage students to have exposure to other areas. This will certainly help students with a variety of options and also encourage them to plunge into entrepreneurship. It will also attract more foreign students to come to India as the Indian education sector will try to accommodate diverse students. Multidisciplinary courses will bring in more flexibility than the traditional formats.
The flexibility to choose subjects from a variety of courses in sciences and humanities with the ability to also learn fine arts will enable and provide students a wide range of subjects to choose from, without having any restrictions that they used to face earlier. This approach will cultivate an intellectual curiosity among students, develop a critical thought process and build sensitivity toward a socio-cultural environment.
The new policy proposes the introduction of vocational streams in schools and vocational education is supposed to start from Class 6 itself. The rigid separation between academic streams is being diluted and students being given avenues to pick up skills from different streams to foster their vocational abilities and performance in the long run. Assessment reforms recommend and include 360-degree holistic progress cards, keeping track of student progress on achievement of specific learning outcomes. This will facilitate the attainment of specific skills in a focused manner and learning outcomes will not be allowed to be left unmet or vague. The objective of goal setting, systematic tracking and holistic assessment will make the learners ready to take on workplace roles with much greater ease.
Another initiative being taken in NEP is regarding setting up of Multidisciplinary Education and Research Universities (MERUs), at par with IITs (Indian Institute of Technology – premier technology institutes of the country), IIMs (Indian Institute of Management – premier management institutes of India), as models of finest multidisciplinary education of global standards in the country. One philosophy behind this initiative is to encourage education which is allowing simultaneous development of skills. According to a report published by BusinessLinc on Campus in August 2020, NEP 2020 targets to achieve a 50 percent Gross Enrollment Ratio in higher education by 2030 from 26.3 percent in 2018, by creating new capacities and consolidating, expanding, and improving existing Higher Education Institutions in India. This exercise of streamlining of building a strong base for higher education will help in removing confusion in the minds of recruiters as the current system of fragmented education industry is filled stand-alone institutions such as IITs, IIMs, and private B-Schools, Central Universities, State Universities, Deemed Universities, and so on.

the digital content is stored and can be used at the convenience of the students. This form of learning does away with the restrictions posed by conventional class schedules and fixed routines. The virtual learning environment (VLE) supports the worldwide exchange of information enabling participation from teachers/facilitators and learners from across the globe to connect effectively and with ease. It is capable of supporting multiple courses and enhances the chances of life-long learning by covering the whole gamut of formal and informal learning. The learners may undergo continuous upskilling and reskilling as per the needs in various life stages due to the flexibility inherent in this approach.

The *Flipped Classroom* methodology is another innovative method of integrating technology into education. It is described as completing all activities outside of the classroom that is normally completed in the classroom (Lage et al. 2000). Instructions, conversations and evaluations in the classroom replace conventional instruction that focuses on tasks to be done at home and revision outside of school/college hours (Wallace 2013). A teacher in a flipped classroom avoids lecturing and asks students to engage in class. These activities will help and ensure to develop a better theoretical and conceptual understanding (Andrews et al. 2011).

The flipped classroom model helps in the development of life skills because students have much more control over the way they study and learn. This approach allows each student to move at their own pace. In this method, there is a shift from the traditional class activities and the focus is on some out-of-class practices with classroom-like setups (Lage et al. 2000). Students are expected to watch recorded lectures on their own time instead of using time for homework and routine lecturing. They will be engaged in active learning like project work, on-the-spot assignments and other experiential learning activities (Rollag and Billsberry 2012), which will lead to skill development through collaborative learning. A large number of students can be reached out to, with the help of the flipped classroom approach, hence it can be seen as a highly inclusive instructional strategy, where video content is accessible, regardless of their geographic location. It can be said that this new-age flipped classroom methodology can lead to skill enhancement by developing life skills because this approach enables the understanding of the learnt concepts and applying them to problems. This will, in turn will help students to succeed not only in class but also far beyond.

Blended learning is a new method of education that brings together the best of virtual interactions as well as traditional classroom methods. It blends online tools, technologies and materials with conventional classroom-based elements of teaching-learning. Blended learning involves collaborative learning wherein the role of the teacher transforms from a knowledge-disseminator to a facilitator mediated by a digital interface. The instructional methods and materials involve digital content along with traditional inputs. It enhances the possibilities of teaching courses in a more practical-oriented way. This is well suited for practice-based disciplines like medical and engineering which have found a special liking for this blended learning approach.

MOOCs for sustainable development is a recent digitally powered technical breakthrough in education. It is emerging as one of the most sought-after Ed-Tech development universalizing remote education. A MOOC is a web-integrated course with diverse participation addressing customized needs. Along with the traditional course materials, MOOC offers a plethora of innovative and interactive course options containing video lectures, self-paced reading options and structured problem sets. Its effectiveness is also rooted in multiple features like affordability, scripted assessments, continuous feedback, real-time interactions, open content, the flexibility of registrations and course completion, etc. Many MOOCs also offer multiple interactive platforms for teaching and student communities via social media forums. MOOCs offerings are frequently regarded as a disruptive innovation with a high potential for displacing traditional learning practices, particularly in higher education. Widespread and effective usage of MOOCs as a new-age knowledge transfer means will expect a certain level of reskilling on the part of the teachers as well as learners. The teaching fraternity needs to come out of the traditional methods of delivery to facilitate the knowledge transfer and absorption by students. On the part of students also, there is a prerequisite of greater awareness concerning learning outcomes. MOOCs open up a plethora of avenues for learners, the success of which depends on the pertinent choices made by the learners. The success in the case of MOOCs and similar other platforms requires a certain comfort level with new-age technological tools.

HEIs in western countries are adopting learning practices such as SCALE-UP. SCALE-UP is an acronym for Student-Centered Activities for Large-Enrollment Undergraduate Programs A SCALE-UP is an innovative classroom that resembles a restaurant, with round tables for the seating arrangement of teams of three to four students. This is an initiative of North Carolina State University. Students sit at round tables that are spread and scattered around the room in this method of learning, allowing the instructor to freely walk around the room and address students as required. In this approach, each team has a laptop to help them learn, as well as easy access to laboratory equipment in the closets around them. Students work out and solve problems on whiteboards and laptops, answer questions based on real-time situations and help each other learn with this method of instruction. This setup allows the instructor to engage all students in lively and enriching dialogues while they work. Students are encouraged to express their opinions, challenge each other and think out of the box. Students can see what others are doing and get involved in more comprehensive ways.

Various technology tools and gadgets are being adopted from the industry. Chatbots and clickers are simple illustrations in this regard. Chatbots are conversational interfaces and indulge in instant communication which responds to queries and enhances the experiences through technology. Using chatbots in HEIs can also promote smart learning. Chatbots perform two important tasks: they analyze a question and frame an answer. Chatbots can benefit HEIs by engaging users through personalized responses and providing information quickly and are available 24 × 7 × 365 which can provide information quickly. Classroom gadgets like *clickers* are common and prominent features of '*smart classrooms*.' These clickers are instructional devices that enable teachers to capture and evaluate student responses to questions in the classroom quickly. Instant feedback on student assignments and comprehension, increased student attendance, focus, and participation, and enhanced exam performance are some of the benefits of using clickers (Keough 2012).

Implementation of the latest outcomes of innovations will facilitate the skill enhancements, upskilling and reskilling initiatives. These new approaches to learning exemplify the potential of new practices in education enabled by digital and technological innovations, as well as the challenges of upskilling and reskilling required to make models successful.

15.4.2 Interface Between Industry 4.0 and Education 4.0

Education 4.0[3] is more than just education, it is a much-needed, desired, and preferred approach to learning that aligns with Industry 4.0 which focuses not just on smart technology but also on AI and robotics. Technology has now completely infiltrated the education process, and both students and teachers have adopted various techniques to utilize technology in basic ways. Education 4.0 is catering to the needs of the education sector in the ‘technological era’ of disruptions. To prepare future graduates for work, HEIs must align their teaching and learning-management processes with technological advancements. Sinlarat (2016) opined that this era’s learning management is a modern learning method that allows learners to develop knowledge and skills throughout their lives. One aspect of Education 4.0 emphasizes lifelong learning which necessitates intermittent skill upgradation on the part of learners, which equips them to face unforeseen challenges posed by continuous industrial evolutions, be it Industry 4.0 or Industry 5.0, and so on. For instance, methodologies like MOOCs and virtual learning will enable learners to remain relevant and productive throughout. An employee who has lost his/her job or a post-retirement employee can enroll for these courses at their choice and convenience and can regain their ability to contribute positively at the workplace. These technologies will enable people who have missed out on formal education to join the mainstream workforce and start their work-life at later stages also. In the absence of such a lifelong learning approach, the learners will be vulnerable to being obsolete and redundant.

Research undertaken by McKinsey & Company (2017) reveals that 60 percent of all occupations will have at least a third of their activities automated as a result of the fourth industrial revolution. In such a scenario being forecast, there is an urgent need for the HEIs to align their strategies and gear up to deliver and equip the students for survival in a technology-driven workplace. For example, HEIs are adopting the tools and technologies embracing Industry 4.0, such as AI-driven smart classrooms, chatbots for instant communication and data analytics. Big data is being used extensively in some of the progressive HEIs for admissions, placements and student-performance tracking and analysis. AI-driven classrooms make the interactions in classrooms more enriching and interactive. Chatbots are helping in resolving issues and queries in daily operations. All these technological adoptions will again demand a high level of reskilling and upskilling on the part of all stakeholders involved in the knowledge transfer process. This will again demand a high level of reskilling and upskilling on

[3] Some salient features of Education 4.0 (Srivastava 2018): (i) Education is demand-led instead of supply-led, (ii) competency-based replacing knowledge-based, (iii) incorporation and implementation of disruptive technologies and skill-sets, (iv) lifelong learning being preferred to front-loaded learning, (v) modular degree instead of one-shot going, (vi) more emphasis on EQ than IQ alone, (vii) focus on purposefulness, (viii) focus on mindfulness leading to overall happiness and wellbeing.

the part of all stakeholders involved in the knowledge transfer process. Universities will ensure that their students are prepared for the fourth industrial revolution by aligning teaching and learning strategies with potential skills requirements.

In 2016, the World Economic Forum produced a report highlighting and exploring the different changes that would be visible due to these disruptions. The report predicted that by 2020, '*more than a third of the desired core skill sets of most occupations will be comprised of skills that are not yet considered crucial to the job today.*' According to the study, social skills, complex problem-solving skills and process skills are among the soft skills that will soon become indispensable.

According to a report published by Quacquarelli Symonds (2019) which is the world's leading provider of services, analytics and insight to the higher education sector globally, another new approach to learning such as project-based learning highlights the need and relevance of studying and understanding a variety and wide set of skills that can then be applied to different scenarios. This enhances the skills of students linking to experiential learning.

The adoption of Education 4.0 will lead to a comprehensive skill-based and competency-based approach to teaching. HEIs are making efforts to gear up to incorporate and implement teaching on a digital platform. Extensive training sessions are being conducted by experts to equip faculty to handle online classes and helping them to use the digital board effectively. More emphasis is being given to modular-based teaching so that the students get to understand and learn the concepts in a better way. This leads to life-long learning wherein greater importance is given to the development of Emotional Quotient instead of Intelligence Quotient.

15.4.3 The Need for Skills Revamp in HEIs

Cambridge dictionary explains **upskilling** as '*the process of learning new skills or of teaching workers new skills,*' while **reskilling** as '*the process of learning new skills so that one can do a different job, or of training people to do a different job.*' To keep India's future talent pool relevant to the evolving industry demands, both upskilling and reskilling are of utmost importance.

Students who are engaged with HEIs will be required to join a workforce where uncertainty and disruptions and are the norms and a multiskilled workforce is a need of the hour and top priority. As a result, HEIs and educators must upgrade their skills to equip and prepare the future workforce. Various new techniques and tools such as mixed reality[4] capabilities are allowing the educators, HEIs, and faculty to offer students experiential learning, as technological disruptions are taking the center stage. Higher education institutions (HEIs) in India must learn to integrate technology into their teaching methods and curricula. Communication, negotiation and relationship-building are all human skills that play a role in career and capability growth and capacity-building.

[4] **Mixed Reality** is an innovative combination and mix of both physical and digital worlds, where human, computer and environment interactions are linked for new outcomes. This new reality is based on advancements in computer vision, graphical processing power, display technology and input systems.

15.4.3.1 Comparing the Demand for Top Skills – 2018 vs 2022

The World Economic Forum, in its *'The Future of Jobs Survey-2018,'* (WEF report 2018) highlights and classifies the future jobs and skills, as well as the rate of transition. It aims to shed light on the disruptions in 2020, placing them within the framework of an evolving economic cycle and the projected outlook for technology adoption, employment and skills in the next five years. A detailed analysis of the various skills that would be declining and skills that are trending to meet the current disruptions due to technology are highlighted in the report. The study illustrates the changing demand patterns of key skills. It depicts a continued decline in demand for manual skills and physical abilities on the one hand and a rise in demand for cognitive skills and abilities on the other.

Analytical thought process and creativity, as well as active learning techniques, will continue to be in demand by 2022. The growing demand for various types of technology competencies found by employers surveyed for this report highlights the dramatically increased value of skills such as technology design and programming. However, proficiency in emerging technology is just one aspect of the 2022 skills equation; 'human' skills, such as ingenuity, originality, initiative, strategic thinking, persuasion and negotiation, as well as attention to detail, persistence, versatility and complex problem-solving, will all maintain or increase in value. Emotional intelligence (EI), leadership, social power as well as service orientation are all experiencing a significant rise in demand compared to their current popularity.

To remain competitive in the face of rapidly evolving workforce capability requirements, companies will need to follow a variety of organizational strategies. To do so, corporate leadership and the role of human resources will need to change to lead the transition successfully. Companies studied in *'The Future of Jobs Survey'* highlight three potential strategies for workers facing changing skills demand: recruiting entirely new permanent staff with skills related to new technologies, attempting to fully automate the job tasks in question and retraining current employees.

According to the report, skills that are quite in-demand presently, such as writing, active listening, math, reading, management of financial and material resources, technology installation, maintenance, management of personnel and quality control will start declining in importance in the coming years. While other skills, such as analytical thinking, innovation, creativity, critical thinking, EI, complex problem-solving and ideation will increase in demand. This depiction of various skills has huge implications for industries, industry leaders, businessmen, educators, faculty and students, policymakers and other direct or indirect stakeholders in the education system.

There is an increasing need to understand and respond to the technological infusion and related skill shifts impacting organizations and workforces through a strategic approach formulated by all stakeholders getting impacted by such changes. Though the demand for new skills will keep on increasing in the future, the traditional skills related to Science, Technology, Engineering and Mathematics (STEM subjects) and soft skills (especially EQ) which enable people to enhance their human capabilities are going to gain more and more significance. Higher levels of technological evolutions will parallelly demand stronger EI. These unique skill requirements will need continuous strategic efforts of upskilling and reskilling to keep the human resources relevant and productive alongside hi-tech machines and technological gadgets.

Some of the strategic interventions which the HEIs and other institutions in leadership situations may think of are discussed in the following segment. These may include strategies to leverage the contemporary technology and innovations, adopting new pedagogies, experiential learning-based education, creation of new knowledge resources, new assessment tools, changes in curricula, focus on a learner-centric approach, and so on.

15.4.4 Strategies for Teaching-Learning Innovations in HEIs

Strategies for higher education must align with Industry 4.0 revolutions to provide and cater to new age personnel. Integration of technology and education is currently seen as a focus and thrust area for strategy formulation as the face of education in the future is set to change due to technology (Vardhan 2014). Digital literacy will replace the traditional benchmark of literacy identification and soon technology-mediated ways of learning will be a new norm. Slowly and steadily, technology-induced disruption in the education system is going to dismantle the conventional educational framework and establish new protocols for teaching and learning interface. This is going to create new quality and control standards along with unforeseen mechanisms of absorption and knowledge dissemination. New skill-set requirements are going to be in demand universally for those involved in teaching, managing and learning processes. Hence adequate strategies for adoption of the contemporary and latest technological innovations and pedagogies will be inevitable for HEIs to sustain.

15.4.4.1 Leveraging Technology and Innovations

Professor Gary Hammel had once famously proclaimed, *'Innovation is the only insurance against irrelevance.'* Today these words are true as gospel truth. The relevance of innovation applies to HEIs and the teaching-learning processes it has to offer. Innovations in higher education systems have deep implications and interrelationships with societal structures demonstrating interdependencies at all levels. It is the ability to innovate continuously and leverage the latest technologies which enables HEIs to remain relevant and contribute positively to social systems in which they operate.

Equitable distribution of technology and innovation outcomes is a prerequisite for HEIs to make a sustained contribution. For such equitable distribution, the prevailing 'digital divide' must be bridged in time. There is a universal occurrence of *digital divide* observable across societies and generations. In contemporary times, that divide is also pervasive among teachers and students. On the one side, there is the generation of youngsters which has been born into a world of technology and gadgets. They represent the group called 'Digital Natives' (they are exposed to digital tools and technological products from early childhood and they develop comfortable relations with them quite easily and quickly). On the other side is the group of 'Digital Immigrant' of the elder generation (who get to use technological gadgets and digital tools much later). Both these groups are on different platforms concerning technology and digital usage. When we take the case of the education sector, most of the teachers fall in the pre-digital group of digital immigrants while the students represent the group of digital natives. The divide creates problems in connecting and both groups face problems as both groups speak in different terms

(Prensky 2001). One big challenge which confronts the education sector is, 'Should the Digital Native students continue to learn the traditional old ways, or should their Digital Immigrant educators adapt and learn the new techniques?' (Prensky 2001). While both sides need to make adjustments to bridge this gap, the prime responsibility of taking the lead can be assigned to the experienced teaching fraternity and educational sector leadership. Teachers are better positioned to take the lead and need to initiate and reskill themselves to connect with the new generation which speaks different languages driven by technology, engage in unfamiliar activities and show novel responses and behaviors (Underwood 2007).

Advances in technologies which is revolutionizing modern workspaces not only need to be understood and analyzed in HEIs but also adopted in their practices too. Technological interventions in teaching-learning and other educational processes through enhanced ICT integration, virtual reality (VR) in teaching (Crosier et al. 2000), web-based instructional modules (Shimamoto 2012), and flipped classrooms (Maheshwari and Seth 2019) is likely to change the whole dynamics of traditional classroom setups. HEIs need to create new delivery mechanisms for the technology-driven teaching-learning process and create a new learning ecosystem. In this regard, infrastructural support will help in the smooth implementation of technology-driven strategies are core necessities. This will enable reskilling through the implementation of the latest gadgets and innovations. HEIs need to incorporate Industry 4.0 technologies such as AI, IoT, robotics, VR, augmented reality (AR), 3D printing, machine learning, chatbots and mixed reality capabilities into their teaching-learning methods. Smart classrooms with the latest tools and gadgets like smartboards are already existing in the educational institutions' premises and the trend needs focused application and to be expedited with greater urgency.

HEIs will need to formulate prudent strategies for promoting learning in the digital and automated world. Higher education's digitalization enhances the versatility of distribution, allowing students to access information at any time and from any place. Automated learning is a new way of learning that involves the use of electronic or mechanical equipment. It contains individualized reading materials, instructional modules, drills, interactive computers and online services that can be used outside of conventional institutions.

HEIs should be able to organize and manage digital learning and tech-activities within a system by focusing more on skill development. This will enrich both the faculty as well as the students to understand the delivery and streamline various tasks digitally, such as online assignments, online exams, lesson plans, course descriptions, messages, syllabus, course materials, etc. (Haghshenas 2019). Thus by becoming technologically competent, HEIs can provide learners with 24-hour access to e-learning platforms such as interactive boards, resulting in many advantages such as enhanced efficacy and quality of learning facilities due to improved communication with teachers and easier access to learning materials. Faculty members' acceptance of e-learning systems and technological disruptions will compel them to enhance the quality of digital interactions leading to skill development. In the evolving educational setup, traditional classrooms will no longer be what they used to be. Traditional classrooms with blackboards and having strict/overbearing/teachers will be replaced by new-age student-friendly 'smart' classrooms. MOOCs, VLE, use of robotics, chatbots, technology-based simulations, use of AI, AR and self-directed learning platforms are going to be the inalienable feature of HEIs in the coming years. Capabilities encompassing AR and VR will be needed to make sense of the world with 'mixed realities.' Integration of technology with education is already imbibed as the core of existence in most leading institutions and others are following the suit. The strategy of technology-supported teaching-learning processes is likely to result in improved academic performance, increased students' learning, higher students' satisfaction as well as greater faculty acceptance (Missildine et al. 2013).

The contemporary HEIs must formulate new strategies to empower the students with the much-needed skills, such as machine learning, big data, AI through virtual, online and digital learning. With the integration of technology in HEIs learning materials should be made available to students and instructors from a variety of operating systems, devices and browsers without the technical glitches that had caused difficulties in the past (Rollag and Billsberry 2012). By adopting active-learning approaches, HEIs must move away from teaching large content and progress toward less content and more holistic learning, thereby encouraging critical thinking and developing problem-solving abilities. Institutions must take care to facilitate experiential learning through increased use of business simulations and activities resulting in more practical and long-lasting learning among students. A pragmatic strategic approach in these areas will enhance the students' perspectives regarding how to be creative, have a multidisciplinary approach, how to innovate and adapt to the changing classroom and work scenarios.

15.4.4.2 Focus on Exploring and Adopting New Pedagogies

In an age when technological disruptions are restructuring the teaching-learning processes, the pedagogies (which are tools facilitating knowledge transfer and absorption by learners) can't remain unaffected. Pedagogies are bound to evolve and respond to the new focus, content and technological support systems being adopted in educational institutions. The pedagogical changes in many ways reflect the paradigm shift in teaching-learning processes wherein the teacher-led teaching and learning practices are giving way to student-centric, self-directed learning approaches. The conventional pedagogies are giving way to new age andragogy and Heutagogy.

Conventional pedagogies represented a state where the learner's maturity and involvement in setting the pace was least and the direction and content were mostly uni-directional from teacher to learner. The teacher's control over the whole knowledge transfer process was overbearing. In traditional systems, the role of technological support systems in the teaching-learning process was rudimentary and their impact was minimal. These conventional pedagogies (dominated by the lecturing method) worked well in such scenarios and served their purpose. But changing times and continuous technological infusion have made them archaic and almost redundant unless used in combination with other evolving pedagogical approaches. Andragogy is one such evolution that is based on the suppositions that grown-up students and adults need different learning inputs and approaches and they can learn under liberal instructor control and nominal guidance when compared to smaller children. While conventional pedagogy was teacher-driven learning, andragogy advocates self-directed learning. Such self-directed learning is hugely dependent on certain basic skill level requirements and also the involvement of the latest technological gadgets and support systems. The better the technological support system enabling the teaching-learning and overall knowledge transfer processes, the more empowered the learners to become to effectively set the pace for self-learning, absorption and effective implementation of knowledge.

The pedagogical evolution sees Heutagogy at the top of the evolutionary tree which gives utmost priority to student-centric learning. Heutagogy is distinct because of its student-led, self-determined learning, which underlines the significance of autonomy, focused approach to capacity building and capability enhancements. It is based predominantly on student-centered instructional strategy and student choice-based content design.

With HEIs giving more importance to a learner-centric approach, heutagogy is now being considered as an option where self-determined learning is the key element (Hase and Kenyon 2007; Hase and Kenyon 2013a; Hase and Kenyon 2013b). Heutagogical pedagogies are considered to have evolved out as an extension to andragogy which is self-directed learning (Blaschke 2012). Heutagogy emphasizes the role of human agency in the learning process. It can be used as an innovative approach for learning, where a learner is considered as 'the major agent in their learning, which occurs as a result of personal experiences' (Hase and Kenyon 2007, p. 112). As opposed to the teacher being the center of the curriculum, heutagogy focuses on the learner being the focus of the learning process. This innovative approach for skill enhancement will lead to skill-based learning.

The possibilities of need-based skill enhancements are multi-fold in such teaching-learning approaches. With the ever-increasing use of technology (digital, virtual, mobile learning, and so on) such self-directed learning approaches (illustrated through Heutagogy) are a natural advancement. Heutagogy encourages the democratic and equitable distribution of knowledge and unrestricted participation of all in the education process.

The goal of heutagogy is to teach lifelong learning and, as Blaschke (2012) wrote, 'to produce "learners" who are well-prepared for the complexities of today's workplace.' The heutagogical method allows students to solve their problems and ask their questions. Rather than simply completing the tasks assigned by their professors, these students look for areas of ambiguity and difficulty in the subjects they are studying. Teachers assist students by providing context for their learning and providing opportunities for them to thoroughly explore subjects. Heutagogical approach will enhance the skills of the learners as per the specific needs they have and it will happen through a self-driven, self-determined learning approach. This will also necessitate the large-scale digital skill requirements among learners as heutagogical approach requires remote/distance learning wherein learner is free to get inputs at their own pace and conveniences mostly through structured digital inputs. The learning inputs can be transferred in capsule-type content inputs to be grasped by the student with limited inputs from faculty. The classroom interactions are more focused on the illustration and application side of those inputs already grasped by the learner. In the heutagogical approach, even the syllabus structure, content and flow can be designed by the learner as per their requirements.

Heutagogical approach is a major step toward the needs of 'customization' in teaching-learning processes catering to very specific up-skilling requirements. It will also throw up challenges for facilitators to be skillful in implementing the pedagogy in its true spirit. This pedagogical approach can be lead to 'smart education' where emerging technologies are considered to be the platform for making distance education work. This pedagogical innovation is likely to serve as a framework for digital age teaching and learning (Anderson 2010). In these learning approaches, the clarity of purpose, understanding of outcomes and coordination among facilitator and learner become important prerequisites for success. Strategies in this direction, for successful adoption and implementation of Heutagogical way or similar other approaches, must ensure that the basic prerequisites are met both in terms of technical support as well as the skill and attitudinal requirements of tutor, learner and leadership of educational institutions.

15.4.4.3 Experiential Learning-Based Education

The teaching-learning process and its approach are drastically changing with the advent of Industry 4.0 and associated technology. Experiential learning refers to the application of theory and academic content to real-world situations, whether in the classroom or the workplace. This helps to improve and enhance course-based learning outcomes that are specifically focused on employability skills. Students can gain knowledge through first-hand experiences instead of reading or being taught.

15.4.4.4 New Bloom's Taxonomy for Upskilling

The six levels of Bloom's Taxonomy of Learning Objectives classify several skills that can be used to teach critical thinking. These include '*remember, understand, apply, analyze, evaluate and create.*' HEIs should try to move away from the conventional lecturing and other rote learning which focus on '*Lower Order Thinking Skills*' (remember, understand and apply) and strive to have pedagogies and content which enhance higher-level learning and development of 'Higher Order Thinking Skills' (analyze, evaluate and create). So, the shift should be more toward facilitating the development of critical thinking and creativity among students through greater emphasis on higher-order skills and application of knowledge.

15.4.4.5 New-Age Assessment Systems

HEIs should focus on redesigning the assessment systems and patterns and move away from merely ranking; instead, they should emphasize competency-based assessments. This will promote better learning and development among students and test higher-order skills, such as critical thinking, analytical skills and conceptual clarity. More focus should be on assessment for learning which will promote students to learn in individual and distinctive ways. The traditional concept of a rigid distinction between curricular, extra-curricular and co-curricular subjects needs to be dismantled. Instead, there should be a provision of multiple entries and exit options for students which will give much-needed flexibility to students to hone their skills and interests. The usage of AI can be helpful in the assessment of students' knowledge and potential. This will lead to a new level of understanding of the students' capabilities.

15.4.4.6 Curriculum Overhaul

There is a strong need to restructure and revamp the entire education system and reconsidering the pattern and purpose of higher education is important (Kaul 2011) and this can be realized through transformational changes in curriculum which must reflect the needs of contemporary times. In the light of radical technological changes and new workplaces, demands relate to new-age skills, the curriculum cannot remain outdated and static. The involvement of major stakeholders, especially the industry representatives and other employers seeking workforces from the graduates coming out of HEIs will help design the curriculum attuned to the skill requirements of contemporary workplaces. The curriculum contents will have to incorporate changes to support capacity building among learners through more dynamic, technologically aligned and comprehensive approach.

A strategy of a comprehensive overhaul of the curriculum breaking the silos between different domains of knowledge is a pressing need of the hour. It is being observed universally that the traditional mode of teacher-dominated interaction and lecture method yielding diminishing returns is soon becoming redundant, and technology is emerging as means of making the interaction more collaborative and learner-oriented. The curriculum and content of HEIs need to be innovative and relevant to address the demands of the present digital economy and meet the unforeseen challenges due to Industry 4.0. This is being well supported by technological interventions to make the shift seamless.

15.4.4.7 New Knowledge Resources

The success of educational institutions owes a lot to the knowledge assets it possesses. Knowledge resources development and management are key concerns for HEIs. Conventional knowledge management through formal libraries may start losing its relevance and effectiveness with the growing technological interventions. Newer technological concepts such as cloud-based storage facilities are opening new dimensions in knowledge management systems. HEIs are increasing their knowledge assets by having digital libraries, stored databases that provide distance access to users. It can be said that HEIs need to be like a springboard for digital innovation and organizational agility and helping students to gain access at any given point of time. Research and consulting activities and training facilities such as conducting Management Development Programs (MDPs) and Faculty Development Programs (FDPs) can become much effective with the support of technology as these can be integrated with various tech-platforms for a wider reach.

New knowledge assets will facilitate the scaling up of upskilling and reskilling activities through greater use of e-content, e-training, digital skill development programs. All this will not only enhance the possibilities of wider reach but will also lead to upskilling in a more focused and customized manner.

15.5 WAY AHEAD

The past few years have witnessed how digitalization has permeated not just in organizations but also in HEIs. Technological advancements will continue to offer new challenges and opportunities for HEIs and they need to continuously ponder over them (see Table 15.2). Continuous assessment and analysis will enable them to respond appropriately. Recent initiatives by all concerned stakeholders are making efforts to leverage the disruptions and channelize them toward skill development via

TABLE 15.2
Points to Ponder

HEIs Scenario in India: Challenges and Dilemmas
i. The Indian higher education system has undergone considerable growth in recent years and is estimated to reach $145bn by the year 2020.
ii. In a short period, India will achieve the distinction of having the world's largest tertiary-age population and the second-largest number of graduates (Heslop 2014).
iii. Due to the huge demand and appetite for education among India's youth, the higher education segment in India is also under tremendous stress and pressure to grow.
iv. The whole education system and the entire teaching-learning process is going to be impacted immensely by Ed-Tech evolutions.
v. In the present state of the education system and the institutions involved in it, there is a wide gap in their approach and effectiveness. They seem to exist at different evolutionary stages – some having reached a stage where they have all inbuilt capabilities to adapt and absorb all the complexities to meet the future, whereas many are still lagging in the evolutionary stage fight to acquire even the necessities. In such a scenario, how do leaders in education systems respond to the challenges of getting the future skill-sets ready?
vi. One moot point to be considered is – 'how will the infrastructural capabilities and the differential growth among leading and laggard HEIs hamper their capabilities response to future requirements?'
vii. Is there a clear picture of what kind of skill-sets (their nature and expanse) will be required? (Since new technology disruptions happening at such frequent intervals are changing the very picture in this regard so frequently.)
viii. Will the existing setups need a complete overhaul or partial adaptations will work allowing the evolution of laggards to happen in a natural organic way?
ix. Will there be a situation where the focus will be on specializations and super specializations only, or generalists will still be relevant in some ways?
x. How will the interface between the traditional human intelligence-based teaching-learning processes and the technology-driven teaching-learning processes change and which of these will have greater relevance in times ahead? Will the 'Emotional Intelligence' (EI) factor retain its prominence in the face of growing importance of Artificial intelligence (AI), machine intelligence, machine learning, and so on? Also, what kind of skill-sets will be required on the part of the instructors and learners in the new systems to balance the traditional human intelligence-based teaching-learning processes vis-à-vis the technology-driven teaching-learning processes?
xi. One major hindrance in the whole scenario is the mindset, inertia and resistance to adapt to new ways with the fear of losing their comfort zone. How can the systems evolve to make a shift?
xii. There are a lot of technological disruptions happening in the industry and corporate world which is at a much faster level when compared to the field of education. Can we imitate them (keeping them as benchmarks) and shape our educational system in terms of content, delivery and style of operating?
xiii. Some institutions, especially many engineering colleges have tried to respond to the dynamic environment by doing away with many core courses and introducing application-oriented courses. Is this a sign of things to come at a larger scale? Is it a workable strategy to follow?
xiv. Technological advancements have given many new ways of communications in the corporate setups (like chatbots, podcasts, blogging, microblogging, etc.; even the likes, dislikes, shares, hashtags may be used to take feedback clues) which are used for various purposes like feedback and all. Can some of these evolutions also be included in the educational set up to facilitate the teacher-learner interfaces? If yes, in what ways? This aspect needs further deliberation.

(continued)

TABLE 15.2 (*continued*)
Points to Ponder

xv. The role of analytics and big data management systems is playing a role in the education field also. The way they have revolutionized the decision processes and operational setups in the educational system at present will demand a continuous effort toward capacity building and skill upgradation of stakeholders involved in knowledge transfer.
xvi. The entire concept of teacher-learner roles will evolve with the likelihood of power distance being reduced and new student-centric approaches, such as one-on-one mentoring (for instance, corporate people becoming mentors to students), coaching style, etc., becoming prominent.

focused and need-based vocational training. *Continuous upskilling and reskilling, learning and unlearning, adaption and adjustments are the new 'mantras of survival and success in a globalized environment.*

There is an urgent need to address the impact of breakthrough technologies on HEIs, organizations and workforces through new education policies and strategies aimed at rapidly raising educational methodologies, deliverables and outcomes resulting in upskilling and reskilling of people enabling them to utilize their optimum capabilities. Relevant intervention must include changes in curricula, pedagogies, comprehensive reinvention of vocational training, teacher training and their skill up-gradation for aligning with the requirements of the age of the Fourth Industrial Revolution.

Some plausible actions which HEIs may contemplate implementing are as follows:

a. Institutions must develop a technology-friendly ecosystem.
b. All stakeholders must be ready to embrace technology and invest appropriately.
c. Universities, education authorities, government bodies and education policymakers must be proactive and supportive to develop a conducive environment, progressive policies, institution-friendly approach and enough flexibilities to modernize the education system.
d. The teaching profession must be prepared to learn and develop their skills continuously. Due to the rapid pace of technological disruption and transformation, it is critical not just for future generations of workers (today's students) but also for faculty to be able to teach students using Ed-Tech.
e. The need for faculty to have research-orientation will help and improve the quality of teaching.
f. HEIs and universities can benefit from research grants or incentives to encourage their faculty to participate in training and enrichment opportunities.
g. Students must be ready to widen their horizons and leverage the tools technology is providing them.
h. Institutions must be ready to explore and experiment with the latest developments like distance learning platforms, VLE, AI, robotics, machine

learning, MOOCs, flip-classroom methods, lecturer captures, small private online courses

i. Possibilities of scaling at a lower cost with uniform quality content and integration of disbursed institutions should be explored. Some leading institutions can play mentors' roles in a cluster of institutions.

15.6 PRACTICAL IMPLICATIONS OF STUDY

The issues discussed in the chapter are very pertinent with ever-growing significance for the people engaged in education, workplaces and policymaking. There are implications for stakeholders and the overall growth of the economy. The leadership of HEIs must keep track of the developments resulting from technological advancements and the much-discussed Industry 4.0 revolution. Discussions of the present study will help the HEIs to understand the intricate connections between Industry 4.0 and Education 4.0. These deliberations will help the HEIs leadership keep track of the evolving technological disruptions unleashed in the phase of the Industry 4.0 revolution. They will appreciate the fact these developments need to be comprehensively understood so that they can first prepare themselves and then transfer the most relevant and updated knowledge to their instructors and student community. They need to grasp the criticality of these disruptive changes and the urgency of adapting rapidly and suitably.

There are significant implications of the study for businesses and organizations. The requirement of high-level technical skills will continue to grow and organizations will need to look out for a workforce possessing relevant skills. Engaging closely with educational institutions and policymakers will help them in putting across their expectations from HEIs and also in positively influencing the educational institutions' future approaches and outcomes. The points discussed in the study concerning certain skills becoming obsolete and some other skills gaining prominence and the necessities of continuous upskilling and reskilling will help the practitioners to chart out their strategies for investing in training and talent development activities. The points explored in the chapter are likely to help future employers in keeping track of what their prospective employees (including students undergoing education in HEIs) are getting trained for and adopt appropriate strategies to acquire the industry fit and future-ready talent.

The discussions in the chapter will provide insights to policymakers at the institutional as well as the government level. The discussion would facilitate a more informed, comprehensive and strategic approach toward policymaking, keeping in mind technological disruptions, industrial adoption of technology and educational institutions' responses toward them. Such a pragmatic approach will help the economies in attaining more sustainable and egalitarian progress. Policymakers would do service to economic and social well-being if they can differentiate among skills and talents becoming imperative or redundant.

15.7 LIMITATIONS AND FUTURE RESEARCH POSSIBILITIES

The present study has focused on highlighting the upskilling and reskilling challenges of HEIs based on secondary data and other pertinent studies already done in the domain. There is a need to extend these studies and further validate the findings with the help of primary data. If researchers in the domain decide to conduct focus group discussions, surveys, face to face in-depth interviews or build case studies related to scope, methodological rigor, generalization possibilities, etc. (which can be taken ahead in future research), the present study may act as a starting point with a good platform to build on. The present study is not based on primary data, which may be seen as a limitation. Validation of the arguments can be strengthened through support from primary data inputs. Future studies may overcome this limitation by way of qualitative or quantitative studies based on empirical facts from the field.

The present study is exploratory research that does not aim at the generalization of its findings. Future surveys and detailed quantitative analysis can be taken up to describe the various dimensions of the upskilling and reskilling-related issues in HEIs discussed here. There are multiple ways of taking the present study forward. For instance, attempts to understand the students' or teachers' or even the administrative perspective on the issues of technological disruptions and the need for upskilling and reskilling in management education can yield some useful outcomes on the issue. There can be future studies on the topic of the 'digital divide' (among digital natives and digital migrants) and its implications for the various stakeholders of HEIs. There are possibilities of examining the alignments among issues of technological disruptions, pedagogical changes and various domains of study in higher education (i.e., engineering, management, medicine, commerce, management, and so on) as a topic of research. The topic is laden with possibilities for future scholars as the impact of technology in education is likely to be relevant for enquiry in times to come. The authors are hopeful that the discussions of the chapter serve a good purpose for future research and scholarly initiatives in this domain of knowledge.

15.8 CONCLUSION

Technology is set to change the face of education in the future (Vardhan 2014). The entire teaching-learning process is undergoing transformational changes due to the impact of Industry 4.0. The contemporary and future education system must be pragmatic and ready with its unique turnaround strategies to survive and prosper.

Rather than treating the new-age technology as a threat, it must be treated as a friendly facilitator enabling the teaching fraternity, students and management to impart quality education in a more focused, equitable and effective way. HEIs urgently need to align and respond to revolutionary developments resulting from Industry 4.0 as they owe to society the responsibility of providing future workforce and leaders to drive the revolution ahead. To make the most out of the Ed-Tech advancements it is vital for educators, students and other stakeholders to move out of the comfort zone of their conventional approaches and upgrade their skill levels to leverage technology in the most fruitful ways. *They must apprehend, adjust and answer the needs of present age learners and the workforce.* Faculty and students alike must own up the responsibilities to develop skill-sets and familiarize themselves with the ever-evolving technological tools to solve new-age problems in the most effective ways. The HEIs need to undergo a major turnaround, beginning with the reskilling and upskilling of educators to meet the requirements of tomorrow's workforce. HEIs need to shift the focus to an alternate teaching pedagogy from traditional teaching methods that will inject and infuse problem-solving and decision-making skills among faculty and students which will prepare them for today's dynamic and competitive business environment.

The dangers of getting swayed away with the flood of technological tools and gadgets and losing the best of human dimensions must be avoided and balance between human sensitivities and technological complexities must always be paramount. Core human qualities like EI, creativity, ethical approach, must be seen as unique advantages bestowed by nature and must be used in sync with technological tools and gadgets. The HEI leadership must invest in technology but at the cost of these unique human dimensions. Sensible and sincere efforts are needed to adjust and respond to these changing realities to continuously evolve and improve the educational processes and institutions.

Overall, the nature of teaching-learning processes, the prevailing culture, ambience of education systems, pedagogical initiatives and styles, skill requirements of students as well as faculty, the role of educators and administrators and policymakers as well as other stakeholders are going to change immensely and all of us need to embrace them with positivity and pragmatism.

REFERENCES

Anderson Terry. 2010. "Theories for Learning with Emerging Technologies". In by G. Veletsianos (Ed.), *Emerging technologies in distance education*, 35–50. Edmonton: Athabasca University Press.

Andrews Tessa M, Leonard Michael J, Colgrove Clinton A. and Kalinowski Steven T. 2011. "Active learning not associated with student learning in a random sample of college biology courses", *CBE-Life Sciences Education*, 10(4): 394–405.

Blaschke Lisa Marie. 2012. "Heutagogy and lifelong learning: a review of heutagogical practice and self-determined learning", *International Review of Research in Open and Distance Learning*, 13(1): 56–71.

Bolton Ruth N., Chapman Randall G. and Mills Adam J. 2019. "Harnessing digital disruption with marketing simulations", *Journal of Marketing Education*, 41(1): 15–31.

Clark Ruth C. and Mayer Richard E. 2016. "E-Learning and the Science of Instruction: Proven Guidelines for Consumers and Designers of Multimedia Learning", Pfeiffer, 4th Edition. https://www.wiley.com/en-us/e+Learning+and+the+Science+of+Instruction%3A+Proven+Guidelines+for+Consumers+and+Designers+of+Multimedia+Learning%2C+4th+Edition-p-9781119158660

Crosier Joanna K., Cobb Sue. V. G. and Wilson John R. 2000. "Experimental comparison of virtual reality with traditional teaching methods for teaching radioactivity", *Education and Information Technologies*, 5(4): 329–343.

Haghshenas Maryam. 2019. "A model for utilizing social softwares in the learning management system of e-learning", *Quarterly Journal of Iranian Distance Education*, 1(4): 25–38.

Hase Stewart and Kenyon Chris. 2007. "Heutagogy: a child of complexity theory", *Complicity: International Journal of Complexity and Education*, 4(1): 111–119.

Hase Stewart and Kenyon Chris. 2013a. *Self-determined learning: Heutagogy in action*. London, UK: Bloomsbury Academic.

Hase Stewart and Kenyon Chris. 2013b. "The Nature of Learning." In S. Hase and C. Kenyon (Eds.), *Self-determined learning: Heutagogy in action*, London, UK: Bloomsbury Academic.

Heslop Lynne. 2014. Understanding India: The future of higher education and opportunities for international cooperation. London: *British Council.* https://www.britishcouncil.in/sites/default/files/understanding_india.pdf

Jakobosne Andra and Sarma Cakula. 2012. "Online Experience-Based Support System for Small Business Development", Proceedings of the 8th WSEAS International Conference on Educational Technologies (EDUTE'12) Porto, ISBN 978-1-61804-104-3, ISSN 2227-4618.

Jarvis, Jeff. 2009. *What would Google do?,* New York, NY: HarperCollins.

Kaul, Vineet. 2011. "Development communication in India: prospect, issues and trends", *Global Media Journal*, 2(2):1–31.

Keough, Shawn M. 2012. "Clickers in the classroom: a review and a replication", *Journal of Management Education*, 36(6): 822–847.

Lage Maureen J., Platt Glenn J. and Treglia Michael. 2000. "Inverting the classroom: a gateway to creating an inclusive learning environment", *The Journal of Economic Education*, 31(1): 30–43.

Maheshwari Prateek and Seth Nitin. 2019. "Effectiveness of flipped classrooms: a case of management education in central India", *International Journal of Educational Management*, 33(5): 860–885.

Maqableh, M. M., Rajab, L., Quteshat, W., Khatib, T., & Karajeh, H. (2015). The Impact of Social Media Networks Websites Usage on Students' Academic Performance. *Communications and Network*, 7 (4), 159–171.

McKinsey & Company (2017). A future that works: Automation, employment and productivity, McKinsey Global Institute. Accessed at: https://www.mckinsey.com/~/media/mckinsey/featured%20insights/Digital%20Disruption/Harnessing%20automation%20for%20a%20future%20that%20works/MGI-A-future-that-works-Executive-summary.ashx

Missildine Kathi, Fountain Rebecca, Summers Lynn and Gosselin Kevin P. 2013. "Flipping the classroom to improve student performance and satisfaction", *Journal of Nursing Education*, 52(10): 597–599.

Prensky Marc. 2001. "Digital natives, digital immigrants", *On the Horizon*, 9(6): 1–6.

Quacquarelli Symonds. 2019. Retrieved from "everything you need to know about education 4.0", *European Journal of Education*, 42 (2): 213–222.

Rollag Keith and Billsberry Jon. 2012. "Technology as the enabler of a new wave of active learning", *Journal of Management Education*, 36(6): 743–752.

Selwyn Neil and Facer Keri. 2014. "The sociology of education and digital technology: past, present and future", *Oxford Review of Education*, 40(4): 82–496.

Shimamoto Dean N. 2012. "Implementing a flipped classroom: An instructional module. Presented at *The Technology, Colleges, and Community (TCC) Worldwide Online Conference.* Retrieved from https://scholarspace.manoa.hawaii.edu/bitstream/10125/22527/1/ETEC690-FinalPaper.pdf

Sinlarat P. 2016. "Education 4.0 is More than Education", *Annual Academic Seminar of the Teacher's Council* on the topic "Research of the Learning Innovation and Sustainable Educational Management", Bangkok: The Secretariat Office of Teacher's Council.

Skog Daniel A., Wimelius Henrik and Sandberg Johan. 2018. "Digital disruption", *Business & Information Systems Engineering*, 60: 431–437. https://doi.org/10.1007/s12599-018-0550-4

Srivastava, V. P., Srivastava, A., & Gambhir, S. (2018). Curricular aspect with special focus on skill development. *ESSENCE Int. J. Env. Rehab. Conserv.* IX (1), 28–34.

Ülker Doğancan and Yücel Yılmaz. 2016. "Learning management systems and comparison of open source learning management systems and proprietary learning management systems", *Journal of Systems Integration*, 7: 18–24.

Underwood Jean D. M. 2007. "Rethinking the digital divide: impacts on student-tutor relationships", *European Journal of Education*, 42 (2): 213–222.
Vardhan J. 2014. *Next Education Disrupting K-12 Education Segment, Bid to Top*, Pearson.
Wallace, Albin. 2013. "Social learning platforms and the flipped classroom", *International Journal of Information and Education Technology*, 4(4): 198–200.
World Economic Forum. 2020. *The Future of Jobs Report*, October 2020. Accessed at: http://www3.weforum.org/docs/WEF_Future_of_Jobs_2020.pdf

16 Strengthening the Retention Rate of Massive Open Online Courses through Emotional Intelligence and Intrinsic Motivation

Richa Chauhan and Nidhi Maheshwari

CONTENTS

16.1 INTRODUCTION

The digital revolution has opened a new vista of ubiquitous learning – the idea that learning can occur anywhere and not necessarily in classrooms (Kalantzis and Cope 2010). The unique feature of anywhere and anytime learning has the potential to enhance the problem-solving and decision-making capabilities of the learner along with enhances learning experiences through using experiential pedagogy (Fallows et al. 2013; de Freitas 2013; Dhawan 2020). Another feature of online learning is

DOI: 10.1201/9781003132097-16

that it got the tremendous potential of enhancing the employability of the learners is much needed in the digitalized era. It seems that the vision of universal higher education, forwarded in the seventeenth century by Comenius, has become achievable in the twentieth century through digital communication (Sadler 2013).

The flexible learning platform has contributed several massive open online courses (MOOCs), which are "online offerings provided at no cost or minimum cost, classically including open access to media-rich online materials and interactions for very large numbers of the audience which removes time-based and spatial boundaries in education" (Stacey 2013; Simonson et al. 2019).

However, the completion rate of MOOCs is not satisfactory (Kolowich 2013; Jordan 2014; Koller et al. 2013; Kizilcec and Schneider 2015). Through the literature review, it was found that the average completion rate is below 7%. Further, a similar result was identified in the study of Breslow et al. (2013) where out of 154,763 students only 5% were able to complete the course. This gives an alarming signal that it is very challenging to keep the learners motivated for completing the course. This instigates that the learners are required to have the right kind of attitude to continue on this platform. Customized MOOCs on learner's learning style and personality attributes may increase the possibility of enhancing learner's engagement and completion of the courses. Further, along with the instructional design of MOOCs the learning style, self-reliance of learners, motivation, and emotional intelligence (EI) are equally important for the successful integration of MOOCs into the curriculum (Romero-Rodríguez et al. 2020). According to Waiswa et al. (2020), EI is the predictor for the academic success of the students. Students need to be equipped with EI skills to allow them to perform more efficiently and effectively. It will help students to balance their emotions and their overall development. EI goes hand to hand with one's personality traits, being the mental ability that refers to an enhancement of the thinking process. Therefore, EI has been described as being a concept that may be referred to in terms of attention to, and discrimination of, one's emotions; it also offers an accurate recognition of one's own and other's emotions, of emotional empathy, control over emotions management, good self-motivation, possession of good social and communication skills, response with appropriate adaptive emotions and behaviors in different life situations, and ultimately achieving a balance between home, school, work, and social life. Thus, EI represents an important personality trait that could influence one's learning process (Viscu et al. 2017).

Therefore, it is suitable for scholars to identify the successful completion of learning with regards to learners motivation toward learning through MOOCS and their levels of EI.

16.1.1 Learner's Motivation and Completion of MOOCs

Successful completion and learning through electronic platforms demands self-reliance, discipline, individual commitment to self-determined goals, and maturity to command emotional instability, and the absence of these characteristics might lead to attrition (Chen et al. 2007). Disengaged behaviors like "carelessness" and "gaming the system" among learners who are not self-regulating themselves appear while continuing on the MOOCs. This disengagement reduces the participation and completion rate of MOOCs. Literature review shows that a learner's achievement is the consequence of self-control, self-efficacy, metacognitive skills, academic motivation, and psychological characteristics, etc. (Semenova 2020).

Further, with specific reference to MOOCs, Downes (2012) explains, "One big difference between a MOOC and a traditional course is that a MOOC is completely voluntary. You decide that you want to participate, you decide how to participate, and when you participate. If you're not motivated, then you're not in the MOOC". This indicates that learner's motivation plays an important role in the completion of MOOCs. Keller (2000) identified that student isolation, self-doubt, and limited learner control are the demotivating factors contributing to the attrition rate. Along these lines, Martinez (2003) found that learners' personality issues are the major factors responsible for high attrition rates. This shows that motivation and self-determination play a big role in the completion of MOOCs. Further, in-depth research disseminated that learner's maturity, self-perception, and self-reliance is the enriching components of learner engagement in the learning process (Bar-On 2006).

Motivation is person-specific and emotionally driven which can affect either positively or negatively (Pintrich 2003). Schunk et al. (2008) defined motivation as "the process whereby goal-directed activity is instigated and sustained". Adults are described to be motivated to learn by internal factors such as "increased job satisfaction, self-esteem, and quality of life" (Cercone 2008).

The research of Lee et al. (2005) identified that intrinsic motivation in terms of perceived enjoyment as well as extrinsic motivation, as skill enhancement, have been found important contributors toward learners' attitudes for online courses.

Motivation also plays a bigger role in how MOOCs are appreciated by the learners even in the presence of heterogeneity of the learner's population (Ricart et al. 2020). The study conducted by Hart (2012) exposed that learners' motivation is not only important for choosing MOOCs as an option for learning but it is also equally important for persistence in online learning environments. Starcher et al. (2011) apprised that the learner engagement and performance primarily depend on intrinsic motivation and its absence can further lead to drop the course or procrastination (Song 2000). Besides motivational components of self-learning, self-determination theory also plays an important role. The self-determination theory recognizes the dynamics of extrinsic, intrinsic, and motivation (a state of no willingness to perform) in a cultural context (Chen et al. 2010).

Ricart et al. (2020) found that individuals who are intrinsic motivation oriented are self-directed people who define a goal for themselves and identify ways to meet the challenges which come in their course of action in achieving goals. The intrinsic motivation ignites them to identify a thoughtful approach to learning which enhances creativity and persistence (Brophy 2013) and this persuasive attitude pays them self-reliance which reduces the extrinsic reinforcement (Ryan et al. 2000). Further, multitasking, which is common in learning through MOOCs, can be handled with intrinsic motivation as it ignites the capacity to do the task simultaneously (Shroff et al. 2009). This instigates that the absence of intrinsic motivation can improve disengagement and experiences concerning online learning (Hartnett et al. 2011). Further on these lines, the association between motivation and engagement is explored by Sun et al.

(2012) and they found that cognitive and emotional engagement depends on an individual's situational interest.

16.1.2 Motivation and Emotional Intelligence

EI is researched a lot in leadership studies and defined as leadership abilities to know the emotions of self and others, the distinction between different emotions, and using this information regarding one's behavior, thinking, and judgment (Serrat 2017; Sukys et al. 2019). EI models in research are explained under trait and ability categories (Petrides et al. 2000). The trait model explains EI as a personality trait that reflects an individual's emotional temperaments and experiences related to emotion perception, expression evaluation, and management (Petrides 2010). The ability model explains EI as "the abilities to perceive, appraise and express emotion; to access and/or generate feelings when they facilitate thought; to understand emotion and emotional knowledge and to regulate emotions to promote emotional and intellectual growth" (Mayer and Salovey 1997). The ability model explanation is directed toward cognitive ability than the trait model. Further, expression, evaluation, and regulation of self and other's emotions and their use in decision-making is emphasized in EI. Consequently, the Ability model explains the role of emotions in generating thoughts and actions.

The self-control and regulation of emotions help individuals in decision-making and enable productive cognitive processes (George 2000). Like, paying attention and regulation of emotions facilitates individuals to direct emotions to encourage themselves to enjoy their achievement aspiration. Focusing on regulating negative emotions help those taking responsibility for their behavior and avoiding emotional obstacles, like mood swings, in their journey of completion of work.

Literature review strengthens that emotions and motivation are closely associated and (Frijda 1994; Zurbriggen et al. 2002) many researchers agreed and accepted that motivation is an important constituent of behavior.

Sabath (2010) identified that motivation arises from EI. Individuals with an advanced level of EI can regulate and inspire themselves. According to Goleman, rational and emotional minds are two important constituents of the human mind. The rational mind tracks logical reasoning while the emotional mind follows emotions and feelings. In this way, EI comprehends individual behavior and encourages decisions. This is further proved in the research of Behnke and Ghiselli (2004) that EI directly influences motivation.

The negative emotional roadblocks in the journey of goal-directed behavior of an individual and stress are restrained by EI. Further, it helps to control emotions and mood swings and keep the individuals in a positive affective state which deters damaging consequences. During perplexing situations, EI enhances problem-solving capacity and helps the individuals to identify themselves as part of the solution, and motivates them to avoid withdrawal intentions. This is true concerning E-learners as well, because the high demand for time management, individual motivation for completion, and self-discipline for online learning need learners to control their emotions.

This is true concerning E-learners as well, because the high demand for time management, individual motivation for completion, and self-discipline for online learning need learners to control their emotions. As Vroom (1964) suggested that performance outcome is important for the learners to keep themselves motivated but equally the enthusiasm, amazement, and obstruction felt by the learner during online learning (Ashforth et al. 1990). At this juncture, it is justified to inspect that how EI contributes to learner's motivation to complete the MOOCs.

16.2 RESEARCH METHODOLOGY

This study is focused on examining the association between learner's motivation of completion of MOOCs and EI. The first is to examine the motivation level of learners toward the completion of self-directed MOOCs. Specifically, it tries to examine the association between EI and the completion of self-directed MOOCs through motivation. Specifically, the objectives are as follows:

1. To investigate the motivation level of learners' toward completion of self-directed MOOCs
2. To measure the EI of registered MOOC learners who participated in the study
3. To know the association between learners' EI and their motivation toward completion of MOOCs in the absence of any personalized direction by the mentors.

16.2.1 Research Questions

a. What is the level of EI of registered Learners with MOOCs?
b. What is the motivation level of learners toward the completion of MOOCs?
c. What is the association between EI and the motivation level of learners toward the completion of MOOCs?

16.2.2 Model to Test

The model exhibits several associations that were described in the reviewed literature but specifically concerning the completion of MOOCs the role of EI is not tested. Consequently, three hypotheses were framed to understand the relevance of EI learner's motivation levels toward completion of MOOCs in the Asian environment. The hypothesis is as follows:

H1: Learner's motivation and completion of MOOCs are positively related.
H2: Completion of MOOCs is positively predicted EI.
H3: EI mediates the relationship between learner's motivation levels toward completion of MOOCs.

As directed by literature review, and through which the model and hypotheses are resulting, this study is trying to establish if any association exist between EI and learner's motivation toward the completion of MOOCs. Starting from these considerations, the study aims to verify the meditational role of the EI in the relationship

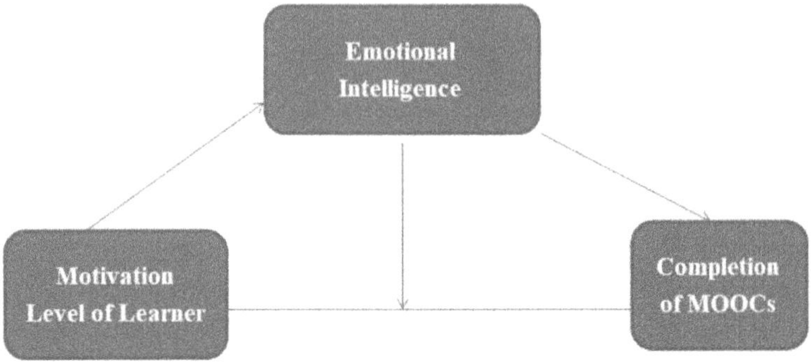

FIGURE 16.1 Relationships between EI and learner's motivation levels toward completion of MOOCs.

between learner's motivation levels toward completion of MOOCs. Given the extent of support among these links, Figure 16.1 serves as the model of our study of the relationships between EI and learner's motivation levels toward the completion of MOOCs.

16.2.3 Method

The study was conducted to have background information and define the terms of the research problem, hence it employed an exploratory research design. For supporting the theoretical perspectives between different variables inter-correlational research approach was applied to know the relationship between completion of MOOCs as the dependent variable and the motivation and EI as independent (predictor) variables.

16.2.3.1 Participants

The population of the study consisted of learners of Banasthali Vidyapith formally enrolled in the number of online courses related to business education available on Alison, Udemy, Coursera.org, NISM, edX portals. These courses are suggested by the faculty mentors as reading electives which are not credit courses but enhance their eligibility for getting jobs. The data was collected using judgmental, convenience, and snowball sampling using a self-administered questionnaire distributed to learners from the postgraduate students of the University. The respondents were regular university students from management courses who are using MOOCs. It is prescribed to them to complete a minimum of two MOOCs in an academic year. Generally, learners complete more than two reading electives from the above-mentioned portal and include these certificates in their curriculum vitae.

During the induction session held at the commencement of the term in July, the students of the management department of the university were briefed about procedures and program level objectives and the learners enrolled in the different online courses were invited to participate in the study voluntarily. The faculty members were approached and explained about the objectives of the study benefit, and

TABLE 16.1
Data Collection

Attempts	Administered	Responded
First Attempt	100	89
Second Attempt	200	172
Third Attempt	100	78

participation guidelines. The questionnaire was then emailed to participants with the promise that confidentiality would be maintained.

Overall, 339 questionnaires were collected (84.75%) with 61 proving to be missing some data, thus not usable (Refer Table 16.1). Only those participants who completed all three sections of the questionnaire i.e. EI, Motivation, and Success of MOOCs, their data was used for analysis. The demographic profile of the respondents includes name, age group, course, contact details. As per the criteria judgment, respondents were targeted. Using Cochran's Formula 339 sample size was determined.

16.2.3.2 Measures

16.2.3.2.1 Emotional Intelligence

For measuring learner's EI of questionnaire (composed of 16 measure items) was taken from Nasir et al. (2011). The questionnaire judged the important dimensions of EI: self-emotion appraisal, use of emotions, emotion appraisal of others, and regulation of emotions.

16.2.3.2.2 Successful Completion of MOOCs

To collect the data regarding the completion of MOOCs, five questions were framed: (1) Do you agree that the opted MOOCs enriched the academic knowledge? (2) How many Courses participants have registered and out of the registered courses, how many courses were they able to finish during July to April of their Master's degree? Those who had taken courses but not finished any one of them was taken as a very poor performer, those completed 25% of courses or modules were labeled as a poor performer, those completed 50% of courses or modules taken as an average performer, for 75% of courses completed labeled as a performer and those you had completed all the enrolled course and earned certificate are tagged as a very good performer. Further, to judge the success of MOOCs, three more questions were included and on the Likert scale, their data was recorded. They were as follows:

(3) Do you agree that the opted MOOCs have given the exposure of global level?
(4) Do you agree that the opted MOOCs helped to understand the industrial perspective of the subject?
(5) Do you agree that the opted MOOCs helped to gain credit toward final placement in the industry?

16.2.3.2.3 Intrinsic Motivation Scale

To assess the motivational level of leaner's the instrument was designed by adapting items from the study. In the study of de Barba et al. (2016), mastery-approach goals, value beliefs, and individual interest were identified as important factors for strong intrinsic motivation. For our research purpose, these three factors were considered to judge the intrinsic motivation level of the learners. For measuring mastery-approach goals, five items were taken from the study of Elliot and McGregor (2001) like "I want to learn in-depth about the concepts from MOOCs", "I need to understand the content of the MOOCs" as thoroughly as possible, "I desire to completely master the material presented in MOOCs". For utility value belief, three items were chosen from the study of Hulleman et al. (2007) and one of them was "The content of MOOC is very meaningful". From the study of Pintrich and García (1994), two items were taken for individual interest and one of them is "I think that what we are learning in MOOCs is interest".

16.2.3.3 Data Analysis

Convergent, discriminant, and nomological validity need to be confirmed for ensuring the construct validity of the instrument. Convergent validity ensures the basic significance of factor loading and reliability analysis. To ensure the face validity, few questions were asked to the panel of experts and respondents in a pilot survey. After analyzing the results researcher found that all the experts and respondents agreed that the questionnaire was measuring what it had to measure. Hence, the face validity of the questionnaire was ensured. Nomological validity ensures when a construct correlates with the other construct in a way as it should do. Observing the correlation matrix, all possible linkages among the constructs found as per the law of nature.

Path coefficient explains the hypnotize relationship among constructs. Their values between +1 and −1 and the Estimated levels of path coefficients are mostly close to plus one it indicates the constructive, positive associations between constructs. The significance of a coefficient is ultimately determined via the calculation of the empirical t-values obtained using bootstrapping. The objective of PLS-SEM is to classify and determine the significant path coefficients in the structural model and also to understand the essential and relevant effects. As per Table 16.2 to check the level of Structural Path Significance in Bootstrapping T statistics values are generated to check the significance level of both the inner and outer model of the structural path, using a statistical procedure called bootstrapping (Hair et al. 2012).

TABLE 16.2
Descriptive Analysis

	Original Sample (O)	Sample Mean (M)	Standard Deviation (STDEV)	T Statistics (|O/STDEV|)	P Values
EI -> MOOC	0.445	0.449	0.077	5.793	0.000
Motivation -> EI	0.772	0.774	0.024	32.444	0.000
Motivation -> MOOC	0.242	0.240	0.078	3.105	0.000

TABLE 16.3
Measurement Model Testing Results

	Cronbach Alpha	Composite Reliability	AVE
EI	0.832	0.881	0.598
MOOC	0.819	0.917	0.847
Motivation	0.740	0.852	0.658

16.2.3.4 Measurement Model

Discriminate validity, average variance extracted (AVE), composite reliability Cronbach alpha were identified and significance of factor loadings recorded to determine the reliability and validity of the model. Items that load high on their respective variables ensure convergent validity. Items loading above the cut-off value of 0.5 are acceptable (Hair et al. 2017). Hence, items with a cut-off value of 0.5 were eliminated from the instrument after the pilot study. The Cronbach alpha coefficient values (α) and composite reliability of all the variables were exceeded the acceptable cut-off limit of 0.70. The AVE also exceed the acceptable cut-off of 0.5 (refer to Table 16.3).

Discriminant validity confirms when the item loads the highest on its variable. The Fornell-Larcker criterium satisfied the criteria for discriminant validity where the item loads highest on its item. Hence, discriminant validity was confirmed for all the factors (refer to Table 16.4).

Chin et al. (1999) and Fornell and Bookstein (1982) recommended the situations of using CB-SEM or PLS-SEM. The purpose of the selection of both lies in the research objective. The primary purpose of the maximum likelihood approach was to examine the construction of the observables. The objective of the PLS-SEM was to envisage the indicators through components extension. Given the statement, Hair et al. (2014) proposed that if the study is exploratory or an expansion of some existing structural theory then applies PLS-SEM. As the study applied exploratory research design PLS-SEM was applied for the data analysis. In a structural equation model, PLS-SEM evaluates the parameters of a set of equations by combining principal components analysis with regression-based path analysis' (Mateos-Aparicio 2011). In management research, PLS is becoming popular as it is based on an iterative blend of principal components analysis and regression, and it explains the variance of the construct in the model (Chin 1998).

TABLE 16.4
Fornell-Larcker Criterium

	EI	MOOC	Motivation
EI	0.774		
MOOC	0.632	0.920	
Motivation	0.772	0.586	0.811

TABLE 16.5
Evaluation of Structural Model

		β	F^2	T Values	P Values	Result
Motivation	EI	0.772	1.479	32.444	0.00	Hypothesis Supported
EI	MOOC	0.445	0.138	5.793	0.00	Hypothesis Supported
Motivation	MOOC	0.242	0.041	3.105	0.00	Hypothesis Supported

In the structural model, the Motivation ($\beta = 0.772$, $p = 0.00$) has a significant positive relationship with EI and also having ($\beta = 0.24$, $p = 0.00$) positive relationship with MOOC. But EI ($\beta = 0.45$, $p = 0.00$) has a strong positive relationship with the MOOC rather than motivation. The results of all the hypotheses are consistent, therefore researcher rejects the null hypothesis. The F^2 value between 0.02 and 0.15 reflects a small effect size, between 0.15 and 0.35 reflects medium effect size, and between 0.35 and above reflects a large effect size. Motivation has a large effect size ($F^2 = 1.479$) on EI and ($F^2 = 0.041$) very small effect or no effect size on MOOC (refer to Table 16.5).

16.2.3.5 Mediation Analysis

The bootstrapping was performed on the sample at the significance level of 0.05 one-tailed distribution to identify the significance of the relationship among different variables. The mediation table shows that the Direct effect between Motivation MOOC (P1) is 0.242 (P-Value – 0.00) and the Indirect effect between Motivation MOOCs is (P2 * P3) 0.344 (P-Value – 0.000). As both the indirect effect of P2 * P3 and the direct effect of P1 are significant, we calculated the product of P2 * P3 * P1 and which is 0.083 and positive which means complementary mediation exists. Further, complementary mediation means motivation does lead to the success of MOOCs but the impact is not higher directly. Higher motivation leads to higher EI which in turn enhances the chances for MOOCs success or successful completion of the MOOCs (Refer Table 16.6).

TABLE 16.6
Effect of Learner's Motivation on Completion of MOOCs through Emotional Intelligence

Path	β	P-Value
The direct effect between Motivation MOOC (P1)	0.242	0.00
The indirect effect between Motivation MOOC is (P2 * P3)	0.344	0.00

Motivation is important for the completion of MOOCs but in the presence of EI, chances are increased for the MOOCs completion.

16.3 FINDINGS AND DISCUSSION

In the structural model, the Motivation ($\beta = 0.772$, $p = 0.000$) has a significant positive relationship with EI, and EI ($\beta = 0.45$, $p = 0.000$) has a significant positive relationship with the MOOC. Excluding insignificant items interest, completely master the content for enhancing understanding and meaningfulness for industry perspective loads high on the EI. High loading of these items shown that Motivation has an impact on the EI of an individual. Specifically, three items of concept learning were identified as more significant and the most significant item was enhancing the understanding level of academic concepts. This may be because learners enhance their level so that the industrial perspective can be added to their learning which will pay them to ensure their placements. Further, the interest item of Value belief is found significant as learners join those MOOCs which are of their interest area. Therefore, it would be useful for course faculty members to reiterate the importance of all components of MOOCs and outline how the particular module relates to industry practice.

In the structural model, all paths are supporting the hypotheses H1, H2, and H3 by showing significant positive cut-off values which indicates that all the constructs have a significant positive relationship among them. Thus, we rejected the corresponding null hypotheses. The analysis shows that the construct Motivation (independent variable) is significantly correlated with EI and MOOC. Therefore, it was confirmed that Motivation and EI together create an impact on MOOC success. Motivation influences MOOC and EI too but Motivation has a large impact on EI and EI has a larger impact on MOOC rather than motivation impact on MOOC success which predicts EI as the main influencer.

Through the analysis, it was found that successful completion of MOOCs is predictable largely through the level of EI of the learner. As it was found that the grouping of EI and intrinsic motivation served as a stronger interpreter of completion of online courses. This finding implies that learner's intrinsic motivation is important for selecting MOOCs as learning pedagogy, but at the same time, EI is equally important in completing MOOCs. As self-discipline plays an important role in the completion of MOOCs it may be inferred that learner's EI nourishes self-discipline through regulation of emotional component. This finding can be utilized while addressing student retention issues related to the completion of MOOCs. While selecting MOOCs as part of credit score, it is desirable that the EI level of learners need to be determined. The EI competencies of learners need to be attributed while doing the profiling of the learners as well as while choosing the teaching pedagogy. The customization of teaching pedagogy based on the level of learner's EI may enhance the probability of learner's successful completion rate of MOOCs. Further, the academic administration can use this finding while determining the most effective platform/pedagogy for learner's success.

In our study, it was identified that items related to self-emotional appraisal, regulation of emotion, and use of emotions dimensions were significant than an emotional appraisal of others. Specifically understanding feelings, managing emotions, setting goals, motivating self, restricting self, control over emotions, task completion determination are significantly associated with the completion of MOOCs.

Theoretically, our results are consistent with earlier explanations of emotionally intelligent individuals (Berenson et al. 2008). It was found that individuals who are helpful and outgoing are more inclined to connect with online success. These personality traits of learners fit with the description of emotionally intelligent individuals who are self-regulators of their emotions and can maintain a strong relationship with others by controlling their emotions (Goleman 2012) as they are self-motivated to take a risk and achieve their goals related with their beliefs.

The research work of Wang and Newlin (2000) promoted that individuals with a high level of EI have a higher level of self-efficacy, internal locus of control, and have the potential to regulate emotions. Self-regulation of emotions is the ability to cope with environmental stressors, controlling impulses, taking responsibility for one's actions. Self-regulated people willingly take the risk to prove one's beliefs, and they are competent in managing conflict and diffusing tense situations.

The non-availability of immediate feedback and face-to-face connection between learner and instructor is a big challenge in completing the MOOCs. In such an environment, self-regulation and self-reliance of learners are very important. In our study, self-regulation was identified as significantly associated with successful completion of MOOC. This finding reinforces the research work of Kemp (2002) where it was identified that those who have high levels of resilience complete the academic degree.

16.3.1 Implication of Research

It is highly appreciated by learners, advisors, academic administrators that a research-based approach needs to be followed for assessing learner's potential concerning online learning. The customized MOOCs can be designed as well as the component of experiential learning can be enhanced while designing MOOCs if a deeper understanding regarding online learner's thinking and learning pattern can be created. As learner's self-reliance and persuasiveness play a major role in the completion of courses role of the course-advisement process is very important. More the matching between personality traits and the courses is there more will be the probability to complete the courses. However, the data was collected from a single academic institution and data were self-reported but it can be generalized for a similar sort of population.

The present study identified the significant relationship between intrinsic motivation, EI, and successful completion of MOOCs but it is not conclusive. The mediation effect of EI provides a basis to further explore the role of EI in the continuously growing online learning ecosystem.

To establish the generalized ability of the identified relationship between EI and successful completion of MOOCs, it is needed that larger samples of experienced students from various universities need to be analyzed. In the present work, only the EI profiling of the learner's was done but learner's, as well as online instructors' EI and intrinsic motivation, could be explored for determining the extent that these constructs play in the online success of learners.

REFERENCES

Ashforth, Blake E., and Raymond T. Lee. 1990. "Defensive behavior in organizations: A preliminary model." *Human Relations* 43 (7): 621–648. https://doi.org/10.1177/001872679004300702

Bar-On, Reuven. 2006. "The Bar-On model of emotional-social intelligence (ESI) 1." *Psicothema: 13-25. Brophy, Jere E. Motivating students to learn.* Routledge.

Behnke, Carl, and Richard Ghiselli. 2004. "A comparison of educational delivery techniques in a foodservice training environment." *Journal of Teaching in Travel & Tourism* 4 (1): 41–56.

Berenson, Robin, Gary Boyles, and Ann Weaver. 2008. "Emotional Intelligence As a Predictor of Success in Online Learning". *The International Review of Research in Open and Distributed Learning* 9 (2). https://doi.org/10.19173/irrodl.v9i2.385.

Breslow, Lori, David E. Pritchard, Jennifer DeBoer, Glenda S. Stump, Andrew D. Ho, and Daniel T. Seaton. 2013. "Studying learning in the worldwide classroom research into edX's first MOOC." *Research & Practice in Assessment* 8: 13–25.

Cercone, Kathleen. 2008. "Characteristics of adult learners with implications for online learning design." *AACE Journal* 16 (2): 137–159.

Chen, Clement C., and Keith T. Jones. 2007. "Blended learning vs. traditional classroom settings: Assessing effectiveness and student perceptions in an MBA accounting course." *Journal of Educators Online* 4 (1): n1.

Chen, Kuan-Chung, and Syh-Jong Jang. 2010. "Motivation in online learning: Testing a model of self-determination theory." *Computers in Human Behavior* 26 (4): 741–752.

Chin, Wynne W. 1998. "The partial least squares approach to structural equation modeling." *Modern Methods for Business Research* 295 (2): 295–336.

Chin, Wynne W., and Peter R. Newsted. 1999. "Structural equation modeling analysis with small samples using partial least squares." *Statistical Strategies for Small Sample Research* 1 (1): 307–341.

de Barba, Paula G., Gregor E. Kennedy, and Mary D. Ainley. 2016. "The role of students' motivation and participation in predicting performance in a MOOC." *Journal of Computer Assisted Learning* 32 (3): 218–231.

de Freitas, Sara. 2013. *Education in computer generated environments.* Routledge.

Dhawan, Shivangi. 2020. "Online learning: A panacea in the time of COVID-19 crisis." *Journal of Educational Technology Systems* 49 (1): 5–22.

Downes, Stephen. 2012. "Connectivism and connective knowledge: Essays on meaning and learning networks."

Elliot, A. J., & McGregor, H. A. (2001). A 2 × 2 achievement goal framework. *Journal of personality and social psychology,* 80(3), 501.

Fallows, Stephen, and Christine Steven. 2013. *Integrating key skills in higher education: Employability, transferable skills and learning for life.* Routledge.

Fornell, Claes, and Fred L. Bookstein. 1982. "Two structural equation models: LISREL and PLS applied to consumer exit-voice theory." *Journal of Marketing Research* 19 (4): 440–452.

Frijda, Nico H. 1994. "Emotions are functional, most of the time."

George, J. M. 2000. "Emotions and leadership: The role of emotional leadership." *Human Relations* 53 (8): 1027–1055.

Goleman, Daniel. 2012. *Emotional intelligence: Why it can matter more than IQ.* Bantam, Hartnett.

Hair, Joe F., Marko Sarstedt, Christian M. Ringle, and Jeannette A. Mena. 2012. "An assessment of the use of partial least squares structural equation modeling in marketing research." *Journal of the Academy of Marketing Science* 40 (3): 414–433.

Hair Jr, J. F., Sarstedt, M., Ringle, C. M., & Gudergan, S. P. (2017). Advanced issues in partial least squares structural equation modeling. saGe publications.

Hair Jr., Joseph F., Tomas G., Hult M., Christian Ringle, and Marko Sarstedt. 2016. *A primer on partial least squares structural equation modeling (PLS-SEM)*. Sage publications.

Hair Jr., Joe F., Marko Sarstedt, Lucas Hopkins, and Volker G. Kuppelwieser. 2014. "Partial least squares structural equation modeling (PLS-SEM): An emerging tool in business research." *European Business Review* 26 (2): 106–121.

Hart, Carolyn. 2012. "Factors associated with student persistence in an online program of study: A review of the literature." *Journal of Interactive Online Learning* 11 (1): 19–42.

Hartnett, Maggie, Alison St. George, and John Dron. 2011. "Examining motivation in online distance learning environments: Complex, multifaceted, and situation-dependent." *International Review of Research in Open and Distributed Learning* 12 (6): 20–38.

Hulleman, M., De Koning J. J., Hettinga F. J., and Foster C. 2007. "The effect of extrinsic motivation on cycle time trial performance." *Medicine & Science in Sports & Exercise* 39(4): 709–715.

Jordan, Katy. 2014. "Initial trends in enrolment and completion of massive open online courses." *International Review of Research in Open and Distributed Learning* 15 (1): 133–160.

Kalantzis, Mary, and Bill Cope. 2010. "The teacher as designer: Pedagogy in the new media age." *E-Learning and Digital Media* 7 (3): 200–222.

Keller, John M. 2000. "How to integrate learner motivation planning into lesson planning: The ARCS model approach." VII Semanario, Santiago, Cuba, pp. 1–13.

Kemp, Sharon. 2002. "The hidden workforce: Volunteers' learning in the Olympics." Journal of European Industrial Training.

Kizilcec, René F., and Emily Schneider. 2015. "Motivation as a lens to understand online learners: Toward data-driven design with the OLEI scale." *ACM Transactions on Computer-Human Interaction (TOCHI)* 22 (2): 1–24.

Koller, Daphne, Andrew Ng, Chuong Do, and Zhenghao Chen. 2013. "Retention and intention in massive open online courses: In depth." *Educause Review* 48 (3): 62–63.

Kolowich, Steve. 2013. "The professors who make the MOOCs." *The Chronicle of Higher Education* 18: 1–12.

Lee, Matthew K. O., Christy M. K. Cheung, and Zhaohui Chen. 2005. "Acceptance of Internet-based learning medium: The role of extrinsic and intrinsic motivation." *Information & Management* 42 (8): 1095–1104.

Martinez, Margaret. 2003. "High attrition rates in e-learning: Challenges, predictors, and solutions." *The eLearning Developers Journal* 2 (2): 1–7.

Mateos-Aparicio, Gregoria. 2011. "Partial least squares (PLS) methods: Origins, evolution, and application to social sciences." *Communications in Statistics-Theory and Methods* 40 (13): 2305–2317.

Mayer, John D., and Peter Salovey. 1997. "What is emotional intelligence." *Emotional Development and Emotional Intelligence: Educational Implications* 3: 31.

Nasir, Farheen, and Seema Munaf. 2011. "Emotional intelligence and academics of adolescents: A correlational and gender comparative study." *Journal of Behavioural Sciences* 21 (2): 93–101.

Petrides, Kostantinos V. 2010. "Trait emotional intelligence theory." *Industrial and Organizational Psychology* 3 (2): 136–139.

Petrides, Konstantine V., and Adrian Furnham. 2000. "On the dimensional structure of emotional intelligence." *Personality and Individual Differences* 29 (2): 313–320.

Pintrich, Paul R. 2003. "A motivational science perspective on the role of student motivation in learning and teaching contexts." *Journal of Educational Psychology* 95 (4): 667.

Pintrich, Paul R., and Teresa Garcia. 1994. "Self-regulated learning in college students: Knowledge, strategies, and motivation." In Student motivation, cognition, and learning: Essays in honor of Wilbert J. McKeachie, pp. 113–133, Routledge

Rasmussen, E. B., and Lawyer, S. R. 2008. *21st century psychology: A reference handbook* (Vol. 1). Sage.

Ricart, Sandra, Rubén A. Villar-Navascués, Salvador Gil-Guirado, María Hernández-Hernández, Antonio M. Rico-Amorós, and Jorge Olcina-Cantos. 2020. "Could MOOC-takers' behavior discuss the meaning of success-dropout rate? Players, auditors, and spectators in a geographical analysis course about natural risks." *Sustainability* 12 (12): 4878.

Romero-Rodríguez, Luis M., María Soledad Ramírez-Montoya, and Jaime Ricardo Valenzuela González. 2020. "Incidence of digital competences in the completion rates of MOOCs: Case study on energy sustainability courses." *IEEE Transactions on Education* 63 (3): 183–189.

Ryan, Richard M., and Edward L. Deci. 2000. "Self-determination theory and the facilitation of intrinsic motivation, social development, and well-being." *American Psychologist* 55 (1): 68.

Sabath, Arpita. 2010. "Emotional Intelligence: Key to Performance Excellence." Souvenir of National Seminar on Emotional Intelligence: A Key to Human Well-being, Amity University, Lucknow, held on 4–5 March.

Sadler, John Edward. 2013. *JA Comenius and the concept of universal education*. Vol. 32. Routledge.

Schunk, D., Meece, J. and Pintrich, P. 2014. *Motivation in education*. Boston: Pearson.

Semenova, Tatiana. 2020. "The role of learners' motivation in MOOC completion." *Open Learning: The Journal of Open, Distance and e-Learning*: 1–15.

Serrat, Olivier. 2017. "Understanding and developing emotional intelligence." In *Knowledge solutions*, pp. 329–339. Springer, Singapore.

Shroff, Ronnie H., and Douglas R. Vogel. 2009. "Assessing the factors deemed to support individual student intrinsic motivation in technology supported online and face-to-face discussions." *Journal of Information Technology Education: Research* 8 (1): 59–85.

Simonson, Michael, Susan M. Zvacek, and Sharon Smaldino. 2019. *Teaching and learning at a distance: Foundations of distance education*, 7th edition, Information Age Publishing, Inc., Charlotte, NC.

Stacey, Paul. 2013. "The pedagogy of MOOCs." edtechfrontier.

Starcher, Keith, and Dennis Proffitt. 2011. "Encouraging students to read: What professors are (and aren't) doing about it." *International Journal of Teaching and Learning in Higher Education* 23 (3): 396–407.

Sun, Jerry Chih-Yuan, and Robert Rueda. 2012. "Situational interest, computer self-efficacy and self-regulation: Their impact on student engagement in distance education." *British Journal of Educational Technology* 43 (2): 191–204.

Sukys, Saulius, Ilona Tilindienė, Vida Janina Cesnaitiene, and Rasa Kreivyte. 2019. "Does emotional intelligence predict athletes' motivation to participate in sports?." *Perceptual and Motor Skills* 126 (2): 305–322.

Song, Sang Ho. 2000. "Research issues of motivation in web-based instruction." *Quarterly Review of Distance Education* 1 (3): 225–229.

Viscu, Loredana Ileana, Cornelia-Ecaterina, Cornean, Roxana, Colojoara, and Ioana-Eva, Cadariu. 2017. The role of emotional intelligence in online learning. The International Symposium of Research and Applications in Psychology, 24th edition, with the theme "Cognitive characteristics of transdisciplinarity. Applications in psychology and psychotherapy," Timisoara, At Timisoara, Volume: SICAP 24 PROCEEDINGS.

Vroom, Victor Harold. 1964. "Work and motivation." San Francisco: Jossey-Bass Publishers.

Waiswa, Helen Christine Amongin, Peter K. Baguma, and Joseph Oonyu. 2020. "Emotional intelligence and academic achievement among university upgrading (Grade V) teacher students." *African Journal of Education and Practice* 6 (4): 39–64.

Wang, Alvin Y., and Michael H. Newlin. 2000. "Characteristics of students who enroll and succeed in psychology web-based classes." *Journal of Educational Psychology* 92 (1): 137.

Zurbriggen, Eileen L., and Ted S. Sturman. 2002. "Linking motives and emotions: A test of McClelland's hypotheses." *Personality and Social Psychology Bulletin* 28 (4): 521–535.

17 Digitalization of Higher Education
Issues and Challenges

Gautam Shandilya and Abhaya Ranjan Srivastava

CONTENTS

17.1 INTRODUCTION

'Education is the passport to the future, for tomorrow belongs to those who prepare for it today'

Malcolm X

The word 'education' is taken from the old archaic Latin word '*educatio*' which means to nurture or to rear through training ("Education" n.d.). Education is the system or process which facilitates learning by primarily acquiring knowledge and skills and subsequently building values, beliefs and habits. Education can be imparted in both formal (physical in this context) and informal (digital or online in this context) settings under the guidance of teachers or educators or by self-learning using methods like teaching, research and experiments, training and development and discussions and discourses. The idea is that it must have a formative effect on how one feels (affective), thinks (cognitive) and acts (behavioral) in a given

DOI: 10.1201/9781003132097-17

situation. This whole gamut or complete range or scope of teaching methodology is called pedagogy.

17.1.1 History

Imparting education is a prehistoric practice, wherein life skills were passed from one generation to other through observation and some oral instructions and legends ("Educate" n.d.). Later on, when clans were converted into society and civilizations were structured, languages with written script got formed, and education got formally structured.

In India, Ashrams were established by various saints and sages (gurus of those times) under the patronage of kings and ruling classes. Formal and informal educations were given to the pupils in the form of 'Diksha'. These institutions also used to have convocation or 'Dikshant' at the end of their courses. Education and its delivery methodology saw a sea change since pre-historic age to ancient era to medieval period to its present modern form. Formal classroom teaching and learning in a set environment with a certain designed curriculum including class size, interaction and assessment methods were formulated and updated time and again.

17.1.2 Contemporary

In the meantime, various educationists and reformists came up with certain new and innovative ideas like alternative schooling, self-learning, home schooling, unschooling, open classroom, play school, charter school, indigenous methods of learning, informal learning, auto didacticism, evidence-based learning and open learning with electronic technology.

Various learning modalities and styles were developed and followed in the last few decades. These learning modalities fit within the above-mentioned methods of learning. Some of the common learning modalities are:

- Visual
- Auditory
- Kinesthetic

Our education system is bound with these learning and teaching methodologies and all the developments in the field of education seen in the past few decades are based on them.

These learning modalities and styles can be understood from the under shown figure.

Figure 17.1 shows the dimensions of preferred learning styles of the receivers (students) and their corresponding styles of teaching adopted by the teachers or educators. This model shows that learners perceive the visual and auditory inputs from their senses and intuitions. They organize these inputs in either inductive or deductive manner, process the information in action or reflection and then understand or behave sequentially or globally. This model also shows that teachers present their

Dimensions of Learning and Teaching Style	
Preferred Learning Style	**Corresponding Teaching Style**
Perception based sensory or intuitive learning	Content could be concrete or abstract
Visual or auditory information	Presentation could be visual or auditory
Inductive or deductive arrangement of information	Inductive or deductive arrangement of presentation
Active or reflective refinement of information	Participation of students could be active or reflective
Sequential or global comprehension of information	Teaching in sequential or global viewpoint

FIGURE 17.1 Learning-teaching styles dimensions. (Adapted from Felder and Silverman (1988).)

abstract or concrete contents using visual or auditory cues. They organize the information in either inductive or deductive manner in sequential or global perspective to make or have active or reflective students' participation.

Learners or students learn from whatsoever content present with them by understanding visually the images, maps and graphics; listening to lectures and discussions; reading and writing verbal texts; and deciphering the signs and symbols. Students learning styles can be understood from the following figure.

Figure 17.2 shows the preferred method of learning styles of students as per their understanding of any subject and the skills which they are good at. The use of stimuli and the sensory responses from and for any situation is the key here in VARK model (Fleming and Mills 1992).

At present, the learners navigate through the contents available in the secondary resources like textbooks, reference books, journals, guides and magazines and do some hands-on practices with projects, case studies, discussions and role plays. Teachers guide their students in classrooms and give them some exercises and experiment assignments to perform the task.

Figure 17.3 accounts both the learning and teaching methodologies adopted by the students as well as the educators focusing on content delivery and the experience and practice. In the learner-centric methodology learners are self-navigated by studying and practicing while in the teacher-centric methodology learners are guided by the teachers through teaching and coaching.

Visual Visual learners learn from graphs, charts, signs and symbols and deduce the meaning by reasoning or logical conclusion.	**Auditory** Auditory learners learn by repeated listening and reciting content. They devise their own mnemonic patterns to learn.
Read & Write These learners learn through a provided text and understand the concept as it is.	**Kinesthetic** Kinesthetic learners learn through physical activities, hands-on practices, practical experiments and tactile movement.

FIGURE 17.2 Students learning styles. (Adapted from "Learning Styles" (n.d.).)

Learner Steering	
Content Focus Students learn by studying books, magazines, journals, guides, periodicals, reports, white papers.	**Hands-on Focus** Students learn by practicing or working on projects, assignments, case studies, role plays, group discussions, forums.
Teacher Steering	
Content Focus Teachers teach students trough classroom lectures, online lectures, video presentations, demonstrations.	**Hands-on Focus** Teachers coach students by making them doing lab experiments, practice and rehearsals, exercise to solve and mentoring.

FIGURE 17.3 Learning modalities. (Adapted from Wenger and Ferguson (2006).)

17.1.3 Future

When the world is changing and evolving at an express pace, our existing beliefs, knowledge, system, process and technology everything become obsolete and inaccurate at a faster rate. We get to see that the products are augmented with newly added features, refined configurations and with a new set of marketing mix. Our new generation wants things at their fingertips and that too quickly or in no time. Transfer of knowledge is not something which is untouched or unaffected by this phenomenon. The transfer of knowledge in a quick and agile way is only possible if the whole education system is driven by technology. This is where the future is calling for digitalization of education.

More so ever, with development in the field of science and technology, it is quite evident that future of the education system is going in a digital mode and this is going to remain for quite a sometime.

17.2 DIGITALIZATION

'Digital transformation is fundamental reality of business today'

Warren Buffett

From time immemorial human beings have been using some knowledge, art, craft, skills, techniques, methods and process to produce or manufacture either products or services. The use of these vocations is called technology which has been of great help in achieving our objectives. Technology is used to improve our productivity by doing our work effectively and more efficiently. The process of converting any information like document, object, sound, image or signal from analog to binary or digital form organized in bits is called digitalization ("Definition of Digitalization" n.d.; "Digitalization/Digitisation" n.d.). Digital information are computer readable and thus can be shared and accessed easily. Several software and computer hardware are used in digitalization which are known to preschool kids also in present-day scenario. Several researchers have found that students gathered more knowledge and information through e-learning (Tunmibi et al. 2015).

17.3 DIGITALIZATION OF EDUCATION

> *'If you want to teach people a new way of thinking, don't bother trying to teach them. Instead, give them a tool, the use of which will lead to new ways of thinking'*
>
> **R. Buckminster Fuller**

With the advancement in technology in the 21st century and the ever-increasing development and use of technological tools, especially in the field of information and communication systems, even education sector could not remain untouched with this technological wave. Educators started using the digital tools and technologies innovatively in their teaching and learning process, and a new form of education i.e., digital education was evolved ("Digital Education" n.d.).

17.3.1 Objectives

Objectives of digital education summed up here from various stakeholders' forums ("Digital Learning Center" 2020):

- To address challenges posed by unequal lives and education
- To support continued higher education through tutoring
- To impart leadership skills and personality development

17.3.2 Digital Education

Digital education is also referred to as electronic learning (e-learning) or technology-enhanced learning (TEL). By using digital technologies, educators can design their course curriculum in such a way that learning opportunities can be deliberated in online courses or programs or in the form of blended learning ("Teaching with Digital Technologies" 2020).

Any form of learning accompanied by the use of technology wherein the technology effectively enables the receivers or learners to follow the instructions and eases the learning process may be called digital learning. A wide range or spectrum of learning and instructional methods like mixed (blended) or hybrid learning and implicit or virtual learning may encompass the application of digitalization in education (Davis 2020).

Digital learning is there to smooth out the learning process, reach out to unexplored space and enhance the learning experience. It is certainly not going to replace the traditional learning–teaching methodologies altogether. Pedagogies are being developed to have an uncanny mix of technology and learning. Some of the common pedagogies used in digital education are:

- Blended or hybrid learning
- Online learning
- Flipped learning
- One to one learning
- Differentiated learning

- Individualized learning
- Personalized or customized learning

Blended learning is a blend or mix of both online and offline learning where a few assignments or tasks are given online which need some exploration and then it could be discussed in classroom. The flipped classroom is an example of blended learning where online assignments or activities are completed outside the classroom and face-to-face in-depth discussions between students and teachers take place inside the classroom. This form of learning helps in developing a quest for learning from the learners' side and then the conceptualization and the exploitation and implementation of the same concept may be guided by the educators.

Online learning gives flexibility to learners as well as educators in terms of time, pace and location. The whole learning process can take place anywhere, at any time and at a suitable pace as per the convenience of both the learners and the educators. Of course, at the design stage of program and course curriculum one needs to plan the digital education approach. A suitable digital education approach in a particular context (for example, theory or sessional courses) needs to be adopted in terms of what digital technology is being used while designing or re-designing the program or course.

Some institutes provide one computer terminal to each student, termed as 1:1 learning, so that every student may have Internet access and learning resources in their comfort zone. An individual student may have his or her own interests and may want to perfect his trade or craft and teachers are expected to adapt to different learning methodologies suited for different learners (e.g. BYJU'S). Sometimes a single teacher is assigned to tutor or mentor a single student (e.g. WhiteHat Junior). This not only helps in tutoring a child but also enables him or her to take the challenge at a very early age and come out as a leader by doing the coding on his or her own. A prodigy today becomes the future leader. These pedagogical styles are covered under individualized learning, differentiated instructions and personalized learning.

Digital learning may be strategized using

- adaptive learning
- badging
- blended learning
- classroom technologies
- e-textbooks
- data analytics
- learning analytics
- learning objects
- mobile learning, e.g., mobile phones, laptops, computers
- machine learning
- personalized learning
- online learning (or e-learning)
- open educational resources (OERs)
- technology-enhanced teaching and learning

- virtual reality
- augmented reality
- artificial intelligence

Extensive usage of cellphone or mobile technology and Internet has certainly helped both the learners and the teachers in a long way.

As per, The Association for Educational Communications and Technology (AECT), the appropriate procreation, utility and management of technological processes and assets or means (resources) using some ethical practices which facilitate learning and thus improve performance is defined as educational technology (Richey, Silber, and Ely 2008). It enhances classroom learning by utilizing technology with face-to-face learning. There is a symbiotic relation between education and technology due to the enormous growth of Internet connectivity and penetration of mobile or smartphones in the market. Roughly 40% of the world population is having Internet access as of now and the percentage is growing northward every day ("The Indian Telecom Services Performance Indicators" 2020).

Open Educational Resources (OER) may be accessed freely at open source by any one and can be remixed and reproduced under some license clauses (Butcher 2011). Educators often do make certain changes in the content of the OER to suit their academic requirements (Bell 2020). OER, like Wikipedia seeks the inputs from its users and makes necessary changes from time to time.

Higher education institutions offer 100% virtual online courses which may exclude MOOCs (massive open online courses). Learners may experience both asynchronous and synchronous elements in their courses. This is a post cursor to the distance education format of education with the World Wide Web on the learners' finger tips. In earlier days, students used to enroll in distance education or distance learning, when they were not able to attend regular physical and could study via correspondence courses, getting course material through post. Online distance learning can be classified into four groups based on the number of participants and the time when they learn (Kaplan 2018):

- MOOCs: unlimited participants learn synchronously at their will and own pace.
- SMOCs (synchronous massive online courses): unrestricted numbers of participants learn synchronously in real time.
- SPOCs (small private online courses): limited participants learn in a choice-based non-sequential or asynchronous manner.
- SSOCs (synchronous small online courses): limited participants learn in real time.

MOOCs, first pioneered in 2008, came with an aim to provide education to unlimited participants with an open access via Internet. Video lectures, reading materials and problem sets are provided along with the traditional course materials. Many MOOCs have discussion forums where students can interact with professors and teaching assistants using social media platforms and may also get immediate feedback on their quizzes and assignments (Kaplan and Michael 2016).

TABLE 17.1
Attributes of MOOC Providers

Attributes of major MOOC providers				
Service providers	Profit making objective	Accessibility	Free Certification (if any)	Institutional credits (affiliation)
edX	No	Partially free	No	Partial
Coursera	Yes	Partially free	No	Partial
Udacity	Yes	Partially free	No	Partial
Udemy	Yes	Partially free	No	Partial
P2PU	No	Free	Yes	No

Source: Li and Powell (2013).

MOOCs may have both open access features like open licensing of content shared by anyone as well closed licensing of contents to protect their intellectual property rights ("Benefits and Challenges of a MOOC" 2020).

Table 17.1 shows the attributes of leading MOOC players in terms of their accessibility, affiliation with institutes to give credit and certification and financial objectives.

17.3.3 LMS and CMS

To run these courses, institutes use certain Learning Management System (LMS). Software-based application like LMS manages the learning or educational resources i.e., they maintain, deliver and track the resources. Institutes of highest reputes may have their own LMS but otherwise most of the institutes opt for outsourced LMSs vendors (Eden, Christopher, and Jacqueline 2014). Some of the prominent LMS vendors are Blackboard (with market share of 31.9%), Moodle (with market share of 19.1%) and Canvas by Instructure (15.3%) as per the report of EDUCAUSE Center for Analysis and Research (ECAR). LMSs are greatly helpful for both faculties and students as they act as a repertoire and depository for course materials, curriculums and syllabuses, learning contents and tools for assessment of learning, etc.

Like it is mentioned above that LMS is a repository of contents, a computer software CMS, i.e., (Content Management System) manages the contents digitally at enterprise level and web level (Ann, Pamela, and Steve 2003). It helps in creation and modification of digital contents including written texts and graphics, pictures, audio-video clippings, maps and programming codes, which can be done by any collaborator with valid access (Martin 2005; Bob 2005). Some of the popular CMSs are WordPress (with market share of 38.7%), Shopify (with market share of 3.1%) and Joomla (with market share of 2.2%) ("W3Techs Content Management Usage" 2020).

After completion of any course on MOOCs platform online credentials (in OERs) are offered. This credential is given instantaneously either after getting the required skills, or by having minimum number of hours of attendance or by passing upon

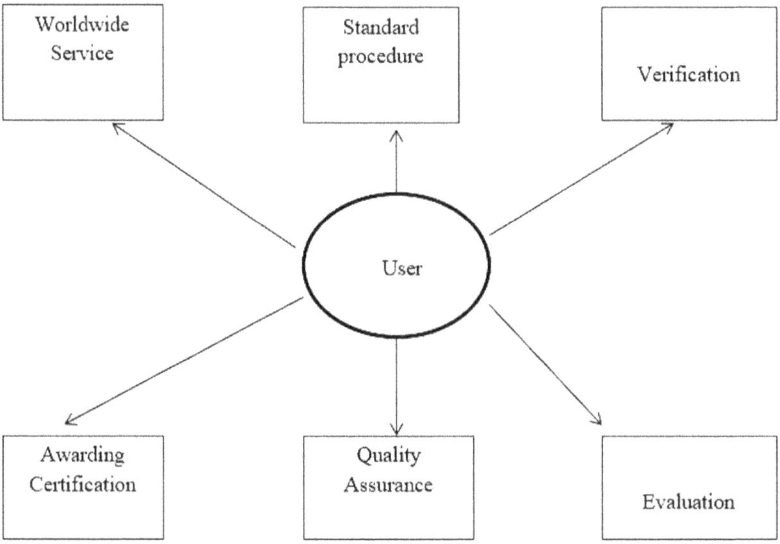

FIGURE 17.4 Digital credential ecosystems. (Adapted from Chakroun and James (2018).)

some assessment tools like quiz, assignment or discussion forum, etc. (Koller 2015). Online badges or online certificates are given to participants on their subsequent educational accomplishment or obtaining the required skills. These are equivalent to paper-based credentials.

Figure 17.4 shows the digital credential ecosystem of participation and involvement of all the stakeholders in certification subsequent upon the completion of any course on MOOC platform. This allows any digital credentials to be treated as valid and equivalent to physical paper-based credential certification.

17.3.4 Digital Learning Tools and Resources

Industries and Institutions are working in tandem day and night to develop new technologies in the field of education so that digital technology may have magnificent impact on our education system. Plethora of digital technologies is present there which are being used by our universities, institutions and schools to deliver the right blend of digital education. Some of the common tools and resources used by modern day educators are (Staff n.d.):

- Google+ Communities
- YouTube Channels
- Cloud-based Word Processors like Google Drive, OneDrive, etc.
- File-sharing platforms like Dropbox
- Digital Pocket
- Video conferencing software like Microsoft Teams, Cisco Webex, Google Meet, Zoom, etc.

Information and communication technology (ICT) tools including email, audio-video conferencing, tele-lessons and FM radio or community radio broadcasts, etc., are used in virtual learning to impart digital education (Sanyal 2001). These days, institutions of higher education and in some places at secondary level also, focus is given on the use of ICT tools in the learning process. Ranchi University, Ranchi, India has come up with Khanchi Radio, wholly managed by students for the same purpose.

17.4 TECHNOLOGY IN VARIOUS COUNTRIES

Technology plays a vital role in imparting education in the form of giving access to underprivileged people living in impoverished quarters of the world. All countries are not economically developed and technologically advanced. There is a great disparity between some of the western countries and their counterpart third-world countries due to lack of advanced technology and poor public distribution system. There are some countries like the USA which can afford to distribute laptops with pre-installed educational software to every child through OLPC (One Laptop per Child) Foundation, an initiative of MIT Media Lab supported by various big corporate houses, wherein, in countries like India the Government had to implement MDM (mid-day meal) scheme to increase the enrolment ratio in schools.

The USA has spent in excess of $13 billion on education technology, wherein students are provided with computers and handheld devices, there are full-time online schools and use digital resources as well ("Use of Technology in Teaching and Learning" n.d.). In Japan, they have prefectural budget to provide expensive facilities like computers and other machines in their institutes (Murata and Stern 1993). China focuses on education technology in their program, research direction, discipline area, institution and enterprise and has surpassed the US in its funding on technology-driven education.

However, developing countries are also taking the cue from their western counterparts and had started investing in education sector off late. They are getting technological support and funding as well, through several United Nations Projects, public foundations, charities and CSR (Corporate Social Responsibility) activities of big corporate houses. NEPAD (New Partnership for Africa's Development) has provided computer peripherals, Internet access and learning materials to more than six lacs schools in Africa through its 'e-school program'("Digital Services for Education in Africa" 2015).

Since 1960 onward, Africa had started using the sustainable development of ICT tools in the form of Program for Education by Television (PETV), an OER project TESSA, Teacher Education in Sub-Saharan Africa, ICT4D initiative By Africa Information Society Initiative (AISI), etc. An Africa Virtual University (AVU) funded by World Bank was established in 1997. With the help of USAID, Amazon distributed e-books on its Kindle platform free of cost to students of Africa ("Information and Communication Tools" n.d.). Just on the line of OLOP, Intel also distributed Classmate PC to the students of Africa.

The under-developed countries like Bangladesh have provided laptops and projectors to more than 20,000 private and state aided educational institutes, books were made available on the website www.ebook.gov.bd and technical training to the teachers were imparted to conduct online classes ("The Ministry of Education" 2009).

In India, the Ministry of Human Resource, Information & Broadcasting, Social Welfare have all come together and drafted a policy through Niti Aayog (replaced Planning Commission) for the betterment of education system in India. The usage of ICT has been mandated and taken up in all the universities and other higher education institutes.

Indian Space Research Organization (ISRO) had also launched a communication satellite EDUSAT in 2004 to provide an outreach access to educational materials in distance learning mode at a reduced cost.

("ASER (Annual Status of Education Report)" 2020) shows that 32.5% school students in India are pursuing education in live online mode with 11% live classes and 21.5% with videos and recorded classes. In total, 74.2% families with an increase of 24% in the last two years have the smartphone ownership. Although the basic structure of digital education is already framed at the household level but the proper eco-system to push for digital education in the country is not developed. This is evident from the fact that several states like Maharashtra, Gujarat, Madhya Pradesh, Himachal Pradesh and Tripura score more than average in smartphone ownership and students enrolled in online classes, while the states like Rajasthan, Uttar Pradesh and Bihar are lagging behind the national average score. A lot of digital content is generated and transmitted through several platforms, but students' engagement and participation are still far reaching and daydream due to poor digital infrastructure.

Uttar Pradesh Government has launched Higher Education Digital Library, having more than 35,000 e-content materials on 134 subjects including science, social science, law, engineering, management, agriculture and foreign languages, etc. These contents were developed by around 1,700 teachers from 23 state universities and have already benefited more than 61,500 underprivileged students (IANS 2020). This shows that even impoverished states like Uttar Pradesh are putting their effort in digitalization of education.

During COVID-19 pandemic, Delhi Government has conducted series of online lessons in more than 1,000 schools on responsible use of social media and to create awareness about various cyber threats. The National Council of Education Research and Training (NCERT) has taken proactive approach in providing guidelines about cyber security and safety (PTI 2020).

National Education Policy '2020 has also highlighted on the various drawbacks in our education system and thus taken a call on having more integrated system of education; hence we see a surge in adoption of blended learning across various universities in India' ("National Education Policy" 2020). The role of technology in education and digitalization of education is going to be a key player in the holistic and integrative development of education in India.

17.5 CHALLENGES AND LIMITATIONS

There are various challenges and limitations we are facing with digital learning which are summed up below:

- Inappropriate distribution of electricity in underdeveloped and developing countries – though countries like India have got 100% electrification, and not a single village is left unwired, there is intermittent and unreliable

TABLE 17.2
Internet Usages in India

Statistic	Figures
Total subscribers	718.74 million
Narrowband subscribers	56.806 million
Broadband subscribers	661.938 million
Wired subscribers	22.386 million
Wireless subscribers	696.36 million
Urban subscribers	450.31 million
Rural subscribers	268.43 million
Overall net penetration	54.29%
Urban net penetration	106.22%
Rural net penetration	29.83%

Source: "The Indian Telecom Services Performance Indicators" (2020) shows the gap in the bandwidth usage, net penetration and inequality in the net subscription geographically.

electricity supply due to electricity loss and uneven distribution ("Access to Clean Cooking Energy and Electricity – Survey of States" 2015).

- Slow Internet and poor bandwidth – India lacks in the provision of Internet i.e., in terms of speed and bandwidth and is nowhere near to the global average. As per ("Speed Test" 2020) Ookla data, the rating and monitoring agency for Internet usage, the average 4G data speed in India is 6.9–9.5 Mbps, whereas the global average 4G data speed is 34–35 Mbps. Some of the government organizations are still working away with 2G data speed thus making it inconvenient to the users. 5G or 6G spectrum is a distant dream for us. Table 17.2 exhibits the internet usages in India.
- Poor infrastructural support – Though India has become a major software hub, the poor infrastructural support is not allowing us to reap the benefits of digital technology ("The Indian Telecom Services Performance Indicators" 2020).
- Purchase disparity in mobile phone users – Although 74.2% household use mobile phones, the percentage of smartphone users is very less. There is again a disparity here due to economic inequality, where many household have multiple numbers of phones while others do not have a single one ("The Indian Telecom Services Performance Indicators" 2020).
- Vast knowledge gap between haves and have-nots – Due to the economic, socio-cultural, political and educational inequality in the total population, there is a huge knowledge gap between the fortunate and unfortunates (Goswami 2014).
- Poor digital literacy – The digital literacy rate at a staggering low level of 10% in countries like India is not going to help in this cause (Bureau 2018).
- Language barrier – With so many regional diversities, where states are formed on the basis of the language spoken in that region, it becomes very

difficult to make learners learn in one language. Less than 30% population understands English language, so it becomes a mounting task for educators to create and deliver the contents in their vernacular languages.

- Lack of training facilities – Both the learners and the teachers need to be trained as they are going through the paradigm shift in education. They need to be acclimatized with the technology beforehand, but due to no or very few training facilities available around, it becomes very difficult for them to be trained.
- Indifferent attitude of the learners and teachers – It is also found that the attitudes of both the learners and teacher are indifferent at times because of the clueless behavioral cues. Unlike the offline class in physical mode where learning occurs face to face, in online or digital mode of education both are unaware of the psyche of each other, making it difficult to achieve the objective.
- Resistance to change – With any change brought into the system, it is inevitable to have resistance to change from certain quarters of people (old school of thought). So the teachers as well as learners who are accustomed with the brick and mortar schooling system shall definitely resist changing into digital education. These stakeholders should be put on board while taking collective decisions ongoing digital way.
- Cyber threats – Students of impressionable age are vulnerable to get distracted and lured by anti-social elements very easily. Constant monitoring and counseling are required to protect them from any kind of untoward cyber threats.

Here many factors are related to the psychological behavioral aspect of a person while few factors are related with the physical facilities or aspects. If the government and agencies pay attention to these physical and material barriers there is not any doubt that digital education cannot mark its presence in the whole educational system.

17.6 MANAGERIAL IMPLICATION

Higher Education Institutes must pay attention to the present need of digital education in their setup. Focus should be on robust LMS and CMS deployed in every University and Higher Education Institutes.

Learning-teaching process should take place in a conducive environment where it is found to be more interactive so that actual learning can happen and it does not remain a mere exercise. Both the learners and teachers should come forward and onboard for better learning.

17.7 CONCLUSION AND SUGGESTION

The inter-networking is revolutionizing, the concept and the conduct of the business. The computer in the classroom can go a long way in helping the quality of education (Sahani and Arshad 2011). Things are moving fast and information and knowledge are available just at the click of the mouse. Physical distance is no more a barrier in the digital world.

All good things come with a baggage of its cons, where we need to be careful. With the greater exposure to the good bad world of Internet and other technologies several incidents of cyber-crime, cyberbullying and fraudulent activities have surfaced off late. Proper safety measures and guidelines need to be charted so that unhindered, uninterrupted and secured digital education can be provided to all.

The importance of digitalization of education is more evident in the present times when the entire world is held standstill and under lockdown amid COVID-19 pandemic. Countries across the world found solace with digitalization and every institute started using some or the other platforms of digital education.

With the little more push from the government agencies in building the appropriate infrastructure for e-education and a ready acceptance of all the stakeholders can surely bring the great revolution in education sector. We just need to have a right blend of our LMS and CMS with the infrastructural support needed to gas up the system. Digital education may provide a consonance to the objective of right to education and education for all. This is where we say future is calling for the digitalization of education.

REFERENCES

Jain, A., S. Ray, K. Ganesan, M. Aklin, C. Cheng, and J. Urpelainen. 2015. "Access to Clean Cooking Energy and Electricity – Survey of States."

Ann, R., K. Pamela, and M. Steve. 2003. *Managing Enterprise Content: A Unified Content Strategy*. New Riders.

ASER (Annual Status of Education Report). 2020. https://www.asercentre.org.

Bell, S. 2020. "Research Guides: Discovering Open Educational Resources (OER)." https://guides.temple.edu/OER.

Benefits and Challenges of a MOOC. 2020. University of Toronto Libraries. https://guides.library.utoronto.ca/digitalpedagogy.

Bob, B. 2005. *Content Management Bible*. 2nd ed. Indianapolis: John Wiley & Sons.

Bureau, F.E. 2018. "A Look at India's Deep Digital Literacy Divide and Why It Needs to Be Bridged." Financial Express, September 24, 2018. https://www.financialexpress.com/education-2.

Butcher, N. (UNESCO). 2011. "A Basic Guide to Open Educational Resources." Vancouver, BC and Paris.

Chakroun, B., and K. James. 2018. "Digital Credentialing: Implications for the Recognition of Learning across Borders." UNESCO, Paris.

Davis, L. 2020. "Digital Learning: Data, Trends, and Strategies You Need to Know." 2020. www.schoology.com.

"Definition of Digitalization." n.d. Accessed October 29, 2020. https://whatis.techtarget.com/definition/digization.

"Digital Education." n.d. University of Edinburgh. Accessed October 29, 2020. www.ed.ac.uk/institute-academic development/learning-teaching/staff/digital-ed.

"Digital Learning Center." 2020. www.planindia.org.

"Digital Services for Education in Africa." 2015. Paris.

"Digitalization/Digitisation." n.d. HarperCollins. Accessed October 29, 2020. https://www.collinsdictionary.com/dictionary/english/digize.

Eden, D., B.D. Christopher, and B. Jacqueline. 2014. "The Current Ecosystem of Learning Management Systems in Higher Education: Student, Faculty, and IT Perspectives." EDUCAUSE, Louisville.
"Educate." n.d. Accessed October 20, 2020. https://www.etymonline.com/word/educate.
"Education." n.d. Accessed October 20, 2020. https://en.wikipedia.org/wiki/Education.
Felder, R.M., and L.K. Silverman. 1988. "Learning and Teaching Styles in Engineering Education." *Engineering Education* 78 (7): 674–81.
Fleming, N.D., and C. Mills. 1992. "Not Another Inventory, Rather a Catalyst for Reflection." *To Improve the Academy* 11: 137–55.
Goswami, C. 2014. "Role of Technology in Indian Education." *IPEDR* 79 (2): 6–10. https://doi.org/10.7763/IPEDR.
IANS. 2020. "UP Launches Higher Education Digital Library." Times of India, October 29, 2020. https://timesofindia.indiatimes.com/home/education/news.
"Information and Communication Tools." n.d. Accessed October 30, 2020. https://bera-journals.onlinelibrary.wiley.com.
Kaplan, A. 2018. "Academia Goes Social Media, MOOC, SPOC, SMOC, and SSOC: The Digital Transformation of Higher Education Institutions and Universities." In *Contemporary Issues in Social Media Marketing*, edited by B. Rishi and S. Bandyopadhyay. Routledge.
Kaplan, A.M., and H. Michael. 2016. "Higher Education and the Digital Revolution: About MOOCs, SPOCs, Social Media, and the Cookie Monster." *Business Horizons* 59 (4): 441–50.
Koller, D. 2015. "An Update on Assessments, Grades and Certification. Blog (Accessed)." 2015. https://blog.coursera.org/an-update-on-assessments-grades-and/.
"Learning Styles." n.d. Accessed October 23, 2020. https://teach.com/what/teachers-know/learning-styles.
Li, Y., and S. Powell. 2013. "MOOCs and Open Education: Implications for Higher Education a White Paper." University of Bolton, CETIS.
Martin, W. 2005. *The Content Management Handbook*. 1st ed. Facet Publishing. London.
Murata, S., and S. Stern. 1993. "Technology Education in Japan." *Journal of Technology Education* 5 (1): 29–37.
"National Education Policy." 2020. New Delhi. https://www.education.gov.in/.
PTI. 2020. "Govt. School Students in Delhi to Get Lessons on Responsible Use of Social Media." The Hindu, November 15, 2020. https://www.thehindu.com/news/cities/Delhi/govt-school-students-in-delhi-to-get-lessons-on-responsible-use-of-social-media/article33102091.ece.
Richey, R.C., K.H. Silber, and D.P Ely. 2008. "Reflections on the 2008 AECT Definitions of the Field." *TechTrends* 52: 24–25.
Sahani, S.B. and Arshad, M. 2011. *Advanced English Essays*. Agra: Sahani Bros.
Sanyal, B.C. 2001. "New Functions of Higher Education and ICT to Achieve Education for All." In *Expert Roundtable on University and Technology-for- Literacy and Education Partnership in Developing Countries*. Paris: International Institute for Educational Planning, UNESCO.
"Speed Test." 2020. https://www.speedtest.net.
Staff, Teach thought. n.d. "9 Digital Learning Tools Every 21st Century Teacher Should Be Able to Use." Accessed October 29, 2020. https://www.teachthought.com/the-future-of-learning.
"Teaching with Digital Technologies." 2020. 2020. www.education.vic.gov.au/school/teachers/teaching resources.
"The Indian Telecom Services Performance Indicators." 2020. 2020. https://www.trai.gov.in/release-publication/reports/performance-indicators-reports.
"The Ministry of Education." 2009. The Ministry of Education. 2009. https://www.moedu.gov.bd/index.

Tunmibi, S., A. Aregbesola, P. Adejobi, and O Ibrahim. 2015. “Impact of E-Learning and Digitalization in Primary and Secondary Schools.” *Journal of Education and Practice* 6 (17): 53–58.

“Use of Technology in Teaching and Learning.” n.d. Accessed October 30, 2020. https://www.ed.gov/oii-news.

“W3Techs Content Management Usage.” 2020. 2020. https://w3techs.com/technologies.

Wenger, M.S., and C. Ferguson. 2006. “A Learning Ecology Model for Blended Learning from Sun Microsoft systems.” In *Handbook of Blended Learning: Global Perspectives, Local Designs*, edited by C. J. Bonk and C. R. Graham. San Francisco, CA: Pfeiffer Publishing.

18 Creating a Sustainable Future with Digitalization in Online Education

Issues and Challenges

Abhishek Srivastava, Lokesh Jindal, and Mukta Goyal

CONTENTS

DOI: 10.1201/9781003132097-18

18.1 INTRODUCTION

Now a day, digital transformation is supposed to be the buzzword in Educational Sectors. With this, the outreach of school services is improving day by day. Online schooling offers new training and learning tools for both teachers and pupils while ensuring increased involvement in the broader phase of learning (Govindarajan and Srivastava 2020). The way education is delivered in schools and colleges has changed since the advent of new innovative learning technologies, such as smart-boards, Massive Open Online Courses (MOOCs), Desktops, and other similar devices (Li et al.2020). One of the most budget forms of educating young people tends to be the Internet of Things (IoT). It is also a strong system for all to implement an outstanding wisdom environment. Edu innovation firms are actively focused on finding new ways to improve access to education for individuals who are currently unable to receive sufficient education facilities (Sharma 2020).

According to online surveys, the number of students from all over the world who have registered after attending classes in the past is growing. The great thing about smart classrooms is that you can only build resources once and use them many times for coming generations (Dasgupta et al. 2020). As a result, a significant amount of money and effort is saved. Simultaneously, immersive curriculum allows educators to customize educational materials to meet the specific needs of individual students. Online learning is also assisting in the advancement of the idea of collaborative learning, where all minds cooperate to create a diverse educational environment that is not constrained by national boundaries. Universities and colleges have launched digital platforms to facilitate online education access in order to promote social equity.

18.2 LITERATURE SURVEY

Bailey and Card (2009) addressed the importance of setting course objectives, learning goals, and expectations. They believe that fostering positive relationships and cooperation between teachers and students is crucial, and that this can be achieved by instructors' sensitivity to students, enthusiasm for teaching, and desire to see people learn. (García-Moya et al. 2020)

These online instructors, understanding the very essence of contact in the online world, recommended that online instructors be attentive, responsive, and timely in responding to emails and text messages. To do the same, they used practical strategies, such as providing timely input on deeds, listening to structured responses, discussing expectations, and informing learners about their absence (Moore et al. 2020). "Giving timely feedback on completed assignments, answering questions directly, communicating requirements, and telling students when they would be gone" were some of their realistic tactics. The importance of involving their students, which can be accomplished by the use of emails and online message boards, prompt responses to discussion questions, and encouraging students to share their backgrounds and work experiences. They recommended online teachers be successful organizers in order to accomplish these goals. In a well-organized course they mentioned, at the beginning of the class, students should be provided with all course content, direct references to the appropriate websites and resources, and detailed guidance on how to complete the course using the university website. They also stated that flexibility is an essential aspect of successful online teaching. Since technology isn't always flawless and reliable, online teachers must be prepared to deal with issues like system delays, app updates, email glitches, and so on. Good online mentors are those who have the knowledge and experience to use and implement new technologies, who are frequently available online, who check for emails and text messages, who respond quickly to questions and complaints (Bailey and Card 2009). The World Wide Web (WWW) was launched in 1991, and it was a watershed moment in the rapid expansion and growth of online teaching and learning.

The study claims that reduced face-to-face contact with learners, lack of time to prepare and offer online courses, and lack of support resources often restrict the participation of the faculty. Rockwell et al. (1999) describe the length of time it took to acquire and upgrade technical capabilities for the faculty; the inadequacy of pay and incentives; and a heavier workload as other online teaching disincentives. Maguire (2005) cites faculty issues as other hurdles over a shortage of expectations, the possibility of fewer jobs, and a reduction in full-time faculty use. Traditionally, teachers rely on face-to-face contact, allowing more opportunities for student responses to be assessed, input received, and pedagogy updated. Choi and Park (2006) find that their online instruction experience was burdensome, and student apathy challenged them. In addition, greater structural help is required, which also discusses adequate pedagogy for the adaptation of online teaching strategies.

Tallent et al. (2006) after reviewing 76 studies on online teaching, studied the course environment and proposed conclusions similar to Wallace (2003) on the benefits of creating an online learning culture that encouraged small discussion groups and effective distribution methods. They emphasized that teachers' involvement was critical in such a process because immediate feedback and teacher responses were critical to students' learning, a factor that contributed to academic performance with their courses. Although they agreed on the significance of teacher-student interactions, they cautioned against extrapolating research results to a wider population due to the small number of participants in most of the studies they looked at.

18.3 ISSUES AND CHALLENGES OF DIGITAL EDUCATION

18.3.1 Unobstructed Ambience

Census 2011 tells us that 71% of families with three or more occupants have two-room residences or fewer (74% in rural areas and 64% in urban areas). In such a case, the issue of how children can benefit from schooling in an undisturbed setting remains immense.

18.3.2 Scarcity of Network Infrastructure

As per July 2017–June 18 National Sample Survey results, only 42% of urban and 15% of rural households had Internet access, and in the past 30 days, only 34% of urban and 11% of rural people had Internet access. These statistics explicitly show that two-thirds of children would be left out of the process of online schooling. Marginalized and disadvantaged communities will be the worst impacted, as usual.

When it comes to online education, it is all about actively connecting with teachers via video calls or attending video tutorials online, and both involve a secure Internet link at high capacity. The whole idea is going to collapse in the absence of sufficient Internet speed. When there are daily protests by the students, we can see the same from the UT of Jammu and Kashmir as they cannot learn in the absence of a proper Internet connection.

18.3.3 Lack of Customary Plan

Online teaching is not about recordings of teachers on the Internet lecturing on blackboards. Relevant networks, technologies, software, interactivity, curation, content, and much more are involved.

18.3.4 Dearth of Solidity in Societies

In terms of social integration and relative equity, public educational systems still play an outstanding role. It is the place where, without one community being required to bow to another, citizens of all sexes, ages, castes, and cultures will interact.

18.3.5 Inadequate Training

Teachers look after children's mental, emotional, and social wellbeing in classrooms. Schooling is intended to take care of children's emotional, social, and mental wellbeing, and is diametrically contrary to social distance. Teachers are not properly qualified to use Internet media to inculcate such information.

18.3.6 Issues with Nurturing Students

Another mission is to keep thousands of kids out of school as their parents return to their workspaces after shutting them down. Who will take responsibility for the welfare and learning of a child at home remains a big concern.

18.4 OBJECTIVES AND RESEARCH QUESTIONS

Most of the previous studies illuminate a variety of issues relevant to the experience of students and teachers in relation with the conventional learning process in the sense of building a viable future through digitalization in online education. However, there are only a few articles that mention the exclusive use of e-learning sites, as it occurred during the pandemic when universities were pressured to use it and incorporate it as a key method in the educational process. The goal of this research is to discuss the future of digital education in India.

In the Indian higher education system, the use of e-learning systems was rare prior to the pandemic: few teachers were using the site, mostly using its simple functions such as online posting of course content. Many teachers who were very inexperienced with online learning sites were shocked by this pandemic, pushing them to switch from conventional learning to purely online learning in a very short amount of time, being the greatest obstacle for them.

In this study, the following research questions were addressed:

- Handling of Required learning objectives during Pandemic with Online Education
- Role and responsibilities of Educators during creation of Online Education Module
- Creating sustainable model, Managing resources, tools, infrastructure, and people for their online activities during imparting Digital Education

18.5 RESEARCH METHODOLOGY

It's an observational analysis that was performed on the basis of online evidence, i.e. secondary in nature. By considering the latest available results, this analysis can be revised and revamped. There is a lot of scope for more analysis on this issue, given other factors that we have not taken into account in our current methodological analysis, it would have been better. The data used for the study was obtained from books, magazines, newspapers, articles or e-journals, books, and databases online. For retrieval of authentic information, numerous government portals have also been used.

18.6 PROMOTING DIGITAL EDUCATION IN INDIA

With the Internet being much more open and accessible, interactive and conventional teaching-learning mediums will merge further. The education sector will see the explosion of Edutech start-ups on a small, medium, and large scale, delivering a number of creative digital goods to academic institutions in the coming days. The government has been taking bold steps to introduce measures that will help the country's digital education sector develop. Attempts are being made around the world to enhance the usability of emerging technology in order to help facilitate the use of innovative educational technologies (Sudevan, 2020).

Digital education, like many other fields, will see significant changes in the way universities and colleges offer education in the coming years that may receive plethora of opportunities to inspire the youth of this country as a result of digital education. With the right information management software, Advance Learning Management Systems (LMS) would help to enhance the design and implementation of educational courses offered in India by universities and colleges. Visual media also helps conquer all obstacles to expression (Agerfalk et al. 2020). Learning tools will also be accessible in national dialects on the Internet. E-learning and M-learning programs promoted by state and private actors provide students and teachers with access to a vast pool of information material. For information, teaching, and research purposes, online education provides publicly available Internet. It enables students to interact openly open on the Internet with a wide variety of study content, thereby creating an ecosystem of self-learning.

When serving on Education technology start-ups, instructors or professionals in the role of instigators face numerous challenges, such as when to begin using it, how to minimize obstacles for learners, and how to improve students' abilities through e-learning technology. Student participation is insufficient and teachers must therefore make concerted efforts to enhance student engagement, sustain their interest, solicit feedback, and evaluate them in a variety of ways. This will result in a positive and meaningful learning environment. Educational technology does not replace a teacher, but it does improve teaching. Ed-Tech companies can be a huge help to students in these difficult times, as COVID-19 has forced schools and universities to close for several weeks due to the severity of the pandemic (Dhawan 2020). According to KPMG and Google forecasts, the Education technology industry will grow rapidly, with revenues expected to reach 2 billion USD by 2021.

Digital platforms like Zoom Classroom, Unacademy, Coursera, BYJU'S, Adda247, Board Infinity, Classplus, GlobalGyan, Vedantu, EduBrisk, Kahoot, Khan Academy, ePathshala, GuruQ are some of the popular Ed-Tech start-ups, and the list is long. The MOOCs and Study Webs of Active-Learning for Young Aspiring Minds (SWAYAM) portals are attention-grabbing learning initiatives launched by the Indian Government so as to attain significant educational policy goals, namely access, equity, and efficiency.

SWAYAM's immediate objective is to bridge the digital divide by incorporating online learning. It provides a wide range of free courses for education, distance learning, graduate, and postgraduate study. SWAYAM has been of immense support to students around the country during the COVID-19 crisis. Indian millennials, living in a globalized economy, are training themselves for the learning needs and challenges of 4.0.0. (Sahasrabuddhe 2020). The education system in the country is bound to undergo significant changes as a result of the digital transition, both in terms of how we learn and what we learn. The youth population of our nation would undoubtedly prove to be a generational dividend if every door move of the country's household is made open to the advantages of technology. In so many ways, online learning would encourage today's youth to learn and engage in the vast sea of information that the digital revolution has made widely available to them. In conclusion, digital education is India's tomorrow, propelling the country to new heights of socio-economic sustainable growth.

18.7 POST COVID-19 ONLINE TRENDS ASSOCIATED WITH DIGITAL EDUCATION

India is experiencing an e-learning surge in the aftermath of the COVID-19 crisis. Across the globe, online learning networks are leveraged by schools, universities, educational institutions, continuing the teaching-learning process. The definition of education has been transformed overnight by this step, and digital learning has arisen as an absolutely essential resource for education.

The great thing is that, with the vast and robust 4G Network and inexpensive data, India is well equipped in practically every part of the country. The mobile usage, which further accelerates the adoption of the learning infrastructure, is even more remarkable. With convenient access to the Internet, modern technology has provided remote learning experiences, interactive classrooms, and access to high-quality education, including in rural and semi-urban areas. Elements, such as location, language, and financial capital are no longer an obstacle to excellent education in the current scenario (Paul et al. 2020). Overall, for India, e-learning holds excellent promise, and it will change the education landscape forever. In India, there are several Ed-tools and strategies for fostering online education that are as follows:

a. **Social Media Applications as Virtual Platforms**: As a learning platform, social networking has come a long way. Both teachers and students use social media as an important part of today's e-learning experience. It is now a critical platform for exchanging information on critical consequences. Apart from being able to exchange knowledge anywhere, at any time, social media properties are also a great source of networking tools for creating social events and potential jobs on a low-cost basis.
b. **Interactive Education-Instruction Plans**: With the introduction of interactive technologies such as flipped classes and smartphone applications, learning is no longer limited to a traditional classroom setting. The learning experience of the millennial generation is being re-wired and thanks to these ground-breaking multimedia supports. Different educators are developing immersive training content with the help of cutting-edge technology.

With collaborative technology, combined learning is being improved. Sitting in one spot, individuals will cooperate and work on activities and projects for others distributed across the globe. Similarly, instruction does not have to be limited to a single subject or locale. It's no longer a regional affair; instead, global trends are being chased. The availability of digital education is critical to the long-term viability and growth of educational institutions (Figure 18.1).

i. **Massive Open Online Courses (MOOCs)**: Platforms for MOOCs make self-learning possible in a decisive manner. An increasing market is the popularity of online courses in India through MOOCs programs. MOOCs support a lot of the country's young people to develop their skills and qualifications. It provides access to a variety of skill-based courses to millions of Indians who do not have access to affordable schooling, allowing them to improve their academic performance. Learning from MOOCs allows students and working professionals to learn from anywhere and at any time at their own convenience. In addition, a number of courses delivered under this platform have a legitimate qualification that is properly recognized as

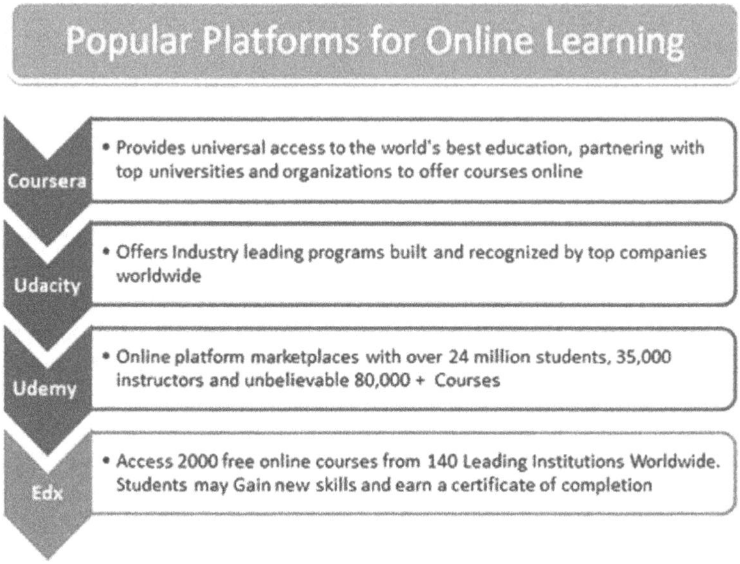

FIGURE 18.1 Digital platforms for online learning acclaimed worldwide.

an effort in the right direction by institutes and businesses. India appears to be a potential market for online courses offered through the MOOC platform, based on recent trends. There is a great demand among students in technical courses, as well as working executives, to continue to enhance their knowledge in order to take advantage of emerging job opportunities in specific fields. For some people, the learning environment through MOOCs is a big boon.

The requirement of the hour is to learn at one's own pace. MOOCs seem to be a successful solution in this respect, providing us with equal access to quality interactive learning material and digital information.

ii. **Visual Learning**: In India, audio-video-based learning is gaining popularity. Video-based educational learning is very popular among students because it combines learning with enjoyment. In the teaching-learning process, such a tool is extremely interactive. This type of classroom instruction is not limited to audio-video but also involves educational software, podcasts, and e-books, among other things. Students are very interested in learning new concepts through these contemporary channels.

iii. **Interactive Learning Packages**: Game-based learning should be the next big thing in India's digital future of education, particularly in the K12 field. It also creates an environment in which students can respond quickly and enjoyably to what they are learning. It will certainly transform the learning offerings of the planet and help create a stronger self-trained population of the future by changing the k-12 market.

18.8 CREATING SUSTAINABLE DIGITAL EDUCATION IN INDIA

The research and design process for Developing Sustainable Digital Education in India, in which we face several difficulties, such as the absence of the design of a suitable e-learning environment, numerous problems, and requirements, as well as the inability to select appropriate educational strategies to enable the best and efficient use of accessible learning platforms that need to foster teaching and learning.

18.8.1 Proposing a Model "MULA" for Creating Sustainable Digital Education in India

By combining different managerial aspects needed for the creation of an Efficient and Sustainable Digital Education in India, we have produced a model as MULA Model holding these problems for considerations. As a consequence, it suits all societies with the development of productive e-learning programs and the process of constructing secure infrastructure, i.e. Teachers and teachers, and taking advantage of technical leverage. To boost performance, we have also added some suggested phases. The flow and dependency of elements applied to the model for further elaboration using the proposed MULA model is shown in Figure 18.2, where prime focus is on Content Management followed by Identity and then Screen Time Management for effective learning.

18.8.1.1 Content Management

Content is simply not a single piece of data but a conglomeration of pieces of data placed together to create a coherent whole. The job of seeking information is like finding a costly thing in deep water because of the exponential growth of information. Content then becomes the foundation of every organization, any contact that takes place across the full spectrum of corporate activity. Content is saved, extracted, changed, upgraded, and managed; therefore, the performance is placed in a different way to minimize the marginal costs over time in each change and development.

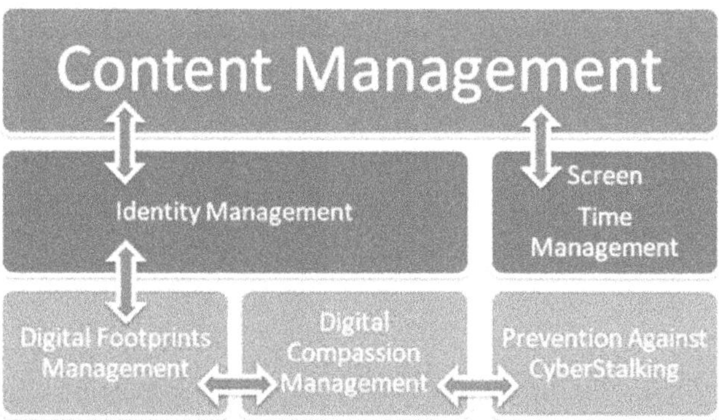

FIGURE 18.2 MULA model for creating sustainable digital education in India.

Content Management is dealing with a variety of issues and concerns. LMS is primarily an Open Source e-learning framework, such as Moodle (Modular Object-Oriented Interactive Learning Environment). Similar kinds of program packages are designed to help teachers produce quality material online. Content Management must enable the production, retrieval, and editing of digital information/knowledge in connection with IT (Information Technology), incorporating partially or fully processed content, such as images/graphics/animation, audio/video, etc., in actual environments as and when necessary.

18.8.1.2 Identity Management

Higher educational institutions, such as colleges and universities, store higher quantities of sensitive analysis and other task-related data. In addition, both institutions store vital data from alumni, faculty, and students (for that matter). There are gold fields for intruders to infiltrate and pose cyber threats to the networks. Here are a few relevant ways hackers target the field of Ed-Tech (Figure 18.3).

18.8.1.3 Screen Time Management

Screen-time rise has impacted students' physical and mental health. In order to reduce risks, some parents plan to allow students to decline a year. Authorities and several schools are seeking to decrease their hours of simulated instruction. There has been a lot of worry for parents. Over sensitivity to an electronic medium creates issues with eye pressure and vision. Mudgal, Specialist, Psychiatry Max Multi Speciality Centre, Panchsheel Park, said that continuous supervision is needed. Experts suggest that providing too much focus to digital screens weakens the capacity of the brain to interpret information, concentrate, make choices, and manage thoughts.

Since children's brains and bodies are still growing, the symptoms of screen addiction are compounded. The act of preparing screen time will lead to a major

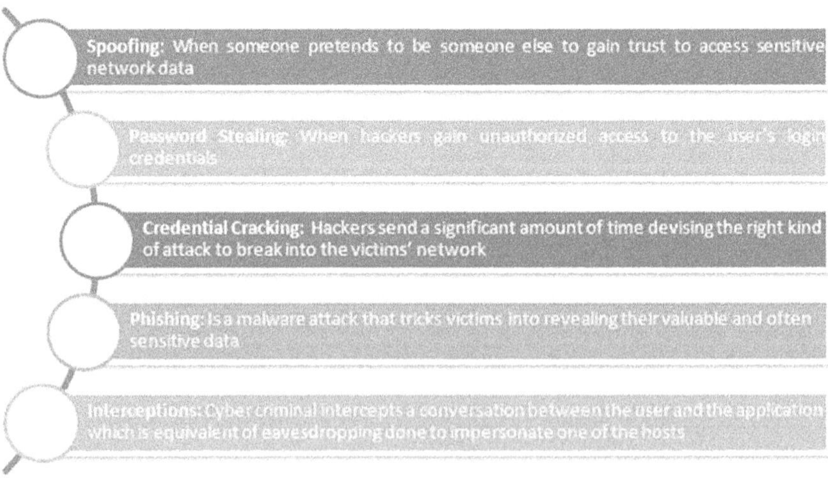

FIGURE 18.3 Ways hackers target the websites of Ed-Tech.

change in attitude. You will gain better mindfulness by marking off blocks of time for playing or working on your tablet and the task will become more purposeful and less instinctual. Screen time instinctually uses compounds which contribute to addiction.

18.8.1.4 Digital Compassion Management

Online networking practices and the resulting online disinhibition phenomena have gradually contributed to an emphasis on "Internet kindness." Digital Compassion can be taught to young people by video production, where, after composing, designing, creating, and screening their videos, the majority of students encountered higher levels of empathy in a project intended to encourage empathy. In order to encourage not just the academic ability of the foster children but also their wellbeing, sense of belonging and media literacy abilities, the practice of cognitive, emotional, and social empathy through digital media has been successful.

18.8.1.5 Digital Footprints Management

Today, in the era of Web 2.0, the essence of personal knowledge is evolving. Internet consumers are becoming more mindful of their digital footprint; 47% have checked online for information about themselves, up from just 22% five years ago. It is, therefore, a much-needed job to spread exposure and thereby create diversity by building a Positive Digital Footprint for being involved on multiple various social media channels. Since social media enables one to highlight imagination and thus to create social links for online credibility.

18.8.1.6 Cyberstalking Management

Maintaining digital hygiene is a recent concept, however, particularly with regard to social networks, it is a very important issue. It helps us to guard from online abuse, cyberbullying, and cyberstalking by ensuring proper digital hygiene. One of the first steps in "cleaning up" accounts is to change privacy settings. Many social media sites and some other types of online accounts can alter who can see and touch other profiles. Holding stuff like timelines, streams, and chat forums free from derogatory feedback is also a smart practice. These can have a major emotional effect, aside from potentially fueling more negativity from others. Website moderators are regularly offered psychiatric assistance since they seriously suffer from reading offensive posts, including those that are not sent to them directly.

18.9 FINDINGS

Since, COVID-19 pandemic has also altered the way many individuals receive and impart education, as per the World Economic Forum, therefore we could bring about some much-needed inventions and improvements in order to find new solutions to such unique problems. In the context of face-to-face seminars, teachers have become acquainted with conventional teaching approaches, and so they refuse to consider any improvement. But we have no choice left in the middle of this crisis, other than to respond to the dynamic situation and embrace the transition. For the education market, it would be helpful and will introduce a lot of unexpected developments.

Students who don't have access to all online technologies should not be overlooked and forgotten. Those students are less wealthy and belong to less tech-savvy families with limitations on financial capital, so as courses take place online, they can drop out. Because of the heavy costs associated with digital gadgets and Internet data packages, they may miss out. The inequalities in disparity will broaden this digital divide. This awful time of fate has taught us that all is uncertain and we need to be ready to meet obstacles. While this pandemic did not give us much time to prepare, we should learn that preparation is the answer. We must not forget that if our plan A fails, we must concentrate on our plan B to execute.

"Before catastrophe hits, the major learning for others might be to follow e-learning solutions!" (Todorova and Bjorn-Andersen 2011). Today, we are compelled to pursue online learning. If we had already learned it, it would have been different. In learning the modes, the time we wasted should have been spent on making more content. There is a need to prioritize all the important and complicated conditions that can arise and prepare accordingly as suggested in our "MULA" Model of creating a sustainable Digital Education in India. This pandemic has also taught us that students need to have those abilities to overcome the epidemic, such as problem-solving skills, analytical reasoning, and, most notably, adaptability. To ensure and prioritize the existence of these skills in their students, educational organizations must have to create resilience in their structures.

18.10 DISCUSSION

Our society and our global perspective might well be transformed by the COVID-19 crisis; it may also tell us about how schooling needs to evolve in order to properly educate our young learners for what the future may bring. Following are few observations:

18.10.1 Compliance Regulations for the Education Sector

Hackers are able to try far harder to hack into educational institutions than most organizations, with significant personal intellectual property at stake. Compliance breaches can be particularly detrimental, particularly with heightened media coverage. The following are a few foreign enforcement laws that, in the unpredictable criminal sense, keep the data of students secure.

- FERPA: Family Educational Rights and Privacy Act guarantees the integrity of information of student education. They will audit, review their information and, if necessary, recommend that their educational record be amended.
- FOIA: The Freedom of Information Act allows US government departments, such as public schools, colleges, and universities to make copies of any documents submitted by the student accessible irrespective of the form or format.
- PPRA: Protection of Pupil Rights Amendment refers to the policies and procedures that control the distribution of a survey, study, or evaluation to students. Marketing surveys and other aspects of student privacy, parental

access to information, and the administration of specific physical examinations to minors are all covered by the PPRA. Under this, parental rights are transferred to a student who is 18 years and above in age.

18.10.2 Educating Citizens in an Interconnected World

COVID-19 is a pandemic that reveals how intertwined we are global – there is no longer such a thing as discrete challenges and behavior. Effective individuals need to be able to grasp this interrelatedness in the coming decades and maneuver across borders to exploit their differences and work in an internationally collaborative manner.

18.10.3 Redefining Educators' Role

For the purposes of education in the twenty-first century, the concept of an educator as a knowledge holder who imparts insight to his students is no longer adequate. Need to redefine the role of the teacher in the classroom and lecture theater by a few clicks on their phones, tablets, and laptops as students gain access to information and develop technical skills. This could mean that educators' position would continue to shift toward promoting the growth of young people as contributing members of society.

18.10.4 Life Skills Teaching – Need of Future

Young people need flexibility and adaptability in an ever-changing global climate, skills that prove to be necessary to move through this pandemic successfully. Looking into the future, imagination, coordination, and cooperation, combined with empathy and emotional intelligence, will be some of the most critical qualities that managers will be looking for professional.

18.10.5 Exposing Technologies for Pushing Online Education

The COVID-19 pandemic has forced universities all over the world to suddenly exploit by using the package of technological resources at their disposal to create flexible learning material for students across all sectors. Schools and teachers all over the world are learning fresh stuff better and more versatile, which will benefit students all over the world in terms of accessibility to school. Most importantly, we anticipate that these memories of isolation and remote learning away from their teammates, educators, and schools will serve as a gentle reminder to Generation Z, Alpha, and future generations of the importance of face-to-face social contact.

18.11 CONCLUSION

The online higher education sector, based on current trends, has an optimistic outlook and can be projected to expand considerably in the coming years. However, not everyone agrees that online education is an equal substitution for conventional schooling, so a hybrid model may gain momentum in the future. If they provide the flexibility of an online approach with the presence of conventional classroom experience, virtual classrooms can also become more common. As many of the current offerings are theoretical in nature, both hybrid models and virtual classrooms can allow a more realistic aspect to be applied to online learning. Due to numerous policy measures and evolving customer tastes, the Ed-Tech industry is still expected to expand rapidly in the coming years. COVID-19 may be an additional impetus for this industry's development and help encourage the acceptance of online education models. In the educational sector, these changing changes may become irreversible. The longer the coronavirus pandemic continues, the greater its effects would be and there is a good likelihood that rather than a temporary measure, online learning will become popular. There will not be a convincing argument to go back to conventional classroom environments until educators and students have access to the remote technologies they need to transition to online approaches.

REFERENCES

Adnan, Muhammad, and Kainat Anwar. 2020. "Online learning amid the COVID-19 pandemic: Students' perspectives." *Online Submission* 2, no. 1: 45–51.

Agarwal, Ms Swati, and Dr Jyoti Dewan. 2020. "An analysis of the effectiveness of online learning in colleges of Uttar Pradesh during the COVID 19 lockdown." Journal of Xi'an University of Architecture & Technology, ISSN 1006-7930: 2957–2963.

Albert, Jim, Mine Cetinkaya-Rundel, and Jingchen Hu. 2020. "Online statistics teaching and learning." *Teaching and Learning Mathematics Online* 99, no. 4: 4–8.

Basilaia, Giorgi, and David Kvavadze. 2020. "Transition to online education in schools during a SARS-CoV-2 coronavirus (COVID-19) pandemic in Georgia." *Pedagogical Research* 5, no. 4: 9–11.

Bevins, Frankki, Jake Bryant, Charag Krishnan, and Jonathan Law. 2020. "Coronavirus: How should US higher education plan for an uncertain future." McKinsey & Company, Public Sector Practice.

Chen, Tinggui, Lijuan Peng, Xiaohua Yin, Jingtao Rong, Jianjun Yang, and Guodong Cong. 2020. "Analysis of user satisfaction with online education platforms in China during the COVID-19 pandemic." *Healthcare*, 8, no. 3: 200.

Code, Jillianne, Rachel Ralph, and Kieran Forde. 2020. "Pandemic designs for the future: Perspectives of technology education teachers during COVID-19." Information and Learning Sciences (Ahead-of-print). doi: 10.1108/ils-04-2020-0112.

Conrad, Dianne. 2004. "University instructors' reflections on their first online teaching experiences." *Journal of Asynchronous Learning Networks* 8, no. 2: 31–44.

Conrad, Dennis, and Joan Pedro. 2009. "Perspectives on online teaching and learning: A report of two novice online educators." *International Journal for the Scholarship of Teaching and Learning* 3, no. 2: n2.

Cook, Kelli Cargile, and Keith Grant-Davis. 2020. *Online education: Global questions, local answers.* Routledge.

Fernández-Chamorro, Vanessa, Sonia Pamplona, and María José Pérez-Fructuoso. 2020. "Assessing prior knowledge of statistics in students attending an online university." *Journal of Computing in Higher Education* 32, no. 1: 182–202.

Franchi, Thomas. 2020. "The impact of the COVID-19 pandemic on current anatomy education and future careers: A student's perspective." *Anatomical Sciences Education* 13, no. 3: 312–315.

Govindarajan, Vijay, and Anup Srivastava. 2020. "What the shift to virtual learning could mean for the future of higher ed." *Harvard Business Review* 31(1): 3–8.

Guo, Lei, Mengting Wu, Zhuoying Zhu, Lei Zhang, Sufang Peng, Wei Li, Han Chen, Fernando Fernández-Aranda, and Jue Chen. 2020. "Effectiveness and influencing factors of online education for caregivers of patients with eating disorders during COVID-19 pandemic in China." *European Eating Disorders Review* 28: 816–825.

Hoffman, Heather J., and Angelo F. Elmi. 2020. "Comparing student performance in a graduate-level introductory biostatistics course using an online versus a traditional in-person learning environment." *Journal of Statistics Education* 29, 105–114. doi: 10.1080/10691898.2020.1841592.

Jahodova Berkova, Andrea, and Radek Nemec. 2010. "Teaching theory of probability and statistics during the COVID-19 emergency." *Symmetry* 12, no. 9: 1577.

Ke, Fengfeng. 2010. "Examining online teaching, cognitive, and social presence for adult students." *Computers & Education* 55, no. 2: 808–820.

Kim, Paul. 2020. "Future implications of the fourth industrial revolution on education and training." In *Anticipating and preparing for emerging skills and jobs*, pp. 17–24. Singapore: Springer.

Kiran, Prabha, Abhishek Srivastava, Satish Chandra Tiwari, and T. Sita Ramaiah. 2020. "Evaluating forces associated with sentient drivers over the purchase intention of organic food products." *Asian Journal of Agriculture and Rural Development* 10, no. 1: 284.

Kulshrestha, Umesh Chana. 2020. "Environmental changes during-COVID-19 lockdown: Future implications." *Current World Environment* 15, no. 1–3.

Kundu, Protiva. 2020. "Indian education can't go online – only 8% of homes with young members have computer with net link." https://scroll.in/article/960939/indianeducation-cant-go-online-only-8-of-homeswith-school-children-havecomputer-with-net-link.

Lau, Joyce, Bin Yang, and Rudrani Dasgupta. 2020. "Will the coronavirus make online education go viral." *Times Higher Education*. https://www.timeshighereducation.com/features/will-coronavirus-make-online-education-go-viral

Li, C., and F. Lalani. 2020. "The COVID-19 pandemic has changed education forever." https://www.weforum.org/agenda/2020/04/coronavirus-educationglobal-covid19-online-digital-learning/

Luthra, Poornima, and Sandy Mackenzie. 2020. "4 ways COVID-19 could change how we educate future generations." World Economic Forum. https://www.weforum.org/agenda/2020/03/4-ways-covid-19-education-future-generations/.

Machado, Renato Assis, Paulo Rogério Ferreti Bonan, Danyel Elias da Cruz Perez, and Hercílio Martelli JÚnior. 2020. "COVID-19 pandemic and the impact on dental education: Discussing current and future perspectives." *Brazilian Oral Research* 34: e083. doi: 10.1590/1807-3107bor-2020.vol34.0083.

Moore, Stephanie, and C. B. Hodges. 2020. "So you want to temporarily teach online." https://www.insidehighered.com/advice/2020/03/11/practical-advice-instructors-faced-abrupt-move-online-teaching-opinion.

Paul, Shuva, Md Sajed Rabbani, Ripon Kumar Kundu, and Sikdar Mohammad Raihan Zaman. 2014. "A review of smart technology (Smart Grid) and its features." In 2014 1st International Conference on Non-Conventional Energy (ICONCE 2014), pp. 200–203. IEEE.

Paulsen, Justin, and Alexander C. McCormick. 2020. "Reassessing disparities in online learner student engagement in higher education." *Educational Researcher* 49, no. 1: 20–29.

Quay, John, Tonia Gray, Glyn Thomas, Sandy Allen-Craig, Morten Asfeldt, Soren Andkjaer, Simon Beames et al. 2020. "What future/s for outdoor and environmental education in a world that has contended with COVID-19?." *Journal of Outdoor and Environmental Education* 23, no. 2: 93–117.

Rowan, Leonie, Terri Bourke, Lyra L'Estrange, Jo Lunn Brownlee, Mary Ryan, Susan Walker, and Peter Churchward. 2021. "How does initial teacher education research frame the challenge of preparing future teachers for student diversity in schools? A systematic review of literature." *Review of Educational Research* 91, no. 1: 112–158.

Shearer, Rick L., Tugce Aldemir, Jana Hitchcock, Jessie Resig, Jessica Driver, and Megan Kohler. 2020. "What students want: A vision of a future online learning experience grounded in distance education theory." *American Journal of Distance Education* 34, no. 1: 36–52.

Srivastava, Abhishek, and K. M. Pandey. 2012. "Social media marketing: An impeccable approach to e-commerce." *Management Insight* 8, no. 2: 99–105.

Sun, Anna, and Xiufang Chen. 2016. "Online education and its effective practice: A research review." *Journal of Information Technology Education* 15: 157–190.

Sun, Litao, Yongming Tang, and Wei Zuo. 2020. "Coronavirus pushes education online." *Nature Materials* 19, no. 6: 157–190.

Tallent-Runnels, Mary K., Julie A. Thomas, William Y. Lan, Sandi Cooper, Terence C. Ahern, Shana M. Shaw, and Xiaoming Liu. 2006. "Teaching courses online: A review of the research." *Review of Educational Research* 76, no. 1: 93–135.

Whipp, Joan L., and Rebecca A. Lorentz. 2009. "Cognitive and social help giving in online teaching: An exploratory study." *Educational Technology Research and Development* 57, no. 2: 169–192.

Zarif Sanaee, Nahid. 2020. "Assessing the criteria for the quality and effectiveness of e-Learning in higher education." *Interdisciplinary Journal of Virtual Learning in Medical Sciences* 1, no. 3: 24–32.

Index

For Product Safety Concerns and Information please contact our EU
representative GPSR@taylorandfrancis.com
Taylor & Francis Verlag GmbH, Kaufingerstraße 24, 80331 München, Germany

www.ingramcontent.com/pod-product-compliance
Ingram Content Group UK Ltd.
Pitfield, Milton Keynes, MK11 3LW, UK
UKHW020020200926
475702UK00014B/1981

9780367676308